The Surface of the Earth 2

Climate, Soils and Vegetation

The SURFACE of the EARTH

Volume 2
Climate, Soils and Vegetation

M J Selby

Senior Lecturer in Earth Sciences
in the University of Waikato

Cassell London

The earth as seen from 35,900 km in space by Applications
Technology Satellite III, at 10.30 a.m. EST, 10 November,
1967 from its position at 47°W longitude on the equator
over Brazil. The photograph shows South and North America,
part of Africa and Europe, as well as the southern part of
Greenland. Antarctica is covered with clouds. *NASA*

CASSELL & COMPANY LTD
35 Red Lion Square, London WC1
Melbourne, Sydney, Toronto, Johannesburg, Auckland

First published 1971

I.S.B.N. 0 304 93671 5

Photoset and printed in Great Britain
by BAS Printers Limited, Wallop, Hampshire
F.770

Preface

Physical geography is usually taught as a combination of geomorphology and climatology, with only limited reference to soils and vegetation, and most textbooks of physical geography reflect this bias. In *The Surface of the Earth* I have attempted to give these four branches of physical geography a more equal treatment than is usual, and also to provide a treatment of the systematic parts of soil geography and biogeography. Some readers may not have a background in the chemistry which is so necessary for an understanding of soil processes, and I have, therefore, attempted to 'explain' the chemistry in the text.

The classification of soils is a particularly difficult problem at the present time, as most countries have their own system and there are few world systems. The USA has recently introduced a new classification which was published in its most complete form in a *7th Approximation*, but as this system is still open to correction and extension, and there is no world map of soils using this classification, I have used the 1949 USA classification with a few modifications.

The writing of a book of this length is a very time-consuming occupation and I am indebted to my wife for her tolerance, support and encouragement throughout this period, and also for her help in drawing many of the figures. Many friends have read parts of the text and made helpful comments and suggested corrections. I am particularly indebted to Dr E. M. Stokes and Dr G. R. McBoyle of the University of Waikato; to Dr J. D. McCraw and Mr W. A. Pullar of the New Zealand Soil Bureau; to Mr P. J. Tonkin of Lincoln College; to Dr C. J. Sparrow and Mr G. R. Cochrane of the University of Auckland; and to Mr R. J. Blong of the University of Sydney. Any errors or misrepresentations which remain are entirely my responsibility. The staff of Cassell and Co. Ltd have always been of great assistance to me, and I am grateful to them for seeing this book through the press.

All photographs not otherwise credited are my own.

Hamilton, New Zealand

Contents

Colour plates: 24 typical soil profiles between pages 192 and 193

1 General Introduction

> I am strongly induced to believe that as in music the person who understands every note will, if he also possesses a proper taste, more thoroughly enjoy the whole, so he who examines each part of a fine view may also thoroughly comprehend the full and combined effect.
>
> —Charles Darwin, *The Voyage of the Beagle*

Geographers are probably the only scientists who have taken it upon themselves to examine 'each part of a fine view' of the earth's surface. Too frequently, however, the result of this examination has been a general description of 'the whole' without an understanding of 'the full and combined effect', which can only come from a disciplined investigation of both the natural environment, physical and biotic, and the human or cultural one. Not only has the investigation been lacking; few geographers have made themselves competent to undertake it, for they have too frequently been untrained in the basic sciences. Most physical geographers achieve competence in geomorphology and climatology and have some understanding of the processes which give rise to the local and world geomorphic and climatic patterns; far too many of them have been content with a general knowledge of world patterns of soils and vegetation. Yet if geography is a study of the relationships between man and his environment, the study of soils and vegetation must frequently take precedence over that of landforms and climate. This is not to deny that climate and relief are major influences in all our lives but to assert that they are only two aspects of the environment.

Traditionally, geographical studies have aimed at a synthesis in which the world has been divided into cultural or physical regions. Within such major physical regions it is assumed that there are related distributions of climate, soils, vegetation and landform types. This kind of correlation is so over-simplified that it has inevitably led to false assumptions and to deterministic attitudes.

The origin of many assumptions is to be found in the search by nineteenth-century naturalists for an understanding of the patterns of distribution of vegetation and climate which their explorations revealed. They sought to 'explain' the limits of vegetation and soils in simple climatic terms. World maps of vegetation were drawn and, once the initial assumed correlations had been made, world climatic maps were also drawn using the vegetation map as a base. Hence the 'correlations' were often the result of circular arguments rather than expressions of reality. In the absence of information about climate and soils this was the best that could be done, but unfortunately the resulting maps and the regions marked on them have become part of a hallowed tradition of regional geography and have been accepted uncritically by students. Figure 1.i is a modern climate-vegetation correlation.

The errors which result from such a method can be seen, for example, in the 'savanna' climate; when it is realised that many of the tropical grasslands, like many of the temperate ones, have been induced through the clearance of forest by natural fires and man's interference, the

Figure 1.1. The classification of world plant formations in relation to climate.
After Holdridge and Tosi

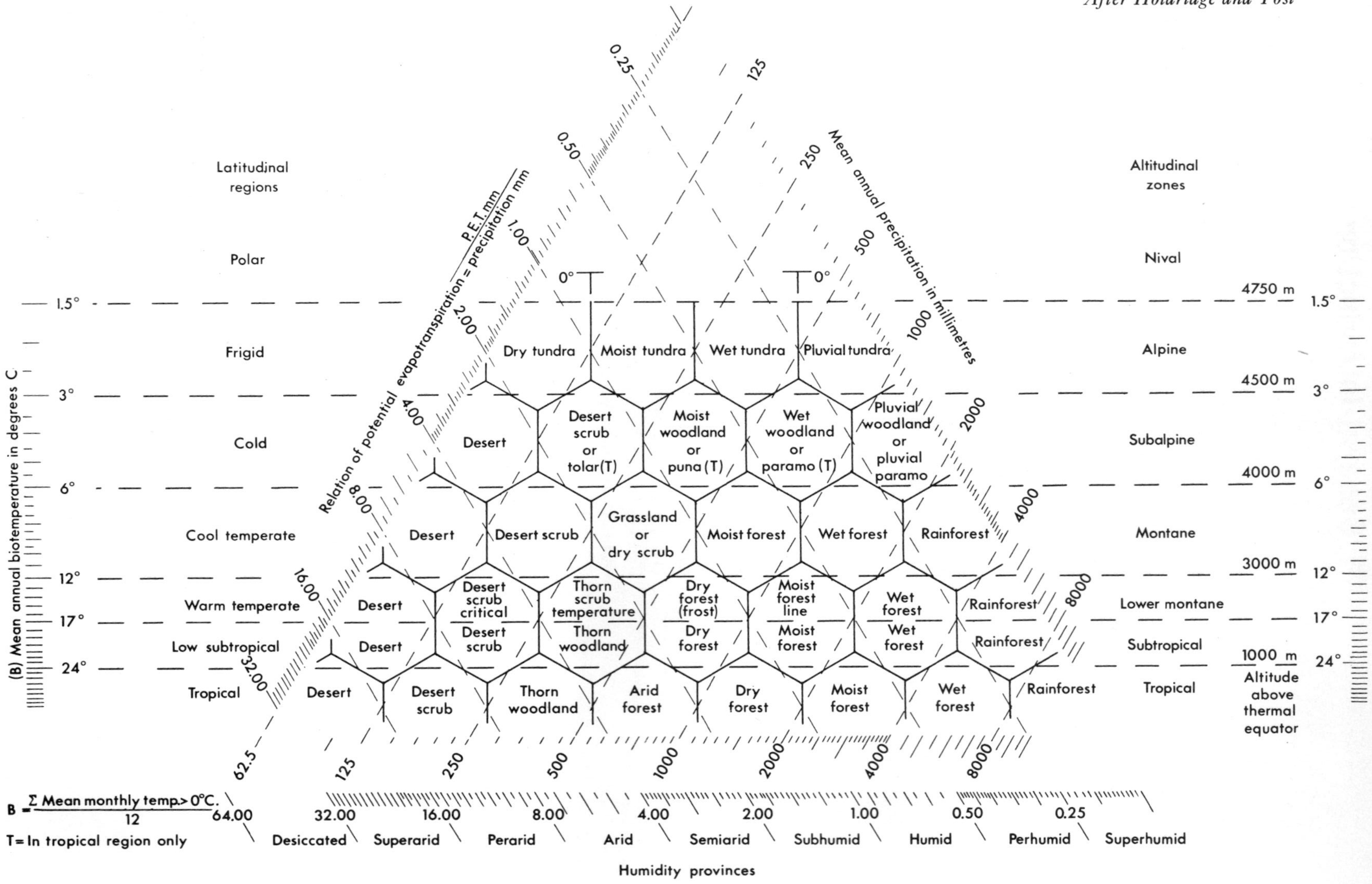

absurdity of a simple climate-soil-vegetation correlation becomes obvious. Climate has only a statistical reality, which we usually express in an excessively simple formula, and it is only one factor in a very complex environment. If the true relationships are to be understood, soils and vegetation must be studied in the field, where the subtle generic relationships, rather than the massive generalisations of world maps, assume their proper importance. The student memorising broad correlations is merely equipping himself with the delusion that he understands, and is stultifying his intellectual processes.

Climatic determinism might have been valid if the earth had had a uniformly flat surface with a homogeneous bedrock; over such a surface each plant species could establish itself wherever climate allowed. However, none of these conditions exists: mountains and oceans form barriers to plant migration; climatic change has caused ice sheets and glaciers to destroy soils and plants over large areas, and their re-colonisation of glaciated landscapes has been inhibited so that only the most effectively dispersed plants have re-entered those areas; man has been a major influence on vegetation not only in the last few hundred years but for several thousand years in some parts of the world.

In the following chapters, therefore, the emphasis is on understanding the processes, composition and structure of the atmosphere, pedosphere and biosphere, and not upon generalised world patterns.

Man and the Environment

An understanding of the physical environment is necessary not only to the geographer but to every man, because for the first time in history man is becoming capable of dominating the environment. It is probable that man exterminated nearly half of the species of the larger mammals in Africa some 50,000 years ago, and that in north Africa, where he arrived much later, he had killed off at least 60% of the species of the larger mammals by 10,000 BC. This 'Pleistocene overkill' is of limited scale compared with the enormous power now available.

Modification of the world's climate by melting the Arctic ice cap, or by bombing hurricanes and typhoons with silver iodide to weaken the airflow and decrease the intensity of the storm are both technically possible now, but what are the implications of such actions? Warming the Arctic might improve the climate of surrounding lands and make them habitable but it would also raise the world sea level, inundate many coastal areas, and have unforeseen effects upon the general circulation of the atmosphere. Destroying typhoons, which are estimated to do annual damage worth 0·45% of the gross national product of the Philippines, 1·43% for Taiwan and 0·72% for Japan, would greatly help those countries economically, but how would this affect the total rainfall and the general circulation? Hong Kong receives 25% of its rainfall from typhoons and the Philippines 67%.

Interferences with the environment can be more subtle than those produced by bombs, and other unintentional disturbances of the environment have resulted in pollution of soil, air and water, and the destruction of flora and fauna. Some losses are inevitable as the world's population expands and increasingly affluent societies demand more of the earth's resources, but if the disastrous waste which has occurred in the past is to be avoided in the future, then a well educated population is the best insurance policy. It is hoped that the following chapters will assist in this education.

PART I   CLIMATOLOGY

2 Introduction to the Atmosphere

Weather is the day-to-day state of the atmosphere and its short-term changes. The science which describes and seeks to explain weather is called meteorology. Climate is the aggregate of weather over a long period of time which takes into account the average and extreme weather conditions. The science which describes and seeks to explain climate is called climatology.

Although observations of atmospheric conditions have been made during a long period of human history, and Aristotle wrote his *Meteorologica* in about 350 BC, the modern study of the atmosphere was not possible until the invention of sophisticated recording instruments and the telegraph in the nineteenth century. Not only were accurate records of temperature, precipitation, cloud cover and other weather elements required, but these had to be collated and mapped at the time of recording so that it was possible to obtain an impression of weather conditions over large areas at a given time, and to study how weather patterns changed with time. Also during the nineteenth century, considerable advances were made in the study of the physics of gases, and Sir Francis Galton even postulated the existence of anticyclones and cyclones. The development of meteorology was hindered by the lack of data on conditions in the upper atmosphere; the invention of the aeroplane made possible great advances in practical meteorology and considerable improvement in the accuracy of weather forecasts. Aeroplane observations have now been replaced by data from balloons, rockets, radar and satellites which extend observations to the full height of the atmosphere, so that a three-dimensional study of the atmosphere is now possible compared with the surface knowledge of only fifty years ago.

The Nature of the Atmosphere

The atmosphere is a blanket of gases which envelops the earth. Without it life could not exist; there would be no weather; the temperature of the earth would soar to over 100°C during the day, and sink to −200°C at night. The gases are probably derived from volcanic eruptions, hot springs, chemical decay of solid matter and contributions from vegetation. The atmosphere is thought to have been relatively stable for about 500 million years. It is constantly receiving additions of gases from plants, decay of rocks and organic matter, and the burning of fuel, but at the same time plants, animals and chemical reactions remove other gases from it. The percentage composition of dry air is given in Table 2.1.

In addition to gases, the atmosphere contains variable quantities of solids and liquids. The solids are mainly particles of smoke, salt, dust or volcanic ash. (1) Smoke may come from volcanoes or forest fires but it is mostly derived from man-made sources related to industrialisation. (2) Salt in the atmosphere is derived from the evaporation of sea spray. (3) Dust comes from areas of bare soil on the land surface and particularly from the deserts. (4) The dust of volcanic ash is thrown into the atmosphere during eruptions and may cause brilliantly red sunsets. Solid particles in the atmosphere scatter light of short wavelength from

Table 2.1. The Percentage Composition of Dry Air

Gas	By volume	By weight
Nitrogen	78·03	75·48
Oxygen	20·99	23·18
Argon	0·94	1·29
Carbon dioxide	0·03	0·04
Hydrogen		
Neon		
Helium	Traces only, being less than 0·001%	
Krypton		
Xenon		

the sun; where there is much dust, so much light is reflected that the sky appears hazy and white. Where there is little dust and especially at great heights, the sky appears blue or even violet.

The only liquid in the atmosphere is water, which may be present either as a gas (water vapour) or as droplets and ice crystals which occur as fog, haze or cloud. The proportion of water in the atmosphere varies greatly from a little over the hot deserts to a maximum of about 4% in humid regions. The study of the variability of water content of the air is one of the chief concerns of the meteorologist.

The Layers of the Atmosphere

The exact thickness of the atmosphere is not known, although it can be calculated at about 32,000 km, the distance at which the earth's gravitational pull approximates the centrifugal force due to the earth's rotation. Air becomes rarer or thinner, that is, the molecules of which it is composed are more widely spaced, with increasing altitude until it loses its characteristics at a considerable height above the surface.

Figure 2.i. The atmosphere has its upper limit at the tropopause beneath which it is about 8 km (5 miles) thick at the poles (P) and about 18 km (11 miles) thick at the equator (E). The scale of the atmosphere is enormously exaggerated.

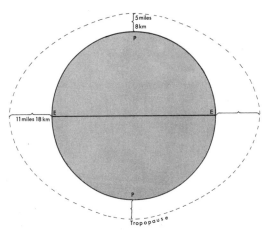

The meteorologist is mainly concerned with the atmosphere within the layer closest to the earth's surface, called the troposphere, which is about 18 km (11 miles) thick at the equator and 8 km (5 miles) thick at the poles (Figure 2.i). Within the troposphere is the realm of weather—storms, clouds and convection currents. Also within it, temperature declines with altitude until a minimum of about −50°C is reached at the

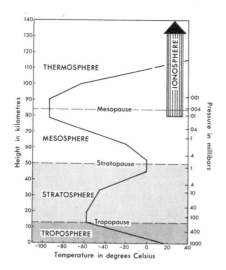

Figure 2.ii. Temperature in the upper atmosphere.

upper boundary, called the tropopause. Above the tropopause is the stratosphere, in which there is no water vapour, no convection and therefore no weather. At about 80 km above the earth's surface begins the ionosphere, which is the layer of the electrical phenomena of the Auroras. The layers of the atmosphere are shown diagrammatically in Figure 2.ii.

3 Heat in the Atmosphere

There are two possible sources for the heat of the atmosphere—the earth and the sun. The earth is a negligible source; it is from solar radiation that the atmosphere gains its heat and therefore its energy. The incoming radiation, or insolation, is balanced by the outgoing radiation lost by the atmosphere, so that the atmosphere's heat remains nearly constant, but during the time that heat is trapped by the atmosphere, it does the work which produces weather.

The incoming radiation is composed of the various rays which make up the solar spectrum. The wavelength of the rays is measured in microns (μ), 1μ being $1/1000$ of a millimetre (see Table 3.1).

Table 3.1. Composition of Solar Radiation

Rays	Wavelength in microns	% of total solar radiation
Shortest: X-rays and gamma rays	$\frac{1}{2000}$ to $\frac{1}{100}$	9
Ultra-violet rays	0·2 to 0·4	
Visible light rays	0·4 to 0·7	41
Infra-red rays	0·7 to 3	50
Longest: Heat rays	3 to 3000	

The ozone gas concentration which extends through the atmosphere as a layer about 24–32 km above sea level (see Figure 2.i) is opaque to solar radiation with a wavelength of less than 0·3 microns; hence the surface of the earth is protected from bombardment by shortwave

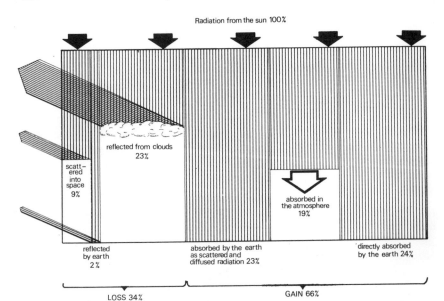

Figure 3.i Gains and losses of solar radiation.

radiation which would otherwise have affected the evolution of life. Much of the energy of incoming radiation is lost in the lower atmosphere, where clouds, light-coloured surfaces, water, snow and vegetation all reflect parts of the solar radiation. The amount of reflection will obviously vary over the surface of the earth, and will also vary throughout the year as cloudiness, snow cover, etc. change.

Of the total radiation which reaches the atmosphere, about 32% is reflected back or scattered into space by clouds and dust in the air, and 2% is reflected back by the earth's surface. This 34% loss of radiation is known as the earth's albedo. Of the remaining 66% of incoming radiation which is effective in heating the earth's surface and atmosphere, about 19% is absorbed by water vapour and dust in the atmosphere, 24% is absorbed directly by the earth and 23% is absorbed by the earth after being scattered by clouds (Figure 3.i). The albedo values of a number of surfaces are indicated in Table 3.2.

Table 3.2. Typical Albedo Values (per cent)

Forests	3–10
Fields (dry, ploughed)	20–25
Grass	15–30
Bare ground	7–20
Sand	15–25
Fresh snow	80
Ice	50–70

It is evident from the table that green vegetation absorbs a very high percentage of radiation, and this explains why there is no glare from it. In a sandy region much of the radiation is reflected and this prevents temperatures rising even higher than they actually do in the hot deserts.

The atmospheric heat balance is maintained because the heat absorbed by the earth is re-radiated out as longwave radiation which warms the atmosphere. This warming of the atmosphere is known as the greenhouse effect because the atmosphere acts like the glass of a greenhouse, letting through incoming shortwave radiation but trapping outgoing longwave radiation.

Variation of Insolation

The insolation received at any particular time and place is controlled by four factors: (1) the distance from the earth to the sun; (2) the transparency of the atmosphere; (3) the daily duration of insolation; (4) the angle at which the sun's rays strike the earth's surface.

The variation in the distance between the sun and the earth is not great and produces only a small effect on insolation received. The transparency of the atmosphere has a particularly important effect on insolation. It varies locally with the amount of dust, cloud and water vapour in the atmosphere, and also with latitude, for at high latitudes the sun's rays have to pass obliquely through the atmosphere so that it is effectively a thicker layer than that in the tropics. This property varies with the seasons, being greatest in the winter when the hemisphere is tilted away from the sun.

The daily duration of insolation varies with latitude and the seasons. At the equator day and night are of equal length throughout the year, but at the poles the daylight period varies from a maximum of 24 hours in summer to nil in the winter.

The angle at which the sun's rays strike the earth varies during the day and with latitude. The significance of this is shown in Figure 3.ii, in which a square sunbeam strikes the earth. When the beam is vertical the energy is not spread, but when the beam is at an oblique angle the energy of the square beam is spread over a rectangular area.

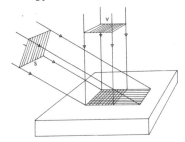

Figure 3.ii. The angle of the sun's rays determines the intensity of insolation at the ground. The slanting ray (S) is spread over a rectangular area larger than the square of the vertical ray (V).
After Strahler

The irregularities in the surface of the earth have so far been ignored, but it should be noted that on sloping ground the energy of the sun will be spread or concentrated, depending on the obliquity of the rays to the surface.

The Measurement of Air Temperature

Temperature is a relative term implying a degree of heat. If the heat of one body flows to another, the originating body is said to have a higher temperature. To measure temperature, scales of reference are employed. The two most common scales are Celsius (or Centigrade) and Fahrenheit. The Celsius scale fixes the boiling point of water at 100°C and the melting point of ice at 0°C. The corresponding values for the Fahrenheit scale are 212°F and 32°F. The Celsius scale is most used in Europe and the Fahrenheit one in the English-speaking world, although in Britain the Celsius scale has now been adopted.

Conversion tables are given in Appendix I, but they can easily be calculated from the two formulae:

$$°F = 32 + \frac{9}{5}°C$$

$$°C = \frac{5}{9}(°F - 32)$$

Records of Temperature

Continuous records of temperature are made by thermographs, but few recording stations are so equipped and daily mean temperatures are usually calculated as the mean of the daily maximum and minimum. The difference between these two is the diurnal (daily) range. The mean monthly temperature is calculated by adding the daily means and dividing by the number of days in the month. The annual range is the difference between the highest and lowest mean monthly temperature. Usually January and July, as the two extreme months, are chosen for comparison purposes. It should be noted that mean temperatures tend to hide the extremes and variability of temperature, which may be very important.

The Horizontal Distribution of Temperature

On an earth with a homogeneous surface and with no surface relief, temperatures would decline evenly from the equator towards the poles. On maps this horizontal distribution of temperature is shown by means of isotherms, an isotherm being a line connecting points with equal temperature values. On a hypothetical earth of constant surface the isotherms would all be parallel to the lines of latitude. In reality, however, the isotherms by no means run parallel to the lines of latitude in the northern hemisphere, and even in the southern hemisphere they have kinks in them where they cross land masses (Figure 3.iii).

The irregularities in the isotherms are caused by the uneven distribution of land and sea. The southern hemisphere is predominantly oceanic, especially polewards of the tropic, and so has relatively straight isotherms, but in the northern hemisphere the irregularities are extreme. They would be even greater, were the isotherms not drawn to eliminate the effects of altitude. This is achieved by not drawing isotherms for the actual observed temperatures, but correcting them so that all temperatures have the approximate value that would be experienced if all land were at sea level. The correction is made by adding about 6·4°C for every 1000 metres of elevation of the recording station. If temperatures were not reduced to the sea level value, the isotherms over hill country would follow the contours almost exactly.

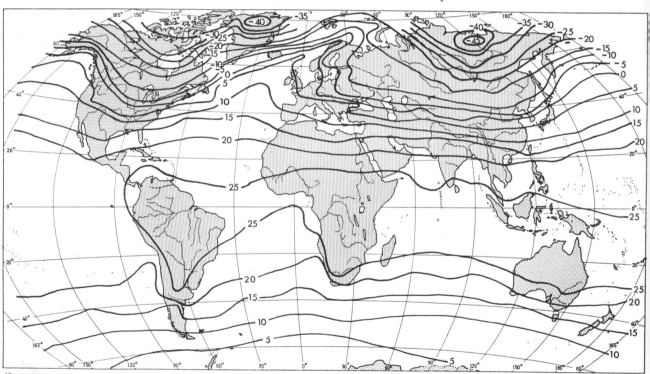

Figure 3.iii. Mean sea level temperatures in °C for January.

There are three main causes of the differences between temperatures over land and sea. (1) The specific heat of water is higher than that of the soil, rocks or plants of the land; in other words, a given mass of water requires approximately three times as much heat to raise it one degree in temperature as does a similar mass of land. Water, therefore, both warms up and cools more slowly than land. (2) Water moves both horizontally and vertically so that mixing distributes the heat received, whereas on land the heat is concentrated at the surface. (3) Water is

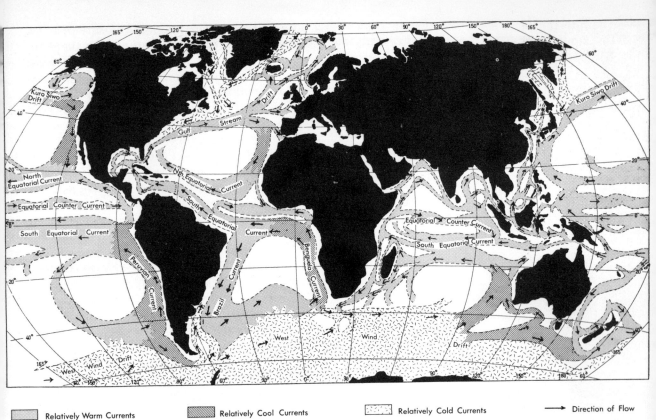

| Relatively Warm Currents | Relatively Cool Currents | Relatively Cold Currents | → Direction of Flow |

Figure 3.iv. Ocean currents in July.

translucent, and radiant energy passes through it to warm water at depths of up to 150 metres or more. On land, however, only the matter close to the surface is affected. Reflection from the surface of land and water is approximately the same, so this does not affect the heating of the two types of material.

Ocean Currents and Temperature

The pattern of temperature distribution over land and sea is mostly controlled by ocean currents. A warm current is one which moves from a warm source towards a cold or polar area, and a cold current moves equatorwards. In moving, such currents act like portable storage heaters, greatly influencing the climates of the oceanic areas they occupy and the temperatures of the coasts around them. Perhaps the most spectacular and best known effect of a current on climate is that produced by the Gulf Stream, which emerges from the Caribbean Sea south of Florida, flows northwards along the coast of the USA, and then travels northeast across the Atlantic as the North Atlantic Drift. In the winter the Drift, at 63°N, warms the Norwegian coast 7·8°C more than the average temperature for that latitude. A similar but less effective current warms the coast of British Columbia in winter.

The most remarkable cold currents are found on the west coasts of South America and Africa where the north-flowing Peruvian and Benguela currents bring coastal temperatures as much as 4°C below the average for the latitudes.

The effect of these water bodies depends upon the wind system and relief of the land. The effect of the North Atlantic Drift is carried across Europe by prevailing westerly winds, but in North America it is

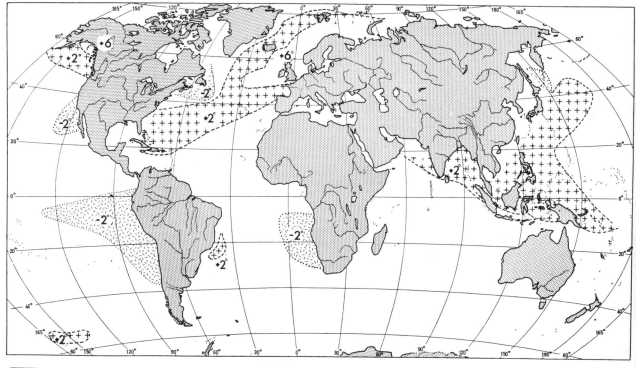

Figure 3.v. Main ocean water annual temperature anomalies. *After Birot*

confined to the coast by the north to south trending ranges of the Appalachian Mountains and the offshore winds.

Continentality

Maps showing thermal anomalies, or differences between the mean temperature at a place and the mean temperature of its latitude, are useful expressions of the effects of ocean currents on climate (Figure 3.iv). They also demonstrate the effects on temperature of a continental location. In Figure 3.v, the greatest anomalies are shown to exist in the northern hemisphere where western Europe and the coasts of British Columbia and Alaska are warmed by the ocean currents, especially during the winter. In the heart of the North American and, more particularly, the Asian continents, temperatures in winter are extremely low. Here there are no oceanic influences; air is descending, the surface is snow-covered, the days are short and temperatures are correspondingly low, making parts of Siberia colder than the poles. In the northern hemisphere summer, the oceans have average temperatures below that of the mean for the latitude, and hence are said to have a negative anomaly, but the land has a positive anomaly. The sea thus has a smaller range of temperature than the land and the temperature extremes of the interior of the continents are great. The largest positive anomalies over land are in the great deserts, where in summer the low humidity and clear skies produce intense heating of the land surface.

In the southern hemisphere most of the surface is water, so the isotherms tend to be parallel to the lines of latitude and anomalies are small. In winter, the Peruvian and Benguela currents produce oceanic negative anomalies in the tropics, and the interior of Australia also

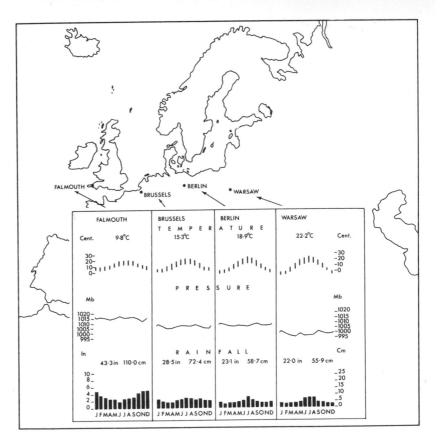

Figure 3.vi. Climate graphs for four stations in Europe. With increasing continentality, rainfall maxima change from winter to summer, temperature ranges increase and temperatures become more extreme.

becomes notably cool, but in the summer, positive anomalies are produced over the three continental land masses and the two ocean currents increase their negative effect.

Vertical Distribution of Temperature

The evidence of ice and snow on mountain tops and the cooler climate of mountain resorts indicates even to the layman that temperature falls with increasing altitude. The main cause for this is that longwave re-radiation from the earth is chiefly responsible for heating the atmosphere, so that air in contact with the surface is heated most. Under normal conditions, temperature declines, or lapses, by about 6·4°C for every 1000 metres rise in altitude. This is the normal lapse rate: the steeper the lapse rate, the more rapid the fall in temperature.

The normal lapse rate is steady up to the tropopause; above this, in the stratosphere, temperature once more rises and then falls again in the ionosphere. The tropopause is highest over the equator, being at about 18 kilometres, but declines towards the middle latitudes to a height of about 11 kilometres and then falls to a minimum of 5–6 kilometres over the poles.

When air in the troposphere is heated so that convection currents are set up in it, the warm air rises. As it rises the air, like any other gas, is under less pressure and it expands. In doing so, it loses some energy with the result that its temperature declines. Conversely, air in a descending current comes under greater pressure and contracts, and the energy of compression is transferred to the gas, raising its temperature.

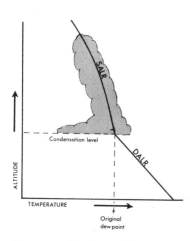

Figure 3.vii. A graph to illustrate adiabatic cooling of rising air, and cloud formation. The original dew point is the temperature to which the air must be cooled before the water vapour present will be condensed.

The amount of change in pressure a gas undergoes is about 1% per 100 metres of altitude. The changes of temperature produced by the expansion or contraction of air are said to be adiabatic.

Dry air, when forced to rise, will expand and cool at the rate of 1°C per 100 metres. This is known as the dry adiabatic lapse rate (DALR) and is the rate at which rising air cools or descending air warms when no heat is transferred from or to the surrounding air.

If the lapse rate for a large body of air were, say, 0·9°C it would be said to have an environmental lapse rate of 0·9°C. In such conditions, if a pocket of air within the main body were caused to rise, after it had risen 300 metres its temperature might be lowered by 3°C and it would then be cooler than its surroundings, and would tend to sink again. Such air conditions are said to be stable.

If the environmental lapse rate were, say, 1·1°C the air forced upwards would have its temperature lowered by only 3°C in 300 metres; it would therefore be warmer than its surroundings and would continue to rise. Such conditions are unstable and produce strong vertical air currents. It is by such vigorous convection currents that much of the heat of the surface of the earth is redistributed.

Rising air which contains moisture will lose its heat at the dry adiabatic lapse rate until the cooling causes condensation. In condensation, however, latent heat is released which will reduce the rate of cooling so that the lapse rate will be less than 1°C per 100 metres. The rate of decrease in temperature in rising saturated air is known as the saturated adiabatic lapse rate. The actual value of the SALR depends upon the amount of moisture condensed, but it is about 0·6°C per 100 metres. The adiabatic cooling of rising air is illustrated in Figure 3.vii.

Temperature Inversion

Under certain conditions, temperature may rise with increasing altitude to form a temperature inversion in which, of course, the lapse rate is also inverted. Temperature inversions just above the ground surface may be produced by the following means.

(1) Cold and hence dense air flows from slopes and valley sides to lower levels, where it collects (Figure 3.viii). This type of inversion is common in hilly and mountain districts and accounts for the cloud or mist in valley bottoms in the early morning, when the upper parts of the valley walls are already free of cloud and sunny. The downward flow of air takes place during the night, and because it displaces the warm air it is very stable, so that when there is little sun to heat the ponded cold air, the inversion may persist for several days. In rural districts the effect is strong enough for orchardists to prefer slope rather than valley bottom sites for their trees in order to avoid frost.

(2) On clear evenings and nights after a warm day, inversions commonly occur over bare ground and land with low vegetation. As the ground rapidly loses heat by radiation in the evening, it cools the air in contact with it and may produce a shallow band of ground mist some centimetres or metres thick. Above the cooled air the atmosphere retains its warmth. This type of inversion is best developed over snow surfaces, and is least likely to occur over water.

(3) When two air masses with different temperature characteristics meet, the boundary between them is called a front. Along the front the

Figure 3.viii. Cold air drains downslope to pond in the floor in the valley, thus creating a temperature inversion. In this diagram the closer the lines the lower the temperature.
Based on Geiger

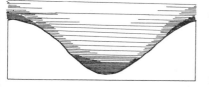

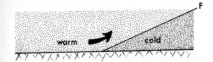

Figure 3.ix. A frontal (F) temperature inversion.

warmer and less dense air tends to ride up over the cooler and more dense air, producing a frontal inversion (Figure 3.ix).

(4) The process of movement of an air mass is called advection. The advection of a warm air mass over a cold surface will create an inversion in the lower layers of the air mass. This process is particularly common when air moves over a snowfield or a cold water surface, when cloud or fog may form.

(5) When a body of air subsides, it is heated dynamically and may spread out over a cooler surface layer of air to produce a subsidence inversion. This type of inversion is most common at considerable altitudes.

Common types of inversion are illustrated in Figure 3.x. A characteristic of all inversions is their stability, for convection currents within them cannot develop readily, and only minor mixing of the air layers is possible.

Figure 3.x. Vertical lapse rate of temperature showing: a surface inversion (S); a normal lapse rate (N) of 6·4°C per 1000 metres; an upper air inversion (U); and an isothermal lapse rate in which there is no change of temperature with altitude (I). *After Critchfield*

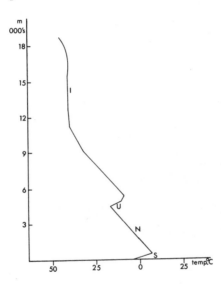

4 Pressure and Motion in the Atmosphere

A column of air reaching from ground level to the top of the atmosphere, with a cross-sectional area of one square centimetre, would weigh about 1·0332 kg. This column exerts on the ground beneath it an atmospheric pressure of 1·0332 kgf per square centimetre = 10·132 newtons per square centimetre. This figure is only approximate because the atmosphere does not have a constant thickness, being thicker over the equator than over the poles, and the density of air can vary locally. The figure of 10·132 N/cm^2 is, however, accepted as being a 'normal' atmospheric pressure at 45°N or 45°S latitude. The following are equivalent measurements: 1·0 standard atmosphere, 1013·2 millibars, 760 mm of mercury, 10·132 N/cm^2. Meteorologists use the millibar as a standard measurement of atmospheric pressure.

Distribution of Pressure

Because the column of air progressively decreases in length and contains air of lower density at higher altitudes, atmospheric pressure becomes less with increasing altitude. The air changes its composition with altitude so the decrease in pressure is not completely regular, but it loses 1/30 of its value for every rise of 274 metres. The decrease in pressure is geometric, so that at 274 metres above sea level pressure is 29/30 of the value at sea level and at 548 metres it is 29/30 of the value at 274 metres.

The meteorologist is interested in the constantly changing pressure in a three-dimensional atmosphere and needs to record the distribution of pressure at frequent intervals. Before the introduction of upper air observations using balloons, aircraft, satellites and electronic devices, the meteorologist had to be content with using only surface pressure records and constructing isobars, or lines joining points with the same pressure on a level surface. The modern method is to draw pressure surface contours, or lines joining points at which a surface of constant pressure has the same height above sea level. By international agreement, the pressures used to define the constant pressure surfaces are standardised at 1000, 850, 700, 500, 400, 300, 200 and 100 millibars.

The 1000 mb surface is almost at sea level, and the 1000 mb contour chart and the sea level isobar chart have a similar appearance. The 500 mb surface is situated about halfway between the tropopause and sea level in mid-latitudes, and the 100 mb surface is mostly in the stratosphere. The difference in height between two surfaces, such as those of 1000 mb and 500 mb, is the thickness of the air layer. Warm air is less dense than cold air and occupies a greater space, so thick air layers are warm and thin ones are cold. Pressure surface charts, therefore, give a picture of the dynamic systems of the atmosphere. Figure 4.ia shows a contour chart of the 500 mb surface and Figure 4.ib another chart showing the thickness of the air layer between the 1000 mb and 500 mb surfaces at the same time.

Isobaric charts showing average pressures over the surface of the earth are usually constructed for winter and summer periods to show the

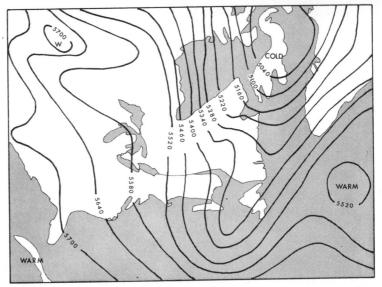

Figure 4.ia. The contours of a 500 mb surface over North America and Greenland. Heights are in metres.
After Sutton

differences in pressure caused by the seasonal variations in temperature. Over some parts of the earth, pressure can be more or less constant for weeks at a time, but in other areas the average pressures are the result of the passage of series of high or low pressure systems. Low pressure systems are called depressions, lows or cyclones—unless they are

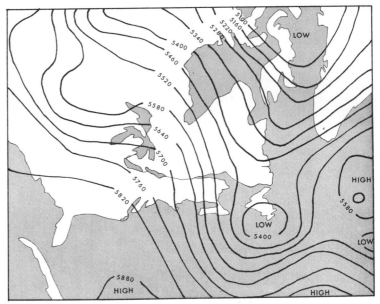

Figure 4.ib. A 1000–500 mb thickness pattern. Thicknesses are in metres.
After Sutton

elongated, when they are called troughs. High pressure systems are called anticyclones or highs. Elongated highs are ridges or wedges.

The causes of differences of pressure in the atmosphere are not fully understood, although at least two—the thermal and dynamic origins of pressure differences—have received the attentions of many research workers.

Wind

Between areas which have different atmospheric pressures, a pressure gradient is said to exist. Air will always flow from an area of high pressure towards one of low pressure and so produce wind. The pressure gradient is said to be steep when the difference of pressures between the two areas is great, and the steeper the gradient, the more rapid the flow of air.

The movement of air is, however, not confined to the horizontal, for in all pressure systems air moves vertically. In high pressure systems air descends and in low pressure systems it ascends, thereby producing a complete circulatory system, with converging air in the low pressure system being forced to rise, and diverging air in the high pressure system resulting from subsidence.

Coriolis Effect

Wind on the surface of the earth, except that which is moving immediately along the equator, is deflected from a straight line course by the effect of the earth's rotation on its axis. All moving bodies in the northern hemisphere are deflected to the right of a straight line course; in the southern hemisphere they are deflected to the left. This phenomenon should be called the Coriolis Effect and not the Coriolis Force, since it is not a force at all. The effect increases from nothing at the equator to a maximum at the poles and it also increases with the velocity of the moving object. A man walking at 6 k.p.h. would be deflected from his intended straight path by 50 metres at the end of 1 km if he were in middle latitudes.

Among physical phenomena the Coriolis Effect is most important in influencing the weather. In the absence of the Effect, wind would rush directly from high to low pressure areas and there would be no opportunity for low and high pressure systems to build up. This is precisely the situation in the tropics, where pressure differences are rapidly smoothed out and relatively calm weather is common for most of the year. Hurricanes or typhoons seldom form within five degrees of the equator.

Away from the equator, the Coriolis Effect causes winds to veer and blow at right angles to the pressure gradient instead of parallel with it; the winds thus blow obliquely across the isobars to produce the typical circular wind systems of depressions and anticyclones which are illustrated in Figure 4.ii. Above heights of 600 metres from the earth's surface the winds do blow parallel with the isobars and are then called the geostrophic wind, but nearer the ground surface friction reduces the velocity of the wind and so reduces the Coriolis Effect, which is proportional to the velocity. The reduction of the Effect prevents air near the ground from flowing parallel with isobars and causes it to cross them.

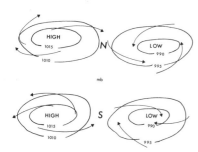

Figure 4.ii. Winds around northern and southern hemisphere pressure systems. Pressures are in millibars.

5 Moisture in the Atmosphere

The Hydrologic Cycle

The moisture in the atmosphere is largely derived from evaporation over the oceans. When air is more or less stationary over the ocean it takes up moisture, and maritime air masses are therefore typically humid, while continental air masses are relatively dry. Evaporation from the land, rivers, lakes and reservoirs, and transpiration from plants do provide some moisture in continental air but the continents obtain nearly all of their precipitation from incursions of maritime air over the land. About 30% of continental precipitation is returned to the oceans by runoff and 70% is returned to the atmosphere by evapotranspiration. Wind can then move much of this moisture back over the sea. The balancing movement of water circulation is called the hydrologic cycle.

Humidity and Condensation

The amount of water in the air is called the absolute humidity; it is defined as the mass of water vapour contained in a unit volume of air. Because air expands and contracts with rises and falls of pressure, absolute humidity is rarely used by climatologists and relative humidity is of greater importance. Relative humidity may be defined as the ratio between the amount of water vapour actually in the air to the amount of vapour the same volume of air could hold, at the same temperature and pressure. The ratio is usually expressed as a percentage, e.g. 75%.

With a rise of temperature the capacity of air to hold moisture increases, but if no more water vapour is available the relative humidity falls. Conversely, with a fall in temperature but no change in the amount of water vapour, relative humidity rises. The average diurnal maximum relative humidity occurs in the early morning when temperatures are lowest, and the minimum relative humidity in the early afternoon. The highest relative humidities occur around the equator, with rather low values to the north and south over the desert areas and high values again in the mid-latitudes of lower temperatures.

If air is cooled so that it is saturated (that is, has a relative humidity of 100%), the water vapour will condense, the temperature at which the change takes place being called the dew point. Depending on the temperature, it may produce dew, white frost, clouds or fog, although these forms of condensation are usually only found near the surface of the ground. By far the most important type of cooling is the adiabatic cooling resulting from a drop in atmospheric pressure, which may be caused by: (1) decrease in pressure at the surface; (2) rising air in a convection current; (3) air rising over an obstacle; (4) rising air caused by convergence of air masses.

Condensation Nuclei

If air is filtered so that it contains no dust or solid particles, and is then cooled below its dew point, condensation of water drops will not occur until the air is supersaturated and has a relative humidity of several hundred per cent. In the atmosphere, however, supersaturation is rarely as great as 1%, so it may be concluded that the dust removed by filtering is of considerable significance in the formation of water droplets.

Experiments in the nineteenth century demonstrated that fogs are made thicker by the introduction of smoke particles, but in spite of much research since then the exact function of such condensation nuclei as dust, smoke particles and salt from sea spray has not been elucidated. Salt is probably a very important nucleus for water droplets, especially near coasts and over the sea, but it is of importance also at considerable distances from its source. Salt particles are hygroscopic, that is, they have an affinity for water. On the other hand carbon, a common component of smoke, is not hygroscopic yet can still act as a nucleus of condensation. Atmospheric pollution by smoke from industrial plant, domestic fires and car engines is no doubt an important cause of the dense fog called smog in Los Angeles and other large cities. Pollution only intensifies fog, for there are always enough natural nuclei about to cause a fog.

Water in ponds and lakes will usually freeze at a temperature a few degrees below 0°C (there is no fixed freezing point of water, only a fixed melting point of ice), but cloud drops do not contain the freezing nuclei of lake and pond water and may remain liquid down to −40°C, at which point all water drops freeze. A few drops will freeze spontaneously at temperatures above −40°C. Condensation is thus greatly speeded by the presence of nuclei.

Rain, Snow and Hail

Cloud drops are exceedingly small and it is one of the problems of cloud physics to ascertain how cloud drops grow to form raindrops. It can be calculated from the rate at which a cloud drop grows by condensation from the air around it, that this is too slow a process to account for raindrops, which may be a million times larger in size. Fine drizzle, however, may be formed this way in a thick cloud.

In large tropical clouds with strong rising convection currents, temperatures are often above 0°C. Observations suggest that most of the cloud drops are very small but some may be 'large' drops as much as several hundredths of a millimetre in diameter—they are probably formed from large hygroscopic nuclei. Large drops in these conditions could grow into raindrops by collision with other drops as they fall after being whirled to the top of the cloud. Some raindrops probably grow so big that they become unstable and break up to form new large nuclei (Figure 5.i).

The collision process probably accounts for much tropical rain, although to produce rain in the tropics the clouds must be thicker than about 2000 metres, and rain is not certain until the cloud is 4000 metres thick. Outside the tropics, rain can fall from thinner cloud than this and the process of raindrop formation is different. The modern theory of rainfall formation was postulated by the Swedish meteorologist Bergeron in 1928. In a cloud composed largely of water drops and few ice crystals, air is supersaturated round the ice but only saturated round the water drops. As a result, the ice gains the condensing water vapour which forms snow crystals; these may break up to form more freezing nuclei. Some of the water drops may even evaporate again but others will freeze on to the ice nuclei. On a cold day with general temperatures below 0°C, snow will fall, but in warmer weather the snowflakes will

Figure 5.i. A falling raindrop will capture cloud droplets on its forward side (A), and will be overtaken by other raindrops coming into its wake (B). *After Petterson*

turn into rain. The Bergeron effect is illustrated in Figure 5.ii.

Hail may occur in any season but it is nearly always associated with large thunderstorm-type clouds. It is thought that in the strong convection currents in these clouds, collisions between ice particles and water drops occur frequently, so that the ice turns into a spherical aggregate rather than a snowflake. Such pellets are soft hail or graupel. Most hail is 1–2 cm in diameter but large, hard hailstones may be the size of a golf ball. Large hailstones are usually composed of alternating layers of opaque and clear ice. It is thought that such hailstones must be repeatedly whirled to the top and then fall to the bottom of the cloud, so that in the cold top of the cloud they gain the opaque skin, and in the warmer bottom they gain the clear skin from liquid accretion. A 13·8 cm diameter stone has been reported from USA with a weight of 680 grammes, and one weighing 3·4 kg has been reported from Hyderabad, India.

Figure 5.ii. Bergeron's precipitation model. When the warm air glides up over the cold wedge, it cools adiabatically and cloud droplets form. Precipitation begins to form in the layer where ice crystals and water droplets are present. The larger elements fall through the cloud and grow further by collision with other drops and droplets.

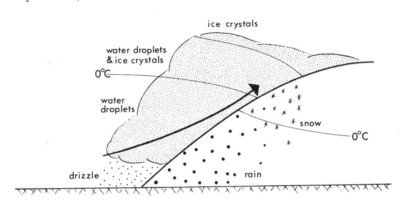

Man-made Rain

One of the great hopes of man is that he may be able to control the weather and so reduce the dangers of drought and flood. In spite of many extravagant claims by commercial operators and others, man has not yet been able to alter the rainfall of any large area significantly. After a comprehensive study of attempts at creating rain in the United States, the Advisory Committee on Weather Control published in its *Final Report* in 1957 the following statements:

(1) The statistical procedures employed indicated that the seeding of winter-type storm clouds in mountainous areas in western USA produced an average increase in precipitation of 10–15% from seeded storms, with heavy odds that this was not the result of natural variations in the amount of rainfall.

(2) In non-mountainous areas, the same statistical procedures did not detect any increase in precipitation that could be attributed to cloud seeding. This does not mean that effects may not have been produced. The greater variability of rainfall patterns in non-mountainous areas made the techniques less sensitive for picking up small changes that may have occurred there than when applied to mountainous regions.

In Australia some success has been obtained on the east coast where suitable cumulus cloud is common, but little success has been obtained in the arid interior.

Artificial stimulation of rain is usually attempted by providing freezing nuclei for condensation in the form of solid carbon dioxide dropped from aircraft or by seeding the clouds with silver iodide 'smoke'. Silver iodide crystals have a form like that of ice and can be placed in clouds either expensively from aircraft or cheaply from ground generators which burn coke impregnated with the silver iodide. The ground method is most commonly used but it often fails because the smoke is not carried into the cloud, or is trapped by an inversion layer, or loses its nucleating power on long exposure to sunlight.

The positive results claimed for mountainous regions in the 1957 report are probably accounted for by high concentrations of silver iodide being carried into the clouds by up-currents of air. It is evident, then, that at best artificial precipitation is unreliable, unless suitable cloud is available, while the chances of breaking a drought are almost nil.

1. A well developed cumulus cloud before (left) and about 30 minutes after being seeded with silver iodide from an aircraft (right).
Photo by E. G. Bowen, CSIRO

Types of Precipitation

The three causes of precipitation from rising air are: (1) thermally induced convection; (2) convergence of air masses or winds to give convergent, frontal and cyclonic rain; (3) orographic lifting. These types of precipitation, however, are not mutually exclusive and most precipitation results from two or more types of lifting in combination.

Convectional precipitation is the result of air being warmed near the ground, expanding and rising. During the rise, if condensation does not take place, no clouds will form and the air will stop rising when its temperature is the same as that of its environment, but if condensation does occur, latent heat of condensation is released, further warming the air so that it will continue to rise and deep clouds of the towering cumulonimbus type will form. Cumulonimbus clouds usually cover only a small area but give intensive rain. As these clouds usually drift fairly quickly, the ground receives the rain as short, sharp showers.

Convectional rain is usually associated with the hot seasons and especially with the hottest parts of each day, so that it tends to be an afternoon phenomenon. In the middle latitudes, especially in the continental interiors, most of the summer precipitation is of the convectional type. Much air that is originally stable may be set rising by passing over a range of hills or by meeting a front, so that the air then continues to rise and produces heavy, and often showery rain. Convectional precipitation is therefore often associated with orographic barriers, convergence and fronts, and is not always merely the product of surface heating.

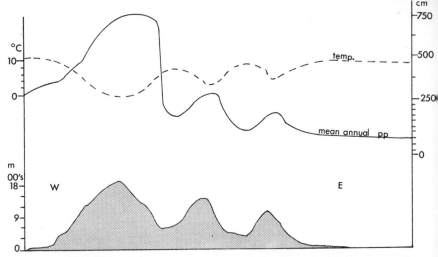

Figure 5.iii. Section from west to east across the South Island of New Zealand showing relief, mean annual precipitation and mean annual temperature.
After Taylor and Pohlen

Orographic precipitation is the result of moist air being forced to rise over landform barriers, cooling as it rises and causing condensation and rain if the uplift results in instability. The windward side of the barrier gets most of the precipitation with some carry-over beyond the summit, but in the lee of the barrier is a rain shadow area. The relationship between precipitation and relief is shown in Figure 5.iii in a cross-section across the Southern Alps of New Zealand. A number of places would show a similar type of pattern.

The world's high rainfall extremes come from areas where topographic barriers lie across the paths of moist air. Cherrapunji, on the southern slopes of the Assam Hills in India, has an average rainfall of 1080 cm, and a record, in 1861, of 2300 cm—930 cm of which fell in July alone. Mount Waialeale, on Kauai Island in the Hawaiian group, has an annual average of 1210 cm. These enormous totals result not only from the orographic barriers, but from the hindering of the passage of fronts, from differential heating on slopes causing convection, and from cooling in contact with cold ground or snow, which reinforce the orographic effect.

Figure 5.iv. Frontal precipitation.

Frontal precipitation results when two air masses of different characteristics meet. The boundary between them is the front. In the tropics the air masses may have similar temperature characteristics, and the lifting at the boundary is more or less vertical and of the convectional type. In mid-latitudes, however, the converging masses have different temperature and moisture characteristics and the warm air rises over the cold. The process is illustrated in Figure 5.iv.

Thunderstorms

Under conditions in which moist, unstable air is made to rise rapidly, a large, dark grey, very thick cloud, called a cumulonimbus cloud can form and produce a thunderstorm. This type of cloud may be the result of intense heating over land or sea, and resulting powerful convection currents, but frequently an additional trigger action is required, such as a front, convergence, or orographic uplift.

The most detailed description available of thunderstorms is that of H. R. Byers, who took part in an intensive study starting in 1945 in which many flights were made into the heart of large storms. A thunder-storm is a short-lived phenomenon seldom lasting for much more than an hour. Three stages of development are distinguished: (1) development, (2) maturity, and (3) decay; they are illustrated in Figure 5.v. In the development stage a cell of warm, unstable air rises, quickly drawing in air at its base so that the wind at the earth's surface blows towards the storm. The updraft of air reaches speeds of over 30 k.p.h. and continues to the top of the cloud at a height of as much as 9000 metres. The base of the cloud may be anything from 1·5 to 10 km across. The development stage only lasts 10–15 minutes but in that time much condensation takes place in the cloud, where raindrops, snowflakes and ice are whirled upwards.

In the development stage, no precipitation occurs but as the drops increase in size they attain terminal velocities which exceed the convectional updraft and fall through the cloud, producing a downdraft and giving rain. Wind speeds at the top of the cloud may reach 110 k.p.h., and the first shower marks the end of the development stage.

Figure 5.v. Stages in the development of a thunderstorm. Heaviest rain at the ground surface occurs in the mature stage (2), and declines as the thunderstorm dissipates (3).
After Byers and Braham

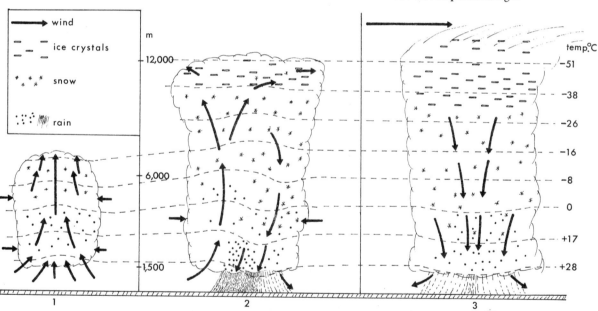

In the mature stage, the updrafts may continue to build the top of the cloud to 12,000 metres or more but the downdrafts gradually take over until nearly all movement in the cloud is downwards. In this stage rain is heavy and may be mixed with hail, and the temperature near the ground is reduced. The descending air meets the ground and spreads out away from the storm, thus reversing the earlier wind direction.

After 15–30 minutes in the mature stage, the thunderstorm begins to decay. The downdrafts become less powerful, the rain becomes light and the top of the cloud begins to dissipate. The descending rain cools the air and the air of the cloud cell takes on the temperature of the environment. This decaying stage generally lasts for about 30 minutes.

2. A big thundercloud— cumulonimbus—with hail falling from the base on the right.
Photo by R. K. Pilsbury

This description of a storm is, of course, idealised and many storms consist of several cells, each of which goes through its own life cycle, so prolonging the storm and making the pattern of wind and rain very complex. In a small storm at least 30,000 metric tonnes of water will be condensed, releasing a great deal of latent heat of condensation. C. E. P. Brooks estimated that about 16 million thunderstorms occur on earth every year and therefore nearly 2000 at any time. Suggestions that man-made explosions can cause changes in climate seem, therefore, to be greatly exaggerated.

The most spectacular feature of a thunderstorm is lightning. A flash of lightning is an immense electrical spark which jumps through a narrow channel in the air. During the discharge, which usually lasts for little more than 1/100 of a second, the air in the channel is heated to about 15,000°C, or about $2\frac{1}{2}$ times the surface temperature of the sun. The expansion of the heated air is so rapid that it is explosive. The rumble of thunder is the result of numerous lightning flashes and echoes. The great heat of the air in the channel of the flash indicates why lightning is so dangerous and why it burns the objects it strikes. Trees are split open by it because the current passes through the tree and turns the sap into a steam which blasts the tree apart.

Lightning is caused because the electrical charge in the base of the cloud is negative while that in the top is positive (Figure 5.vi). Lightning

Figure 5.vi. (A) the distribution of the electrical charges in an idealised thunderstorm. (B) the possible distribution of charges in a real thunderstorm. (C) the observer at O will hear the noise of thunder created by the lightning discharge at Y long before he hears that from X even though the two discharges occurred at the same time, hence the thunder sounds like a long rumble.
After Dobson

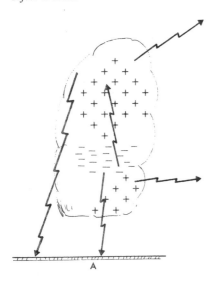

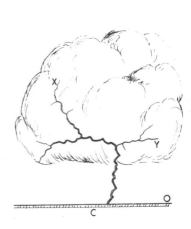

is the spark which jumps the gap between them. The problem is to explain the separation of the charges. There is no simple answer, nor are the theories which follow fully accepted. It is probable that in the convection cell the heaviest rain and hail drops with their negative charge collect near the base of the cloud, and the smaller, positively charged ice crystals collect at the top.

There are two main processes at work which result in this separation. (1) When a water drop is formed by condensation from vapour, it has an attraction for negative ions and becomes negatively charged, leaving the surrounding air with an excess of positive ions (an ion is a molecule which has become electrically charged by losing or gaining an electron); falling drops with negative charges accumulate at the bottom of the cloud. (2) If ice particles rub against each other, they become negatively charged by friction and as they fall to the base of the cloud they leave behind an excess of positive charge in the form of ions, small drops and tiny ice particles. By these two processes, positively charged drops will accumulate at the top and negatively charged drops at the base of the cloud.

Thunderstorms can seldom be heard more than 10 km from the source but as each lightning flash sends out radio waves—which produce the static noises heard on a radio receiver—they can be detected at a range of some thousands of kilometres. The most thundery places on the earth are the areas of great convectional instability—Java, central and southern Africa, Mexico, Panama and Brazil. The cooler and drier the climate and the fewer the triggers, the less likely is thunder. For this reason it is hardly ever heard polewards of the Arctic or Antarctic Circles.

Clouds

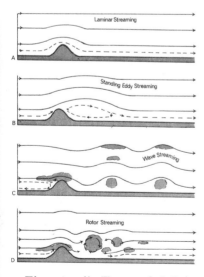

Figure 5.vii. Types of airflow and resulting cloud over a long ridge. *After Förchtgott*

A cloud is an aggregate of particles, chiefly of water in either a liquid or a frozen state, which is formed by the cooling of moist air to a temperature below its dew point. The cooling may be the result of lifting and consequent expansion of the air, or it may be caused by cooling of air near the ground by radiational loss of heat.

Clouds may be classified either according to the physical processes by which they are produced or according to their form. The modern tendency is to classify by process but the traditional, and still more widespread, method is to classify by form. Three main groups of cloud producing processes are recognised:

(1) Orographic clouds are the result of air being forced to cross hills and mountains.
(2) Layer clouds are formed by the mixing of air masses of different characteristics.
(3) Convection or cumuliform clouds are the result of strong upward air currents.

The formation of orographic clouds is illustrated in Figure 5.vii. When the wind-speed is low, the air moves as a series of layers, and on crossing an obstacle the airstreams are merely bent upwards (A). As the wind-speed increases, the laminar streaming is interrupted in the lowest layer by a standing eddy in the lee of the hill (B). With still stronger winds, the lee eddy is replaced by a series of lee waves (C), and in even higher wind-speed conditions or very high mountains, the wave

3. Lee-wave cloud over
Sicily.
Photo by R. K. Pilsbury

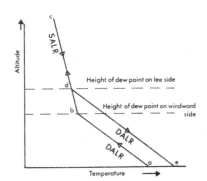

Figure 5.viii. The föhn effect.
The precipitation above the
condensation level results in
loss of moisture and a higher
dew point on the leeward
side. Moist air cools as
indicated graphically by the
line abc, but on descending
it warms as shown by cde so
that on the leeward side
temperature is higher by ae
than on the windward side.

streaming breaks down into rotor streaming (D). The size and form of
the clouds will depend upon the local conditions of air temperature and
humidity, but the diagram indicates their predominant lenticular form,
or lens shape. They can be produced at heights of 30 km by large
orographic barriers.

Air which is forced to rise over a range of hills will cool at the dry
adiabatic lapse rate until the level of the dew point is reached and
condensation occurs. As the air continues to rise, the cooling will be at
the saturated adiabatic lapse rate. If rain falls, the air will pass over the
mountain and descend on the leeward side containing less moisture than
before it started its rise on the windward side. The air will warm up
during its descent at the saturated adiabatic lapse rate until the level of
the dew point is reached again, but because there is less moisture
present the dew point will be at a lower temperature and hence a higher
altitude on the leeward than on the windward side (Figure 5.viii). High
ranges of mountains may produce a warming of some 8°C on the leeward
side. These warm, dry and often dusty mountain winds are given local
names, but because the effect was first named in the European Alps as
the föhn (or foehn), the phrase 'föhn effect' is in common use. In
Canada a similar wind in the Rockies is called the Chinook. In the New
Zealand Alps it is called the Nor'wester.

Layer clouds depend largely on horizontal movements in the air and
are most commonly produced by the mixing of air of different temperatures and humidity. Advection fog, described below, is a type of layer
cloud, but the most usual cause is the mixing along fronts in depressions.
Such cloud systems may cover thousands of square kilometres and
contain an enormous weight of water. Their occurrence is more fully
described in Chapter 8 in the section on depressions.

Convection clouds are the result of strong heating and consequent
rising of air, in some cases triggered by the passage of the air over
rough land surfaces. The rising air occurs in discrete cells and the
clouds are therefore separated by clear sky in which the air may be
descending. Such clouds are cumuliform, that is, thick and dome-
shaped and constantly changing.

The traditional classification of clouds is based on their form and
altitude. Ten main types are recognised, and these are listed in Table 5.1
with their characteristics. (See also Figure 5.ix.) Any cloud may be
further described by the addition of another classifying word of which
the following three are the most important:

4. **Large lee-wave cloud over Central Otago in the lee of the Southern Alps, New Zealand.**
Whites Aviation photograph

Fog

Castellatus is a turreted structure, indicative of upper atmosphere instability and often associated with thunderstorms.
Fracto is a prefix used to distinguish broken or fragmented clouds.
Lenticularis indicates lens-shaped cloud.

To the meteorologist fog is ground level cloud in which visibility is less than 1000 metres. Over hills the fog may just be orographic cloud but fog is usually distinguishable from cloud because the cooling processes in its formation do not involve ascent and consequent adiabatic cooling of the rising air. The condensation of water vapour which gives rise to fog is produced almost entirely by the direct effect of cooling from a cold surface. Two main types of fog are recognised: (1) fogs resulting from evaporation; and (2) fogs resulting from cooling.

5. **Hooked cirrus (or mares' tails).**
Photo by R. K. Pilsbury

Table 5.1. Cloud Characteristics

Name of Cloud	Usual abbrevi-ation	Range of height of cloud base in middle latitudes (metres)	Vertical thickness of the cloud	Description
Cirrus	Ci	6000–12,000	A few hundred metres	High, feathery clouds with a delicate wispy form. Entirely composed of ice crystals. They are the first signs of an approaching front. Never give precipitation.
Cirrocumulus	Cc	6000–12,000	Thin	Small high patches of cloud, often grouped in lines or ripples. No precipitation.
Cirrostratus	Cs	6000–12,000	A few hundred metres	A thin white sheet which covers the sky and gives it a milky appearance. When seen through it the sun and moon have haloes. Often indicates an approaching front. No precipitation.
Altocumulus	Ac	2500–6000	A few hundred metres	Layers or patches of cloud flakes or flat globular masses. The Ac castellatus are often associated with thunder. No precipitation.
Altostratus	As	2500–6000	Thick—up to 3500 metres	A uniform grey sheet of cloud. It may give precipitation but this does not necessarily reach the ground, being evaporated in its fall.
Nimbostratus	Ns	100–600	Thick—may be up to 4500 metres	Dark grey, shapeless cloud giving rain.
Stratus	St	150–600	Thin—30–300 metres	A uniform fog-like cloud, but above ground level unless capping high ground.
Stratocumulus	Sc	300–1400	Thin—150–900 metres	Large globular masses, fairly flat, often with a regular pattern. No precipitation.
Cumulus	Cu	300–1500	May be thick—1500–4500 metres	Dome-shaped with cauliflower tops and horizontal base. Indicative of stable conditions and strong vertical currents.
Cumulonimbus	Cb	600–1500	Very thick—3000–9000 metres	Heavy masses of dark grey clouds. The upper parts may become fibrous and spread out in an anvil. Heavy showers and often thunder and hail.

(1) Fogs resulting from evaporation are steam fog and frontal fog. Steam fog occurs when intense evaporation of water takes place from relatively warm water into colder air. It is most common in high latitudes and in the Arctic it builds to heights of 1200–1500 metres. Sailors often call it 'sea smoke'. It is most common over water but may form over warm land after rain, especially over warm roads.

Frontal fog is the result of warm rain evaporating as it falls through drier, colder air beneath it so that saturation and condensation occur. If the fog forms above ground level, it is called stratus cloud.

(2) Fogs resulting from cooling include radiation, advection, upslope, mixing and barometric fog. Radiation or ground fog is produced when still moist air is in contact with ground which has been cooled by

	10,500 m
Ci	
	9,000
Cc	
	7,500
Cs	
	6,000
	Cb
Ac	4,500
As	
	3,000
Sc	
	1,500
Ns	
Cu	
St	

Figure 5.ix. Main types of cloud.

night-time radiation. A cloudless sky is usually necessary for radiation fog, and descending air in anticyclonic conditions encourages its development. Low-lying valley bottoms and marshy areas are particularly susceptible, and in industrial areas where smoke produces suitable hygroscopic nuclei such fogs can be very thick. The former London 'pea-souper' was a good example.

6. Cumulus clouds characteristic of fair weather. Note the flat bases of the clouds.
Photo by R. K. Pilsbury

Advection fog is produced when moist air is transported over a cold surface. It is especially common at sea when warm, moist air passes over the surface of cold currents. Warm air in a southwesterly airstream blowing over the cold water of the Labrador current produces the infamous fogs of the Grand Banks of Newfoundland. Cold upwelling currents along the western margins of continents produce the summer fogs of Southwest Africa, California and Chile.

Upslope fog is the result of a very gradual rise of moist air up a slope without lifting off the ground to form cloud. Cooling in this case is usually adiabatic.

7. Stratus cloud enveloping the top of Benbulbin, Ireland. Such clouds are shapeless.
Photo by R. K. Pilsbury

Mixing fog occurs when cool moist air comes in contact with warm moist air so that the temperature of the mixture is low enough to produce saturation and condensation. It is usually formed along a warm front which is the boundary between two suitable types of air.

Barometric fog is extremely rare, as it requires a layer of moist air to experience a lowering of barometric pressure and hence adiabatic cooling which can lead to condensation. It is most common in a basin or valley of stagnant air.

All fog develops best in slowly moving air, for strong wind breaks it up and turbulence may lift it above ground level, mixing the air.

6 Air Masses, Convergences and Fronts

Air which is more or less stationary over any homogeneous surface gradually acquires characteristics related to the nature of that surface. Thus air becomes cold or warm, and moist or dry. Air masses can therefore be classified according to the characteristics of their source region. The simplest classification divides air into continental polar (cP); maritime polar (mP); continental tropical (cT); and maritime tropical (mT) types. These four types are significant because a source region must be a large, even surface with calm conditions; an uneven surface or turbulent air would prevent the development of deep, homogeneous air layers. The subtropical and polar high pressure zones are the most important source regions.

Air masses usually move away from their source regions and become modified as they do so. Dry continental air (c) will absorb moisture as soon as it moves over the sea and maritime air (m) will become drier as it moves over land, by precipitating its moisture.

The two principal types of air mass modification are thermodynamic and mechanical alteration. Thermodynamic changes occur when air moves over a surface which is warmer than its own lower layers. These will be warmed, the lapse rate will be increased, the heated air will

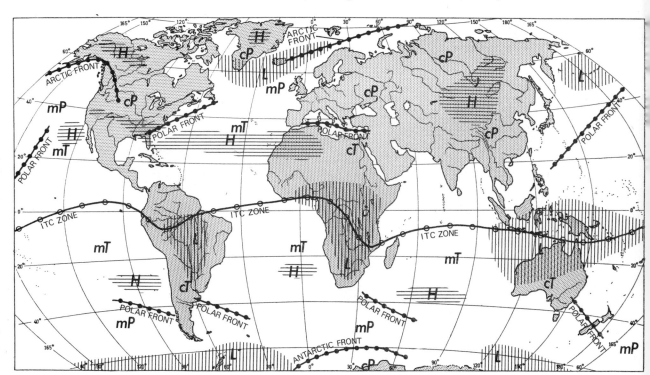

Figure 6.i. January. Main pressure centres, fronts and air masses.

become unstable, and cloud and precipitation may result. Conversely, air passing over a cooler surface will become chilled in its lower layers, a surface temperature inversion will result, and the air will become very stable. These modifications are noted by the letters W (warm) and K (kalt or cold). Usually polar air passes over warmer surfaces and is therefore K, whilst tropical air is W, or warmer than the surface beneath it.

Other factors of importance in producing the characteristics of an air mass are the mechanical effects of convergence or divergence of air. Diverging air, as in an anticyclone, is descending and increasingly stable air; converging air, as in a low pressure system, causes instability, clouds and maybe precipitation. To denote these characteristics a fourth letter is added to the classification: s = stable air aloft; u = unstable air aloft.

Using this system, therefore, it is possible to categorise 16 types of air mass:

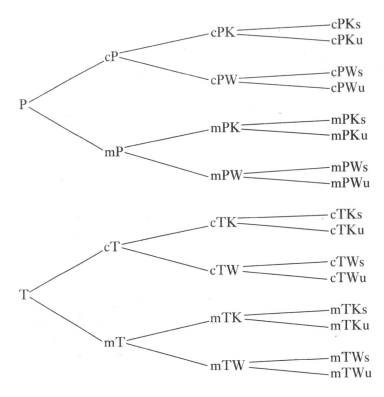

It is also possible to recognise equatorial (E) and Arctic and Antarctic air (A), but these may be more conveniently regarded as modified tropical and polar types.

The World Pattern of Air Masses

Figures 6.i and 6.ii attempt to show the mean positions of the air masses and the fronts and convergences which represent the lines along which contrasting types of air come into contact. There are three major zones of contact: the intertropical convergence, the polar front and the Antarctic and Arctic fronts.

Figure 6.ii. July. Main pressure centres, fronts and air masses.

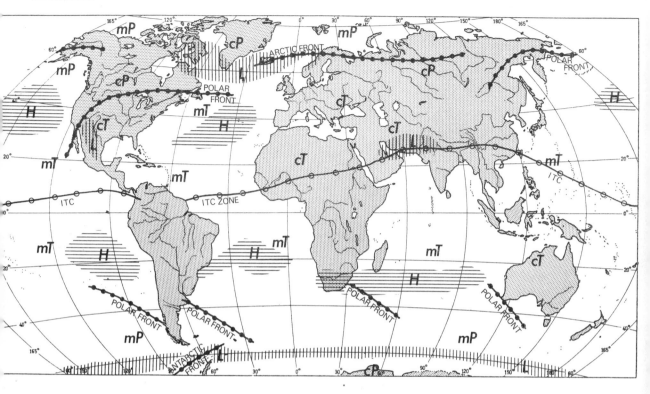

The intertropical convergence (ITC) is formed in the zones where the trade winds from the northern and southern hemispheres meet. The tropical air masses along this convergence have very similar characteristics, so density differences seldom exist. There is still a great deal of doubt about the nature of the ITC. Some climatologists think there are two convergences, a northern and a southern (NITC and SITC); others say it is a single phenomenon; and others regard it as a broken zone in which evaporation of the rain falling along the convergence may cause it to become a source of cool descending air, so that more ITC's are formed north and south of it.

The polar front is formed between the polar and tropical air masses and is the zone of greatest air mass temperature contrasts, and hence of the strongest frontal cyclogenesis—as will be discussed later. The Arctic-Antarctic fronts are formed between the cold polar air masses direct from the polar anticyclones and easterly wind zone, and the modified polar air on the polar edge of the zonal westerlies. Temperature contrasts along the Arctic-Antarctic fronts are therefore small and cyclogenesis is correspondingly weak. All of the frontal zones, together with the air mass source regions, move with the seasons, with the single exception of the Arctic front, which is drawn southward over Eurasia during the summer by the warmth of the heated land, and driven northward in winter by the circulation around the Asian High, while spring and autumn positions are intermediate between them. The main positions are indicated in Table 6.1.

Table 6.1. Air Mass Source Regions

	January	*July*
cP	*N. hemisphere:* In this winter season the anticyclonic circulation over all the large land masses—N. America, Greenland, Eurasia—produces enormous volumes of cP air. Polar front at about 30°N. *S. hemisphere:* Only the Antarctic continent is large and polar enough to produce cP air.	*N. hemisphere:* The polar front is now between 45° and 60°N and only the northern halves of N. America and Eurasia produce this type of air. *S. hemisphere:* Again only the Antarctic can act as a source for cP air.
mP	*N. hemisphere and S. hemisphere:* All ocean areas polewards of the polar front. Air gaining its characteristics in a zone of low pressure disturbances.	*N. hemisphere and S. hemisphere:* Source as for January except that in the N. hemisphere the source region is smaller and in the S. hemisphere it is larger.
cT	*N. hemisphere:* Formed only over northern Africa. *S. hemisphere:* Australia and a small area of Argentina.	*N. hemisphere:* southern N. America, Eurasia and N. Africa. *S. hemisphere:* Australia only.
mT	At all seasons and in both hemispheres this air is formed in the subtropical highs which dominate the oceans, being larger and stronger in summer than in winter.	

7 The Primary Circulation of the Atmosphere

It is convenient to classify the movements in the atmosphere according to their scale, ranging from world size to local situations:

(1) The primary or planetary circulation, usually called the general circulation, is the worldwide wind system which tends to equalise the heat in the atmosphere by transporting it between the tropics and the poles.

(2) The secondary circulations are the movements of cells of high and low pressure, anticyclones and depressions, which are largely responsible for day-to-day weather.

(3) The tertiary circulations are localised cells of the type represented by such phenomena as land and sea breezes, föhn winds, thunderstorms, tornadoes and hurricanes. Secondary and tertiary circulations are discussed in Chapters 8 and 9.

The Primary Circulation

On a non-rotating earth with a homogeneous surface the average temperature would vary only with latitude, so that there would be a gradual decrease of temperature from the equator to the poles. Warm air would expand and rise over the equator and move at height towards the poles as cold polar air moved along the surface towards the equator. Such a simple circulation is known as a Hadley cell, after the Englishman George Hadley who proposed in 1735 that this was the usual circulation in the atmosphere. He accounted for the known average wind directions by pointing out that because of the earth's rotation, air flowing towards the heat equator would be caused to lag behind the movement of the surface and would have an easterly component, while compensating air masses moving polewards must have a westerly component.

A simple Hadley cell does explain the main wind directions within the tropics, but outside the tropics differential heating of land and sea, surface frictional drag on the air, and turbulence from the earth's rotation complicate the pattern of the general circulation to such an extent that it is not possible to find evidence for the existence of a single cell.

On a rotating earth of uniform surface the average wind system would be similar to that shown in Figure 7.i with the following features:

(1) At the equator is a belt of uniform low pressure, characterised by calms and light variable winds with occasional squalls and thunderstorms, originally known as the doldrums, and now referred to as the intertropical convergence zone.

(2) Polewards of the equatorial doldrums is a zone of steady easterly winds blowing towards the doldrums from the subtropical high pressure zones.

(3) The subtropical high pressure zones occur round about latitudes 30° in the northern and southern hemispheres. In them, air is

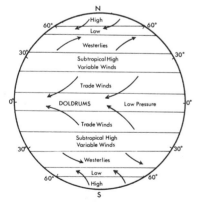

Figure 7.i. The basic pattern of surface wind and pressure fields over an earth of uniform surface.
After Sutton

descending, rainfall is low and winds are light. They used to be called the horse latitudes.

(4) A zone of westerly winds between latitudes 35° and 60° is characterised by variable winds and pressure systems with a low pressure zone.

(5) Polewards of the main zone of the westerlies, the subpolar low pressure area receives air moving away from the polar high pressure.

(6) The polar high pressure zone is one of descending air and mainly easterly winds.

The above scheme is grossly oversimplified and Figure 7.ii is an attempt to provide a more realistic model of the general circulation. It will be noted that at the surface the two diagrams are similar, with the eddy nature of the westerlies being emphasised. The circulation in the upper troposphere, however, is quite different from that within a simple Hadley cell. In the tropics, a simple Hadley cell does exist with very little north and south transportation of air and a large east and west movement. In the middle latitudes, turbulent movements with wind direction depending upon pressure systems are characteristic, and in the upper atmosphere jet streams are important features. Little is known about the apparently complex polar winds.

An outstanding feature of the model in Figure 7.ii which requires explanation is the net transport of air northwards in the westerlies, which is hardly what would be expected, for it implies that warm tropical air moves polewards and cold polar air rises. Such a system must use up energy in maintaining itself. Where, then, does this energy come from? Over a period of many years checking weather charts and building laboratory and mathematical models, Starr and his fellow workers concluded that the energy comes from the great rotating air masses of the westerlies. The main principle seems to be that high pressure systems are usually masses of cold air which move equatorwards from polar regions and that lows are usually masses of warm air which move polewards. The cold air masses moving into the warm air descend, and warm lows moving to cooler regions rise. In so doing the air masses convert potential energy into kinetic energy and provide the energy for the momentum of the westerlies. The tropical easterlies

Figure 7.ii. A highly schematic pattern of the general circulation.

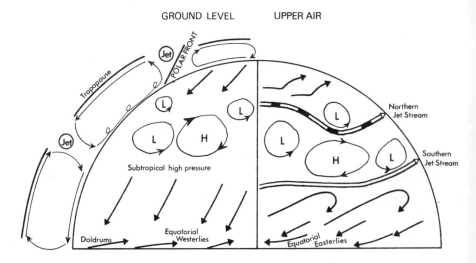

Figure 7.iii. Rossby waves. The sequence of development is known as an index cycle. North of the jet stream lies cold polar air and south of it warm tropical air. The deepening waves carry cold air equatorwards and tropical air polewards until the waves break up and leave cells of cold air and warm air separated from their source. *After Namias*

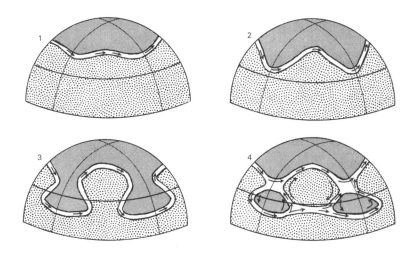

might be regarded as the result of lows carrying westerly momentum into the mid-latitudes and leaving behind a deficit in the tropics—an easterly flow.

The most important recent advances in understanding the general circulation have been the result of considerable progress in the study of the upper atmosphere. The simplest picture of the circulation is obtained from the upper part of the atmosphere at about the 500 mb surface. At this level the circulation is almost entirely westerly, but imposed on the westerlies are a number of wave-shaped disturbances, varying in number from 3 to 7, which move slowly eastwards while constantly changing shape. These waves, called Long or Rossby waves, carry cold polar air equatorwards while between them warm air moves polewards.

The amplitude of Rossby waves varies, but is of the order of several thousand kilometres. The form of the waves varies from a gentle curve to a deep finger-like form. The gently curved version is characterised by a well developed broad westerly flow of air in mid-latitudes which is described as zonal circulation with 'high' index. The extended long waves produce an air flow which is mainly north and south with compression of the westerlies into a jet stream of high velocity winds along the polar front. This type of flow is said to be meridional with 'low' index. The index cycle of development from the high to low index is illustrated in Figure 7.iii and typical zonal and meridional circulations in Figures 7.iva and 7.ivb. A typical cycle seems to end as the meandering waves finally break up into closed cyclonic and anticyclonic cells at the same time as a new zonal circulation begins to develop near the poles (see also Figure 8.x).

There is a tendency for stationary waves to develop in the lee of large mountain systems like the Rockies and Andes which run across the direction of the general circulation. As is shown in Figure 7.v, an air column moving over a range contracts vertically, to expand vertically and horizontally on the leeward side, at the same time undergoing an increase in absolute vorticity or rotation. As a result, a permanent wave occurs in the lee of the range and this affects the general circulation; hence there is a tendency for cold air to be drawn southwards into central North America while a ridge occurs over the Atlantic and

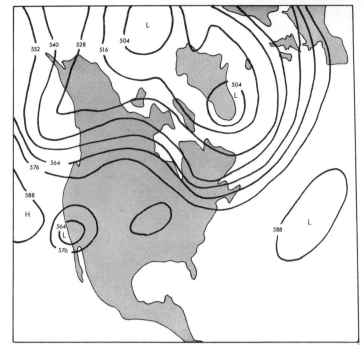

Figure 7.iva. Flow at 500 mb on 17 October, 1963 showing contours of the 500 mb surface in 10 metre units. The waves are widely spaced and have little north to south extent.
After Riehl

another trough over eastern Europe. The moving of large bodies of air north and south thus has the effect of increasing the continentality of the North American and European climates, and carrying warming influences northwards in the Atlantic.

Equatorwards of the irregular westerly circulation of mid-latitudes, upper air movements in the trade wind belt are more stable and symmetrical, with southwesterly winds predominating at high levels, although they do have occasional easterly airstreams embedded in them. At the surface, northeasterly winds flow towards the heat equator or intertropical convergence zone.

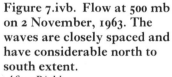

Figure 7.ivb. Flow at 500 mb on 2 November, 1963. The waves are closely spaced and have considerable north to south extent.
After Riehl

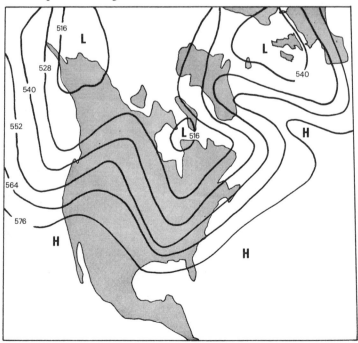

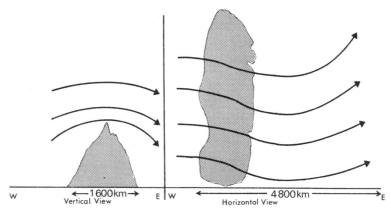

Figure 7.v. Large scale air-flow over a mountain range showing changes in the vertical and horizontal extent of the air columns crossing the range.

The Jet Stream

8. A jet stream area is indicated by the shadow of high cirrostratus cloud upon the cloud layer beneath. The jet stream is located from Alabama to Lake Erie, USA. The photo was taken by satellite TIROS V, orbit 1534, 4 October, 1962. *NASA*

The first knowledge of the jet stream was obtained in the early 1940's by the Japanese meteorologists, who discovered that a constant high wind blew from their islands across the Pacific to North America. The speed of the wind was as high as 500 k.p.h. Now that more is known about the jet streams, aircraft are able to avoid them when flying from east to west and to use them to increase speeds in the opposite direction.

The World Meteorological Organization recommends the following definition:

> A jet stream is a more or less horizontal, flattened, tubular current, close to the tropopause, with its axis on a line of maximum wind-speed and characterised not only by high wind-speeds but also by strong transverse wind shears. Generally speaking, a jet stream is some thousands of kilometres in length, hundreds of kilometres in width, and some kilometres high; the minimum wind-speed is 30 metres per second at every point on its axis.

As indicated in the discussion of the index cycle, the jet stream above the polar front migrates with the seasons, just as the westerlies do. In winter it moves towards the equator and is centred around 30–35°, and in summer it is around 40–45°. It is strongest in winter, when it may attain velocities of over 500 k.p.h. It may break down into several streams separated by gaps—especially during the summer.

The origin of the jet stream has been suggested as being the result of convergence in the westerlies above fronts where tropical and polar air meet, and although this is no doubt largely true there are as yet many features of the streams which are not fully understood.

The Effects of Land, Sea and Seasons on Wind and Pressure Distributions

Figures 7.vi and 7.vii show the complication of the general circulation produced by the distribution of land and sea in July and January. During July the northern continental area is warmer than the surrounding oceans, so that over land pressure is low and over the ocean the subtropical high persists. The northeast trade winds are therefore interrupted at this season and replaced by winds blowing round the high and low pressure cells. In the equatorial region, air crosses the equator from the south and changes its direction from southeast to southwest as a result of the Coriolis Effect. In the southern hemisphere, the land is colder than the oceans but the area of land is so small that it hardly affects the subtropical high, and the southeast trades are persistent.

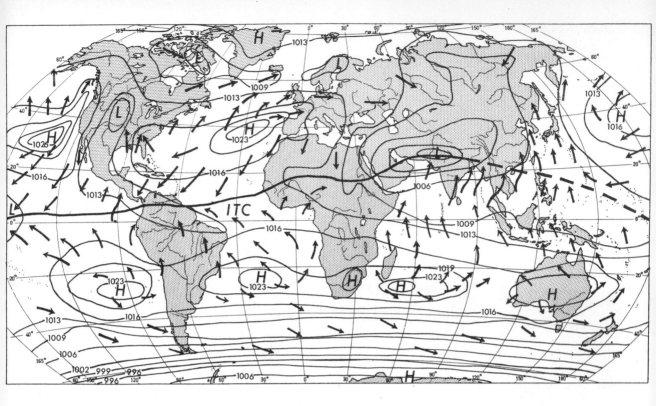

Figure 7.vi. Mean sea level pressures and winds in July.

During January the conditions are reversed, although modified by the asymmetrical shape of the continents. In the north the continents are very much colder than the oceans and a strong high develops. The centres of low pressure over the oceans are greatly strengthened; depressions are many and fast-moving, and displaced northwards. In the southern hemisphere, pressure over the continents is low and the doldrums extend southwards, although not nearly as prominently as the development of the northern hemisphere winter low. The interruption of the subtropical high is slight.

The actual mean sea level pressure for July is shown in Figure 7.vi. In July the northern continental lows extend from the tropics northwards over the continents, confining the subtropical high to the Pacific and Atlantic (called the Azores) cells. Over the north Atlantic the polar lows are less intense and pushed far to the north. In the southern hemisphere, on the other hand, the subtropical high is continuous and the polar almost continuous. Associated with this pressure distribution, the southeast trades of the southern Indian Ocean blow across the equator into the Indian low as the monsoon. The northeast trades, therefore, do not exist over the Indian Ocean.

In January Asia and North America are covered by the continental high pressure systems which are almost continuous with the subtropical highs. Over the oceans, the Pacific and Icelandic lows are large and vigorous and bring stormy weather to the mid-latitude oceans. In the southern hemisphere, the subtropical highs are interrupted over each of the continents by lows. The northeast trades are almost uninterrupted in the northern hemisphere, and the winter monsoon produces strong northwesterly winds blowing out from the Asiatic high to the Pacific low.

44

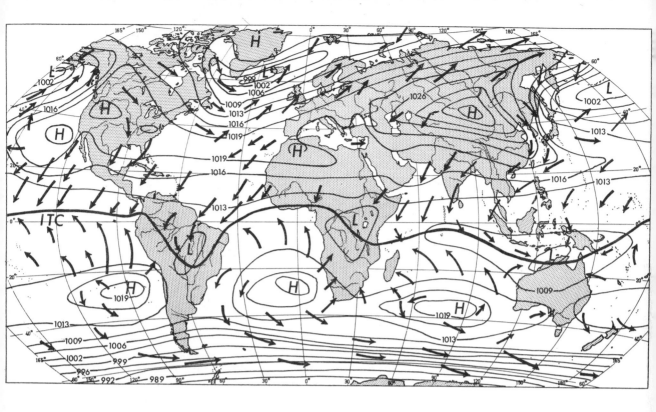

Figure 7.vii. Mean sea level
pressures and winds in
January.

8 Secondary Circulations of the Atmosphere—Monsoons and Pressure Systems

The causes of secondary circulations are of two main types—thermal and dynamic. The thermal effects are the result of heating and cooling of the earth's surface and for the most part are well understood. Dynamic effects, by contrast, are not so well understood and are far more complex.

The largest secondary circulation is that of the monsoons—traditionally, and in part misleadingly, described as a thermally induced effect which operates during hot and cool seasons as a result of the large temperature contrasts between land and sea areas in low latitudes. The monsoon effect is best developed over the enormous land mass of Asia, although a similar effect occurs over the other large continental masses of North America and Africa. It is most important on the eastern side of continents, where the influence of moist westerly airstreams is at a minimum.

The Monsoons

The word monsoon is thought to be derived from the Arabic word *mausin*, meaning 'season'. It originally referred to the seasonal wind conditions of the Arabian Sea, which blow from the southwest for the cooler six months of the year and from the northeast for the hotter six months. The first reasonable explanation of the monsoon was put forward by Halley in a paper read before the Royal Society in London during 1686. He suggested that during the summer the interior of Asia becomes so hot and the resulting low pressure centre so intense that air is drawn into it from the Indian Ocean and west Pacific, and that this influx of moist tropical air causes heavy rain over south and southeast Asia. In winter the reverse applies, as the vast interior of the continent cools and a high pressure centre with outblowing cold air develops. The winter or reverse monsoon, therefore, gives cold dry conditions to all land areas except those which, like northern Australia, Japan, Indonesia and southern India, lie on the far side of sea areas from which the winds could gain moisture.

This simple and attractive hypothesis is now thought to be inadequate, as a result of upper atmosphere observations, but the whole truth is still not known because long period records are not yet available. The modern ideas of the monsoon are largely derived from the work of the German Flohn, who described the significance of upper air phenomena, and the distinctiveness of the south and east Asian monsoons. The importance of Flohn's work is that he postulates that the monsoons are not fundamentally thermal systems, but complex modifications of the general planetary circulation. This modification is caused by the great extent of the eastern hemisphere landmass which, because of the greater heating and cooling effects of a landmass compared with an ocean, causes the surface equatorial trough to oscillate in position over

30° of latitude, whilst in the more oceanic western hemisphere the shift is only 5° of latitude. It is true, then, that heating and cooling of the landmass is important in producing the monsoons—but only indirectly—by the effects on the position of the equatorial trough. This is not the same as attributing the monsoon to an independent circulation, as Halley did.

There are important differences between the northeast Asian (China and Japan) and the southeast Asian (India-Pakistan and Indo-China) monsoons. The northeastern occurs in temperate latitudes in which the winter circulation is strongest, so that the outblowing or return monsoon is strongest. It is cut off from India by the high barrier of the Himalayas, although not from southeast Asia. The southeastern monsoons occur in tropical latitudes and are strongest and most constant during their advance in summer, when they blow with speeds of over 16 k.p.h. but during the winter speeds of 3 k.p.h. are more common.

The India-Pakistan Monsoon

During the period from March to May the northern part of the Indian subcontinent heats up and a hot drought period prevails. Centred approximately over the Thar Desert, the pressure average is low, not because a permanent low forms but because numerous cyclonic disturbances pass through the area giving the 'statistical' low. Far to the south of India, tropical maritime air is drawn northwards as the intertropical convergence and equatorial westerlies migrate northwards.

The dry period over India ends abruptly with the arrival of the mT air and the ITC (Figure 8.i). The monsoon rains begin over Burma in May but are delayed over India and spread swiftly northwards during June. The 'burst' of the monsoon is associated with turbulent conditions and heavy rain, but after the burst the turbulence decreases. The advance of the monsoon is not regular over India, probably because the jet stream which flows both south and north of the Himalayas repeatedly retreats beyond the mountains and then reappears south of them during the advance, thus alternatively drawing on and blocking the monsoon. Over southeast Asia, on the other hand, the jet stream does not occur and there are no blocks to retard the advance. In the upper atmosphere the airstream is always westerly during the summer and it is only near the surface that the air is drawn northwards.

With the onset of the winter cool season, the jet stream reappears south of the Himalayas and pushes back the monsoon. A deep anticyclone develops over northern India and the winds round it produce a northeasterly airstream at the surface which is reinforced as the easterly trade winds dominate the upper atmosphere. Around the southern edge of the Himalayas, the jet stream conducts a succession of depressions beneath it along the Ganges Valley, bringing some precipitation which may be extended across Burma into southern China. Over southeastern India and Ceylon, the easterly trades bring rain after crossing the Bay of Bengal and convective disturbances occur along the intertropical convergence. Elsewhere over India the winter is dry.

According to the Köppen classification, two-thirds of the Indian subcontinent is arid or semiarid. The wettest areas are the mountain regions of the Western Ghats and the northern mountain chains. The driest areas are those of central India, which are in the rain shadow of

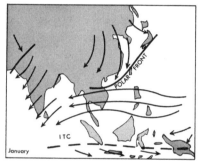
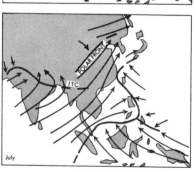

Figure 8.i. Low level circulation over eastern and southern Asia in the cold (January) and warm (July) seasons.
After Thompson, Watts, Flohn and Trewartha

the Ghats during the advancing monsoon and covered by continental air during the retreat, and the dry region centring on the Punjab and the Thar Desert. The latter is at the northern end of a tropical sea; yet it is enclosed on the north by mountains which might be expected to give rise to convection; it is the centre of low pressure systems in winter; and in summer it has the highest temperatures of the subcontinent—all factors which might be expected to produce convergence. All this is in sharp contrast to the similarly placed Bengal, which is north of the Bay on the east of the subcontinent. The explanation for the desert is to be found in its position in the belt of subtropical highs. The summer thermal low is very shallow and such mT air as it does draw in can carry relatively little moisture. Above this low, however, is a deep anticyclone containing stable dry continental air, some of which subsides into the surface air. The situation is illustrated in Figure 8.ii, which shows how the ITC forms a nose bringing aridity to the Indus lowlands. The fluctuation of the ITC can thus control the precipitation so that when the continental lobe remains over much of the Indus area, the monsoon may be said to have failed.

Figure 8.ii. Vertical structure of the air over the north-western part of the Indian subcontinent in summer.
After Sawyer

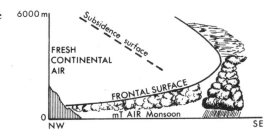

The Monsoon of Tropical Southeast Asia

In this region, useful climate data from a close network of recording stations is only available from Java, Malaya and the Philippines so generalisations about the area may well be inaccurate. A further cause of possible inaccuracy is the extremely varied relief, and the pattern of land and sea.

Watts has defined three airstreams which dominate the area: (1) the northeast is covered by the north Pacific trades; (2) the southeast by the south Pacific trades; (3) the west is covered by a wedge of equatorial westerlies. The boundaries between these three airstreams are zones of convergence. This system moves north and south with the sun to give seasonal weather changes. Most of the precipitation is therefore regarded by Watts as being attributable to the development of cumulonimbus along the lines of convergence, to the effects of relief which produce minor convergences in local wind systems among the islands and mountains, and to the daily effects of land and sea breezes. Little has been said about tropical storms because of the paucity of reliable data, but it is known that disturbances, probably guided by the jet stream, frequently cross southeast Asia in winter and bring rain, while typhoons and other storms move westwards with the north Pacific trades over the China Sea. The precipitation pattern resulting from movements of the airstream is one of permanently high rainfall throughout the area of northern Indonesia and the southern Philippines between about 7°N and 5°S, while to the north and south of this zone a dry season occurs in the period of low sun, and maximum precipitation in the hottest season.

In essence, then, the monsoons of tropical Asia are an unusually great northward displacement of the ITC, partly caused by intense thermally induced low pressures over northern India and Pakistan, and partly controlled in the case of India by the northward retreat of the jet stream beyond the Himalayas. The main westerly airstream, according to Flohn, is an expanded equatorial westerly system which usually lies embedded in the tropical easterlies. The rain-bringing storms are disturbances occurring along the ITC. The winter circulation is not the result of cooling but the re-establishment of the trade wind circulation as the ITC and the heat equator retreat southwards.

The Monsoon of Mid-latitude China and Japan

Most of eastern Asia is covered by a westerly airstream of cold continental (cP) air during winter, and by maritime air during summer. The traditional explanation of these monsoon winds is that in winter the Siberian High is the origin of the cold air, and that in summer a continental low forms in central Asia which draws in moist air from the Pacific. This explanation has now been refuted by a greater understanding of upper air circulation and by the realisation that much of the maritime air which reaches China in summer is from the Indian Ocean (Figure 8.i). Furthermore, the surface circulation of eastern Asia is far too weak to cause the summer monsoon rain.

During the warm season, northeastern Asia is covered by a westerly airstream in the upper atmosphere which is concentrated into a jet stream at around 45°N. This air is relatively dry and stable, but to the south of it is a warm, moist airstream (mT) flowing from the Indian Ocean and Bay of Bengal. Along the line of convergence heavy rain and cyclonic disturbances occur, often guided by the jet, to bring heavy rain to northern China and Manchuria (Figure 8.iii). The mT air from the Indian Ocean is a very deep layer which gives abundant precipitation over more southerly areas. The third important airstream is the easterly stream from the Pacific, and although this is often described as the moist monsoon airstream, it is actually too shallow to give much

Figure 8.iii. Characteristic features of airflow at about 3000 metres above southern Asia, from November to March. (A) the Tibet lee convergence zone; (B) the polar front; (C) the ITC. The Tibetan plateau is shown by heavy dots.
After Thompson

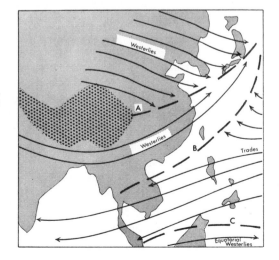

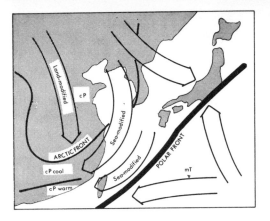

Figure 8.iv. Air masses and fronts over eastern Asia in winter. China is affected by continental polar air, some of it modified by being passed over the sea.
After Lu

precipitation and is only of significance where it is drawn into convergence with the westerlies. Even over Japan, Pacific air is relatively unimportant in producing monsoon rainfall except perhaps in the south.

In winter the jet stream migrates south and bifurcates round the Himalayas and Tibetan plateau. The southern jet and associated westerlies pass over southern China and Japan, converging at height with the colder westerly streams to the north. The surface air, however, has a more complicated pattern (Figure 8.iv). Siberian cP air moves west and south so that some of it has a purely land trajectory and some of it crosses cold seas before reaching either Japan or southern China. Convergence of the sea-modified air may produce cloudy and even showery weather. The cP air with a land trajectory brings to China the coldest weather and lowest average winter temperatures for the latitudes anywhere on earth.

Another factor which is often disregarded in the traditional explanation of the monsoons is the fact that both in summer and winter most precipitation is the result not of convection but of convergence along fronts and in travelling cyclonic storms; this precipitation is often triggered by orographic effects.

Thermally Induced Circulations

Figure 8.v. A typical disturbance of the easterly wave type showing the areas of high (HP) and low (LP) pressure.
After Riehl

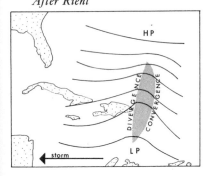

Tropical Storms and Hurricanes

A much smaller disturbance of the atmosphere than that of the monsoon is the tropical hurricane. This storm has a variety of names—typhoon in the west Pacific, cyclone in the Indian Ocean, hurricane in the east Atlantic. In the humid tropics, the weather at the surface is usually described as monotonous, with constant winds, regular afternoon showers and even temperatures, but although this may be largely true at the surface it is certainly not true aloft. In contrast with mid-latitudes, above about 9000 metres variable winds and pressures are usual.

In the trade wind belts the surface air is warm and moist; convection produces patches of cumulus cloud of variable thickness, which seldom grows above 3000 metres, for at about this level there is the trade wind inversion separating the surface air from the dry upper air. Most cells of moist surface air which break through the inversion are rapidly dissipated in the dry upper air unless they are constantly supplied by strong rising cells, in which case they can produce the well known chimney clouds.

In the lower atmosphere, wave-like disturbances quite frequently sweep round the equatorial edges of the subtropical highs (Figure 8.v) and bring stormy weather. The wave is like a dent causing a shift in the trade winds. Ahead of the wave the weather is clear but in it banks of deep cumulus and cumulonimbus bring heavy rain and thunder. Such waves are common, but few develop into hurricanes—why? The answer to this question could not be given until much data from upper air observations had been collected.

The occurrence that turns a wave into a more intense storm with a vortex at the centre was found to be the passage of an upper air anti-cyclone over the surface wave. In that situation, air at high levels is moved out from the centre of the storm and has to be replaced at low levels by incoming air so that the surface wave becomes an intense low pressure system with a powerful vortex and high inblowing winds (Figure 8.vi). Even at this stage there are further requirements before this storm can become a hurricane.

The energy that drives most tropical storms is derived from the latent heat of vaporisation as the moist air condenses into clouds, but most of the heat is dissipated by mixing with cooler air. When, however, the warmed air is funnelled upward without mixing, it breaks through the temperature inversion; but even at this stage a hurricane will not develop unless a very powerful vortex causes an 'eye' to form. It is still not known how or why an eye is produced, but when it is, air in it begins to descend, and as it does so it is warmed by adiabatic compression from 5° to 8°C above its surroundings. It seems that the air is continually drawn in at the top of the hurricane to descend into the calm eye, which may be 15–60 km in diameter, but at the bottom of the eye air is drawn outwards into the vortex, where its heat helps to evaporate more water from the sea and draw it up to condense in the clouds and release more heat to make the storm self-perpetuating. The rising air travels round the eye at speeds of 45–200 knots (83–370 k.p.h.). A hurricane can exist only as long as it can draw up water vapour from the surface; as soon as it passes over land or a cold sea it must die out.

Figure 8.vi. (1) An easterly wave caused by deflection of the trade winds showing the isobar pattern (top) and a section through the cloud band (bottom). (2) The easterly wave has deepened and formed a vortex with winds up to 70 k.p.h.; the ascending clouds have broken through the inversion layer and built up to thunderheads. (3) The vortex has developed fully and is now a hurricane with a calm eye and wind speeds in the vortex exceeding 200 k.p.h.

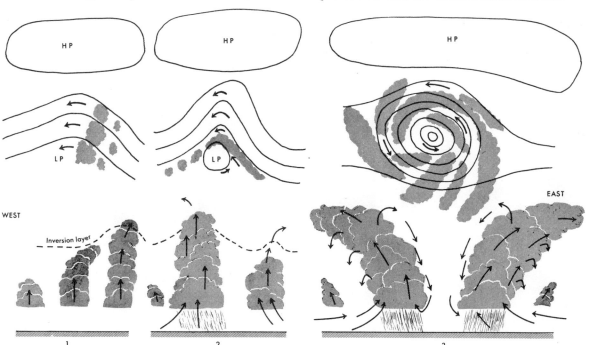

It is evident that the chances of any particular wave disturbance turning into a hurricane are small, but even so the hurricane is a dangerous feature of some tropical seas. The distribution of hurricanes is shown in Figure 8.vii. They are usually late summer and autumn features although they have been recorded in midwinter. They travel along the equatorial edge of the subtropical highs at about 15–20 knots (28–37 k.p.h.) affecting an area rarely more than some hundreds of kilometres in diameter. Their tracks are, however, seldom straight, for they recurve away from the equator and move polewards round the western ends of the anticyclones. The danger of the storm is that it produces high and variable winds, mountainous seas, and piles up the sea to as much as 9 metres above normal along the coasts. Over land, floods often cause more damage than the wind and torrential rain—117 cm of rain in 24 hours have been recorded in the Philippines. Radar and more recently weather satellites are now able to spot hurricanes by their cloud patterns, but in earlier times a rapidly falling barometer was the warning. Pressure in the eye becomes exceedingly low, usually about 960 mb—although a reading of 914·6 mb was reported in 1932 by a ship in the Caribbean.

Attempts to regulate hurricanes in the past have involved seeding the wall clouds round the eye to retard the rate of air movement up through the inner vortex, thereby causing the low pressure zone to fill. Future

9. A series of 6 satellite photos of tropical storms, typhoons and a hurricane. *NASA*

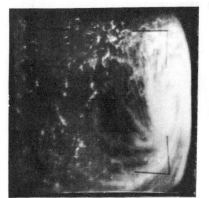

TYPHOON KAREN (140kts)
TIROS VI, 754/754, fr 27
0138 GMT NOV 9, 1962

TYPHOON RUTH (100kts)
TIROS V, 812/811, fr 16
0400 GMT AUG 15, 1962

TYPHOON SARAH (70kts)
TIROS V, 827/826, fr 14
0500 GMT AUG 16, 1962

HURRICANE ALMA (80kts)
TIROS V, 1019 DIRECT, fr 12
1431 GMT AUG 29, 1962

TROPICAL STORM RUTH (60kts)
TIROS V, 798/798, fr 16
0430 GMT AUG 14, 1962

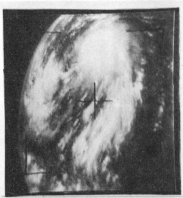

TROPICAL STORM AMY (45 kts)
TIROS V, 1024/1024, fr 22
2250 GMT AUG 29, 1962

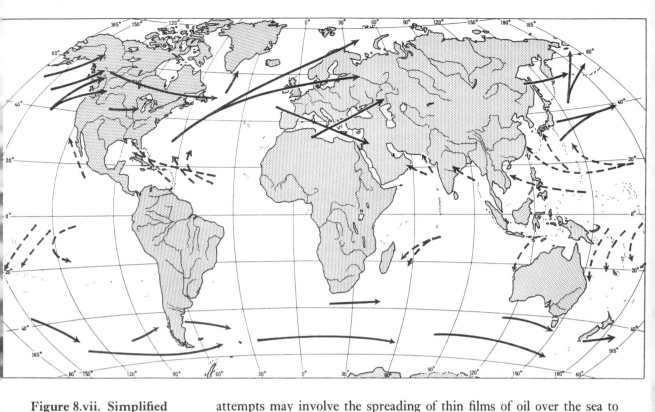

Figure 8.vii. Simplified diagram of the main tracks of mid-latitude (solid arrows) and tropical low pressure systems.
After Petterssen

attempts may involve the spreading of thin films of oil over the sea to prevent the evaporation of water vapour, and thus deny embryonic hurricanes their indispensable supply of energy in the form of latent heat of condensation. If such schemes are to work, the detailed knowledge of the location of the hurricanes, which satellite photographs provide, will be essential. The long-term effects of such controls, are, however, difficult to predict, for hurricanes and other tropical storms may be important systems in the transfer of heat energy from tropical to cooler latitudes. If part of the system is disrupted, what will be the result?

Secondary Circulations of the Dynamic Type

Scientific observers and men whose activities were conditioned by the weather have been aware for several centuries that there is a relationship between changes of barometric pressure, wind direction and weather in the mid-latitudes of the northern hemisphere. Admiral Fitzroy, the first Director of the British Meteorological Office, suggested as early as 1863 that depressions mark the meeting of two great airstreams with different properties—one from the tropics and the other from polar regions. Other British workers, notably Abercromby, Napier Shaw and Lempfert, refined Fitzroy's ideas but the development of the air mass concept made little more progress until the period of 1910–20, when a group of mathematicians and physicists at Bergen, in Norway, under V. and J. Bjerknes (father and son), evolved the polar front or Bergen theory of the life history of depressions.

Before the Bergen school published its work abrupt changes in wind and temperature were regarded as unreal, but the new frontal concepts of the Norwegians were based on discontinuities as the significant features of the weather. Along the polar front, cold polar air flowing from the east and mid-latitude mT air from the west are in close contact (Figure 8.viii). In this condition, the front is likely to develop waves, just as air blowing over a lake produces waves. The waves may die out, but they may grow and produce a true depression. J. Bjerknes and H. Solberg published a famous diagram in 1921 which represented the development of an idealised northern hemisphere mid-latitude depression. However, this diagram, upon which Figure 8.ix is based, represents only an ideal, and many depressions do not develop in precisely this way. In the southern hemisphere the same principles apply, but the centre of the depression is to the south of the warm sector.

As a result of a strong wind shear along the front, the warm air develops a large bulge called the warm sector and the polar front becomes divided into a cold front and a warm front. As the whole system advances from west to east, any point ahead of the depression will first experience cold polar air and then, as the front passes, mT air of the warm sector. The easternmost front is therefore called the warm front. As the warm sector passes, the point experiences polar air again, and the western front is therefore the cold front.

In a depression, the air of the warm sector always tends to rise over the cold air so that both the fronts slope. Warm fronts have low slopes of about 1/80 to 1/200 but cold fronts are steeper—1/40 to 1/80. As the depression grows older, the cold front begins to catch up with the warm front and the warm air is lifted off the ground, when it is said to be occluded. The occlusion starts at the centre and gradually works equatorwards until all of the warm sector is occluded and the depression degenerates into a gently rotating mass of homogeneous air.

Figure 8.viii. Life cycle of a northern hemisphere mid-latitude depression. Pressures are in millibars; the shading indicates rain. The upper front is shown by broken lines.

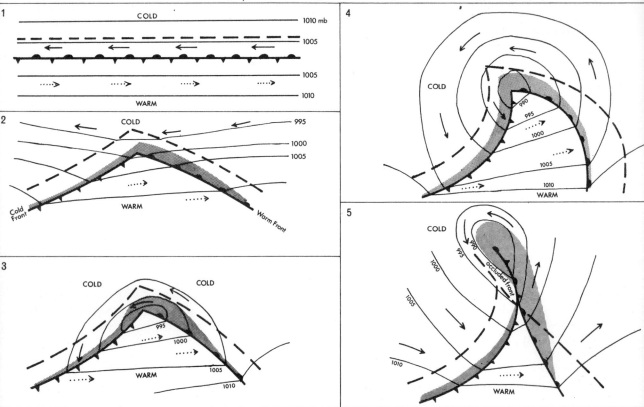

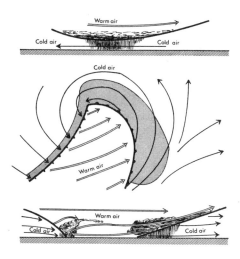

Figure 8.ix. Bjerknes's depression model.

Weather in Depressions

The first warning that an observer to the east of a depression would receive is the appearance of a belt of high cirrus cloud far ahead of the low. As the warm front approaches, the clouds gradually thicken and become lower, usually passing through the sequence of cirrostratus, altostratus, nimbostratus and cumulonimbus. There is rain or snow for some miles in advance of the front, for the mT air is forced to rise up the front. The extent and type of precipitation will depend upon the stability of the air, for if it is stable cumulonimbus will not develop and precipitation will be light; unstable air will, however, give rise to heavy rain, which will produce fog and low cloud as the warm rain evaporates in the cold air beneath the front.

As the front passes overhead, the observer will be in the warm sector. The sky will clear, temperature will rise, the wind direction will shift, and showers will alternate with sunny periods. The advancing cold front undercuts and lifts the air of the warm sector to form cumulonimbus clouds, which may form a squall line of black, menacing clouds and torrential rain. The cold front brings a fall of temperature and rapidly veering wind. Most cold front weather is of the showery type indicative of instability.

The cold air ahead and behind a depression usually has a sufficiently varied life history for slight differences of temperature and humidity to exist between the two cold masses. As a result, an occluded front can frequently act like a minor cold or warm front depending upon whether the easterly air is warmer or cooler than the westerly air. Precipitation beneath an occlusion is much like that beneath ordinary warm and cold fronts, but late in the life of the occlusion the warm air is able to spread out horizontally over the cold air beneath it; further lifting ceases and precipitation is less. For this last reason western Europe, which experiences many occluded depressions, has a great deal of low cloud with only modest precipitation and small temperature fluctuations.

Depressions vary greatly in size but a typical winter one would have a long axis of about 2000 km (in the northern hemisphere this would be aligned NE-SW) and a short axis of about 1000 km (NW-SE). It would advance at about 50 k.p.h., although a summer depression will seldom exceed 30 k.p.h.

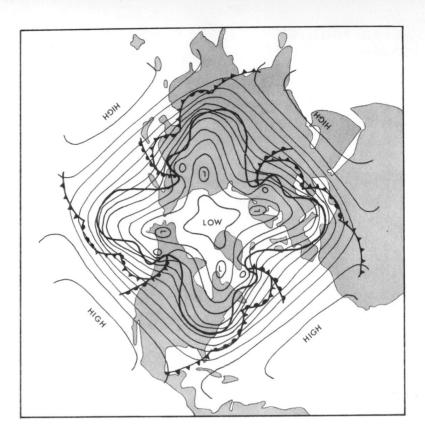

Figure 8.x. Schematic circum-polar chart for the 500 mb level showing four long waves. The heavy line represents the 500 mb polar fronts and the surface fronts are represented by the usual symbols.
Based on Rossby

The relationship between surface and upper air conditions has not been fully worked out but it appears probable that depressions form along the forward edges of Rossby waves where the jet stream is strongest; the relationship between the front and the jet stream is shown in Figure 7.iii. In the occluded stage, new depressions can form behind the leading decaying depression so that a family of cyclones results, with the youngest being closest to the equator. Each depression is probably a minor wave-like perturbation along the leading edge of a Rossby wave (Figure 8.x).

Depressions can form for reasons which are not associated with frontal cyclogenesis. The case of a depression forming in the lee of mountain ranges has already been cited. A second case may involve an island or a large lake, since land and water respond differently to incoming solar radiation. Low pressure may form over an island during summer or over a large lake during winter, and cyclonic circulation develops as air flows into the surface low pressure region. A third case is the retardation of air flow over a coastline caused by the increased frictional drag as air passes over from sea to the land and the change in velocity causes a change in direction so that cyclonic circulation again develops.

Anticyclones

An anticyclone is a pressure system characterised by high pressure and outflowing or diverging air at the surface. The outflowing air is deflected by the effect of the earth's rotation so that in the northern hemisphere it develops a clockwise course round the high and in the southern hemisphere a counter-clockwise course. The isobars in highs are usually widely spaced, so wind speeds are low and near the centres calms are common. The most active highs occur in the mid-latitude belt of westerlies as migratory systems moving more slowly than the lows, and producing more settled weather. Along their leading edges they move air from polar regions and on the trailing or westerly edges they move it from tropical areas, so that although frontal systems do not usually develop in highs, weather does change as a high passes over a region.

Tracks of Migrating Pressure Systems

Figure 8.vii indicates the main tracks of migrating depressions. The main lows occur in the mid-latitude zones of the westerlies but they frequently invade more polar and tropical areas. The most vigorous circulation occurs in the southern hemisphere, where the very stable and cold Antarctic anticyclone produces cold air which gives greater vigour to the depressions which circle the continent. The depressions mostly follow the trough of the subpolar low but they extend southwards to overlap the continent and northwards to affect southern Argentina and Chile, Tasmania and New Zealand, especially in the winter.

In the northern hemisphere, summer circulation is less vigorous than that of the southern hemisphere as the Arctic and subarctic anticyclones are relatively weak and the depressions move along northerly tracks. In winter, on the other hand, the Arctic and subarctic anticyclones are much better developed and the circulation is both more vigorous and located further south.

Mid-latitude depressions usually originate as waves along the Arctic, Antarctic and polar fronts where there are marked temperature contrasts in the air masses. In the northern hemisphere, therefore, the most common regions of cyclogenesis are the troughs and zones of convergence along the east coast of the continents of Asia and North America, where the cold air from the continental winter highs comes into contact with the warm maritime air of the subtropics. A less important area of cyclogenesis is the interior of Canada and northern USA where polar and tropical air converges over the plains. Minor disturbances also develop in the Mediterranean area during winter. The depressions from the east coasts of Asia and North America travel across the Pacific and north Atlantic and reach the eastern sides of those oceans as occlusions, so the most vigorous weather occurs over the ocean.

9 Tertiary Circulations in the Atmosphere

The tertiary circulations are the result of local thermal and dynamic disturbances of the atmosphere. The main types are classified below, but as many of them have been discussed before, only land and sea breezes, mountain and valley winds, and tornadoes are dealt with here.

Classification of Tertiary Circulations

(1) Thermal circulations produced by adjacent local cooling and heating:
   (a) land and sea breezes;
   (b) mountain and valley winds.
(2) Thermal circulations produced by heating:
   (a) cumulus convection and thunderstorms;
   (b) dry convection (sand and dust whirls);
   (c) valley breezes.
(3) Thermal circulations produced by cooling:
   (a) gravity winds;
   (b) glacier winds.
(4) Dynamic circulations:
   (a) föhn winds;
   (b) large scale vertical eddies (lee eddies);
   (c) orographic and frontal showers and thunderstorms;
   (d) tornadoes.

Thermal Circulations

One of the best known types of local wind is the *land and sea breeze*. Along most coasts equatorwards of about 50°, but especially in the tropics, the land is warmed more than the sea during the day. As daytime heating begins, the air over the land expands and rises, producing a pressure gradient directed from sea to land at low levels, and from land to sea aloft. The result is that air flows from the sea towards the land as a sea breeze that in the tropics may have a speed of 16–30 k.p.h. The breeze usually lasts from mid-morning to mid-afternoon, when the sinking sun and increased cloudiness allow the temperature differences between land and sea to diminish. The situation is reversed at night, when the land becomes cold so that air descends over it and flows towards the sea which is warmer, and over which air is still expanding and rising. The land breeze is usually weak and seldom exceeds 3–8 k.p.h.

In tropical regions the Coriolis Effect upon the sea breeze is slight and it may penetrate 80 km inland, but in temperate latitudes the Effect can be strong enough to deflect the breeze so that it tends to flow parallel to the coast and hardly penetrates inland. Another feature of the land and sea breezes is that strength may be determined by the trade winds (Figure 9.i).

Mountain and valley winds are one of the characteristic features of mountain climates. During the day, especially under the clear skies and weak general circulation of anticyclonic conditions, mountain slopes facing the sun are intensely heated while the air over the middle of the

10. Cold air and thin stratus cloud pouring over the edge of the Antarctic plateau is warmed as it descends and the cloud dissipates.
NZ Dept of Scientific and Industrial Research, Antarctic Division

valley remains cooler. Consequently the air in contact with the slopes expands and rises, creating an upslope wind that frequently causes cumulus clouds to form round peaks in the afternoon. At night the air on the slopes is cooled and so contracts and sinks, producing a downslope wind.

Slope winds can occur on any slope, whether it be part of a hill and valley system or not, because in the clearer atmosphere of the upper slopes daytime heating and night-time cooling are always more intense than at the base of the slope. The resulting daytime upslope movement

Figure 9.i. The effect of the trade winds on land and sea breezes.

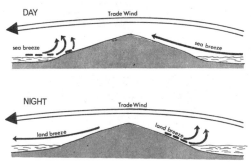

of warm, low-density air is called an anabatic wind and the downslope movement of cold, dense air is called a katabatic or *gravity wind*. A very cold, strong, outblowing wind is a feature of the Antarctic and Greenland ice caps, and there are many similar winds from different mountain regions in the world. The Bora flows from the plateau of Yugoslavia down the Adriatic; the Mistral from the European Alps flows down the Rhone Valley to chill the Riviera; and the Harmattan from the cooler Sahara brings dry air to the Guinea Coast of Africa. Most of these winds are strengthened by a pressure gradient from their source towards the area of dispersal.

Tornadoes

Tornadoes are perhaps the most destructive of meteorological phenomena, for although the damage caused by floods and high winds associated with hurricanes may be greater, the destruction in the path of a tornado is usually complete. The damage is caused by: (1) the very high winds which, although no anemometers have survived a tornado, are estimated as reaching 800 k.p.h.; and (2) the very low pressure at the centre of the 'twister' which causes buildings to explode.

The typical tornado is a vortex with a funnel-shaped cloud which extends as a narrow spout from the ground into a thundercloud aloft. The spout averages 360 metres in width and travels at 40–65 k.p.h. The tornado approaches with an enormous roar and passes rapidly, leaving anything in its track wrecked. Its life span seldom exceeds one hour and its average distance travelled is about 25 km. In 1957, 924 tornadoes were reported in the USA, killing 191 people, injuring 2343, and causing damage estimated at $73 million.

The meteorological conditions favouring the development of tornadoes are particularly common in the USA and India, where a surface stream of southerly mT air is drawn into a depression in which there is a stream of dry cP air aloft with a temperature inversion at 1500–1800 metres. Along a front or a squall line in which the isobars are very close so that a pressure jump exists, a parcel of air which starts to rise may develop a rapid lift, hastened by the heating produced as the moisture in the parcel condenses. The air that breaks through the inversion may produce a chimney of rising air which spins as a result of convergence.

Most tornadoes occur in spring and summer during the afternoon. They cannot be forecast accurately as yet, but when the above conditions occur, it is possible to predict that certain areas will have them.

11. The funnel-shaped cloud of a tornado.
US Information Service

10 Climate and Human Activity

Man, in common with other mammals and birds, is a homoiotherm, that is, he is able to keep the temperature inside his body relatively constant when he is exposed to a wide range of external temperatures. Such an ability represents a notable evolutionary advance over the primitive organisms, or poikilotherms, which have no such ability to regulate their internal temperature. Poikilotherms have body temperatures which rise and fall with the environmental temperature, so that in cold conditions their metabolic processes slow down until they become sluggish, and in hot conditions their body heat rises until it becomes incompatible with life and the animal dies. Poikilotherms are therefore very much at the mercy of the environment. Homoiotherms are affected by extreme temperatures, but much less so.

Homoiotherms prevent a fall in body temperatures in five ways:

(1) by varying the amount of muscular activity;
(2) by the involuntary muscular activity we call shivering;
(3) by non-shivering thermogenesis, which means creation of heat by basal metabolism without muscular movement;
(4) by reducing blood flow to the skin and so reducing heat loss (this carries with it the risk of frostbite and clumsy fingers);
(5) by fluffing out hair or fur or feathers to trap more insulating air next to the skin (goose pimples on man are vestigial remnants of this process).

Heat loss is effected:

(1) by increased blood flow to the skin;
(2) by sweating. Man can lose water in sweating at the rate of a litre an hour for several hours, although this places a severe strain on the body and the work output is reduced in high temperatures.

Racial differences in adaptations to heat appear to be less significant than the differences between individuals within each race. It is true that a dark skin protects against solar radiation, but it has the disadvantage of absorbing more solar heat than a light-coloured skin. Of much greater importance is behavioural adaptation to high temperatures.

Primitive peoples chronically exposed to cold do seem to have developed physiological temperature regulating mechanisms, in that they can allow their body temperatures to sink lower and they appear to experience less discomfort than a European would in similar circumstances, but it is not clear whether or not this effect is indicative of ethnic differences.

In dry climates sweat will evaporate readily, and the desert dweller such as the Bedouin wears thick clothes to give insulation from the heat of day and to prevent the evaporating effect of wind on his skin. He avoids hard work during the day and often moves by night. In humid climates, clothing has most value in protecting the body from sunburn. It should be light in weight, admit air freely and reflect the sun's heat. Overheating of the body by exercise tends to produce a reaction by slowing down all activities, and the person acclimatising to a hot, humid climate has to live quietly for a few weeks until his body has adjusted so that he develops a lower pulse rate and loses less salt in his sweat. People of European stock can live healthily and productively in the humid tropics as long as they are willing to adjust themselves psychologically to their environment. Europeans in the cane fields of Queensland or the wars of southeast Asia have proved this beyond doubt.

Nearly everyone regards temperatures of over 30°C as hot, but whether or not they are regarded as unpleasant depends upon individual conditions. High relative humidity—over about 70%—unsuitable clothing, hard physical exercise and lack of wind will all increase feelings of discomfort. The first response of the body to heat is vasco-dilation, a process in which the blood vessels of the skin dilate so that body heat will more readily be lost to the surrounding air. Next the two million or so sweat glands start working and the evaporation of this body water produces a cooling effect. A man sitting in the shade under desert conditions with an air temperature of 38°C will lose a quarter-litre of sweat each hour, and walking at a steady pace will lose a litre an hour. This enormous loss of water—which may be 9 litres a day—must be replaced or the pulse rate rises, the blood becomes thicker and the heart becomes less able to force it through the veins and arteries. If the water loss exceeds about 12% of the body weight the man collapses, his body temperature soars and he dies—cooked to death by his own body heat.

Normal body temperature is about 36·7°C, but because of internal heat, which will be maintained as long as there is a sufficient food supply, the human body will maintain this temperature even when the air temperature falls well below this. The first reaction of the body to cold is vasoconstriction: the blood vessels near the skin contract to reduce heat loss, then muscle tone changes, and shivering starts and continues until the body temperature is back to normal. Death from cold occurs as soon as the body temperature falls below about 27°C.

Low humidity encourages rapid evaporation from the skin, so that although temperatures of 18°C are only regarded as cool in a relative humidity of 70%, if the humidity falls to 30% the air may feel very cold. In other words, it is the sensible temperature which makes us feel hot or cold, and not the air temperature.

In cold climates clothing should be designed to insulate the body and to retain as much warm air as possible trapped in the layers of clothing. Low temperatures seem to act as a spur to mental and physical activity, but extremes are of course most unpleasant. Once body tissues freeze, they usually go gangrenous unless circulation can be restored rapidly and frostbite is an ever present danger to people living in sub-zero temperatures. Some acclimatisation to cold does occur very rapidly with the onset of cold conditions. The blood volume decreases and sweating becomes less effective. It is possible also that some glandular changes also occur. The main reaction to cold, however, must be a cultural one in which suitable clothing, diet and housing are related to the climate.

Climate and Agriculture

During several thousand years of experiment, man has selected and cultivated plants for his use. Because all plants are adapted to certain climatic conditions under which they achieve optimum growth, it is seldom possible to move plants into an alien environment and to acclimatise them successfully, but by careful plant breeding and selection it is possible to raise strains of plants adapted to the required climatic conditions. This type of breeding has in the past been carried out largely in field experiments and often under conditions in which chance was important. A more modern technique is artificially to create the required atmosphere in a laboratory similar to that of the 'phytotron' at the California Institute of Technology.

In the phytotron 54 different environments can be created in which day temperature, night temperature, light intensity, duration of daily illumination, light quality, relative humidity of the air, wind, rain and gas content of the air can all be varied as required. In this world free from insects, dust or fumes, the effects of climatic variables on plants can be studied in detail.

Very little can be done to improve the selection of climate for the great staple crops of the world, for as a result of experience, economic forces, and the development of suitable local species, crops such as wheat, rice and cotton are already grown in the most favourable localities and extension of their areas of growth is usually prevented by a decrease in profitability. Specialised crops, however, can be grown almost anywhere, provided that labour and expense are not significant factors.

By providing suitable soil, drainage, irrigation, shelter or shade, the farmer can help to provide a suitable environment for his crops and the agroclimatologist can assist him by providing warning of damaging weather. In Holland the Meteorological Institute warns farmers when climatic factors threaten the spread of potato blight so that crops can be sprayed in time. In many orchard areas, frost warnings are essential to avoid damage to the fruit. The orchardist who expects frost can light smudge pots to combat radiation temperature inversions by creating some heat at tree level and forming a dense pall of smoke which spreads over the orchard, reducing loss of heat by radiation in the same way that a cloud does. In some citrus orchards in California, large fans are switched on to mix the air when there is a danger of frost, but this of course cannot be effective if the low temperatures are caused by an invading cold air mass. Another technique is to spray the orchards with water which releases latent heat when it freezes on the blossoms, and so protects them.

All types of farming practice depend ultimately upon the suitability of specified crops to prevailing temperature and moisture conditions. Temperatures must be suitable at each stage of growth during a sufficiently long and dependable growing season. Moisture must be available in the right amounts and form, so that soils are not parched or waterlogged, nor crops beaten down by torrential rain or hail. The farmer can seldom do anything about unseasonable temperatures or protect his crops against them, nor can he usually counter exceptional floods, but the amelioration of drought and wind conditions is a common feature of farming practice in many parts of the world.

Drought may be defined as the condition in which the amount of water needed exceeds the amount available in the soil. Three types of drought may be distinguished: (1) the permanent drought associated with arid climates; (2) seasonal drought which occurs in climates with a marked dry season; and (3) drought caused by precipitation variability. High winds, some soil conditions, low humidity and high temperatures contribute to drought conditions as well as lack of precipitation. Whether or not a drought is damaging to crops depends upon its occurrence in relation to the growing season. Some of the worst famines of southern Asia have resulted from the failure of the summer monsoon to develop as early as usual.

To combat drought many farming techniques have been developed. In semiarid regions, crops with low water demands such as the sorghums—kaffir-corn, feterita and milo—are preferred to wheat or maize. Deep-rooted crops like alfalfa can withstand drought by tapping moisture in the subsoil, and soils can be cultivated to improve their water retention capacity and to limit runoff. The addition of some minerals and humus to the soil and the destruction of weeds which increase losses by transpiration are common practice. In semiarid and subhumid climates, dry farming methods aim at using the precipitation from two or even three years for one crop. By tilling the soil and keeping it free from growing weeds during the fallow period, enough soil moisture can be accumulated for one cropping season.

The most productive means of avoiding drought is the provision of irrigation water. This encourages larger yields and a greater variety of crops, and in some arid regions it is essential for any form of production. The chief limitations on irrigation are the availability of water and the cost of getting it to the crops.

Irrigation has the advantage that it can be regulated to meet the demands of different crops under any drought conditions. On well controlled irrigation schemes, a constant check is kept on the soil water available to the plant so that deficits can be rapidly made good. Where water is made available, temperature becomes the dominant climatic factor affecting crop distribution and yields.

Wind is chiefly damaging to crops by breaking of plants, but indirectly soil erosion, wind-blown sand, wind distributed disease, spores or seeds of weeds are harmful to production. In areas commonly afflicted by winds, crops have to be planted in naturally or artificially sheltered areas.

Climate and Transport

Sea travel has historically been more affected by weather conditions than other forms of transport. In the days of sail, the general wind systems dominated the choice of routes so that, for example, the arrival of Columbus and his successors in the Americas was influenced by the trade winds of the central Atlantic and the return journeys to Europe were commonly made in the westerlies of higher latitudes. The steamship is not directly dependent upon wind systems but, even so, long ocean voyages are still planned to make use of helpful winds and currents and to avoid head winds, in order to conserve fuel and reduce running costs. The largest modern ships are still partly dependent upon weather, for even with modern radar and other navigation aids, heavy fogs bring shipping to a halt in coastal waters and ports. Major storms, particularly hurricanes, generate large cross swells which are dangerous even to large ships, and routes have to be planned to avoid them. Another major climatic hazard is ice in high latitudes, which creates blockages as in the Great Lakes and St Lawrence Seaway for 3–4 months a year, and either blocks Arctic and Antarctic routes or produces danger from drifting icebergs.

Aircraft are particularly susceptible to the vagaries of weather. Even the largest airliners, which may be able to fly above many storms, still have to land and take off in the weather zone. Winds, storms and visibility are the three factors which most closely affect air travel. Winds are particularly important because most airports have to be built so that the runways are aligned with the major winds and cross winds are avoided. In flight, tail winds can greatly increase the speed and hence the economy of a flight, whilst head winds will slow it and increase the cost. Before taking off, therefore, pilots and navigators study weather reports to plan the most advantageous routes. On the very busy north Atlantic route, east to west flights are planned to avoid the jet streams but west to east flights make use of them. In pressure systems, the flight paths have to be plotted so that an aircraft is not only flying along the best course but also at the right altitude, for wind speeds and directions vary with height. If it is not possible to fly over or around a major thunderstorm, it is often necessary for an aircraft to turn back.

Poor visibility still causes more disruption of flight schedules than any other factor. The minimum requirements for takeoff are usually regarded as being visibility of 5 km and a cloud base not lower than 300 metres. It is possible to make instrument-controlled takeoffs and landings but few aircraft or airports are suitably equipped; aircraft may have to be grounded for several days, and those in flight diverted to open airports, during bad fogs or periods of low cloud.

The most dramatic interruptions of land traffic are floods and snowstorms. Floods wash out bridges, cause landslides and block roads. Snowstorms do most damage by blocking roads and railways, but they also bring down telephone and power lines. In some mountain areas, it has been found economic to protect railways and even roads by enclosing them in tunnels or building snow breaks beside them to prevent drifts. It is technically feasible to prevent snow settling or water freezing on roads by heating the roads but, apart from exceptional installations like the Chiswick flyover in London, such a scheme is seldom economically possible.

Air Pollution

Writing in 1661 in a tract entitled 'Fumifigium, or the inconvenience of the Aer and Smoak of London', John Evelyn said:

> Whilst these [chimneys] are belching forth their sooty jaws, the city of London resembles the face rather of Mount Etna, the court of Vulcan, Stromboli, or the suburbs of hell, than any assembly of rational creatures, and the Imperial seat of our Incomparable Monarch.

Air pollution, then, is not a recent phenomenon but has been in existence ever since men started using fossil fuels in large quantities. Complaints of London pollution were recorded as early as the sixteenth century. Pollution is not a result of weather or climate but it is strongly influenced by them, for the diffusion of contaminating gases, solids and liquids is the result of atmospheric conditions. The haze and smoke cap which sits over many modern cities is the result of smoke, dust and gases from domestic fires, heating systems, industrial furnaces, incinerators, motor vehicle exhausts and other wastes which together cause injuries to fabrics, buildings, health and plants, and irritation of eyes, noses and throats. The Ministry of Technology estimated that the cost of air pollution in Britain in 1966, through material damage and ill health, was £350 million.

Many contaminants released into the air by combustion are hygroscopic, and act as nuclei of condensation to create a haze or fog, often known as smog. The word 'smog' was originally coined from 'smoke and fog', but it is now used to denote almost any polluted air, whether it contains smoke or not. Smog droplets are often very stable; because their oily substances form a protective coating round the droplets, they do not readily evaporate on being heated. Smog reduces the solar radiation which can reach the ground, thus reducing the daytime temperatures, and increasing the night temperatures by impeding outgoing radiation. It also filters the sunlight so that the quality as well as the quantity of light is reduced.

In Britain as a whole, the tonnage of smoke emitted each year is declining as a result of the Clean Air Act of 1956. Tall chimneys and smoke cleaners have reduced industrial pollution, so that although some 2·7 million metric tonnes of smoke were emitted in 1938, less than half this amount was emitted in 1966. By contrast, the weight of carbon monoxide from motor vehicles each year is 6·1 million metric tonnes and this amount is increasing.

In general, valley sites, lack of wind, high pressures, temperature inversions, and a lack of precipitation to clean the air are the main factors favouring air pollution in large cities.

Urban Climates

Although air pollution is the most outstanding characteristic of London's climate, it also has other features which may be taken to illustrate some characteristics which are common to many cities, although local conditions of weather and topography make each city distinctive. (The following paragraphs are based on T. J. Chandler, *The Climate of London*, published by Hutchinson in 1965.)

Perhaps the best known feature of London's climate has always been its fogs, which were once related to the city's position in a tidal estuary. Now the most usual pattern for fogs is for them to form in the evening in the cold humid air over the fields around London, so that the urban area becomes surrounded by a fog belt. The fog gradually invades the outer suburbs and by dawn the whole of the urban area and its rural surrounds are covered. Soon after dawn the fog in the rural district is dispersed by the rising temperatures, but in the central zone the temperatures rise more slowly, wind-speeds are less and the fog is more persistent.

The decreased visibility in the centre of London is one effect of the smoke haze and greater daytime incidence of winter fog. A further effect is the reduced amount of average sunshine hours; the outer suburbs receive 16 minutes per day less than the rural Green Belt around the metropolis, the inner suburbs 23 minutes less and the centre 44 minutes less. These losses tend to be least in the summer and greatest in the winter.

Figure 10.i. Under the influence of a high pressure system covering England on the night of 10–11 July 1959 a 6°C heat-island developed above London with a displacement towards the eastern districts of the city and a smaller centre of local heating in the Richmond district of western London. A tongue of cooler air covered the more openly settled areas of Elstree and Hampstead in northern London.
After Chandler

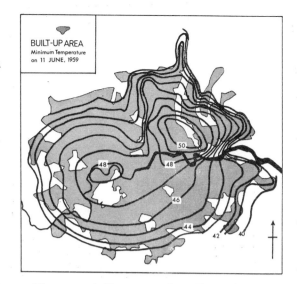

BUILT-UP AREA
Minimum Temperature
on 11 JUNE, 1959

Figure 10.i illustrates the effect of the built-up area on minimum temperatures. The warm air which rests over the central area at night is known as the heat-island; on calm clear nights in summer, after sunny days, it may be as much as 8°C warmer than the rural periphery, but if the night is windy, humid or overcast, the contrasts will be small or non-existent. Most of this heat comes from the buildings, which store it from the daytime and release it gradually at night. Peculiar local variations over very short distances can be observed as a result of the buildings channelling wind, and as a result of parks and open spaces cooling more quickly than buildings. Contrasts between the built-up areas and open spaces during the day are small.

The effects of urban areas on precipitation are difficult to assess because of variations which would be caused by topography in any case. It is probable that the main effect is to reduce the likelihood of snow and certainly to reduce the time it lies on the ground. Whether or not the increased cloudiness over the city has any effect on total precipitation is not known, although there does seem to be a rather high intensity of summer thunderstorms around the higher areas of north London.

The Rural Microclimate

Sensational and often irresponsible claims have frequently been made that man can now control the climate—by which is usually meant the macroclimate. It is known that climate varies over short and long periods, but such variations are not the work of man. Claims that hurricanes and tornadoes can be prevented by explosions, that rain can be readily induced by seeding clouds or by gunfire, and that sunshine hours can be increased by dissipating the cloud cover, have little basis in fact. Even the much venerated beliefs that precipitation will be reduced by draining areas of marsh, or that by planting trees it will be increased, have no evidence in their support. It is, however, indisputable that man can modify the microclimate—usually defined as the climate within 2 metres of the ground, or beneath the vegetation canopy, or in the case of urban areas beneath the general roof levels.

The significance of the microclimate is indicated by temperature differences in the air layer nearest the ground. The highest screen temperature ever recorded (1·2 metres above the ground) is 58°C at

Azizia in Libya, but in a desert soil at 4 mm below the surface 71°C has been recorded, and it is possible that in hot tropical sun ground temperatures might reach 90°C. Even in England, 60°C has been recorded at the surface of a macadam road. Not only higher but also lower temperatures are recorded at ground level. Insects and seedlings must, therefore, live in a climate which is far more extreme than that to which men, large animals and mature plants are accustomed. The gardener frequently has to protect young plants from temperature extremes which would not occur at head height.

The temperatures of soils vary greatly because of their colour and porosity—which affects their water-holding capacity and air content. Sandy soils are pale in colour and so reflect more of the sun's radiation than dark clay soils. Sands are also porous, and the air in the pore spaces of the soils is a poor conductor of heat. In calm weather the air over sands is cooler (3·5°C or more) than that over clay. Sandy soils are therefore more subject to frost. A loose sandy surface also prevents heat penetrating into the ground, so that some centimetres below the surface on a hot day the temperature may be 20°C below that of the surface. Sandy soils, however, often warm up more quickly than clays in the spring, for they are free-draining, whereas clays retain their moisture, which has to evaporate and so keeps the soil temperature low.

A gardener must consider not only the thermal regime of his soil but also the effects of air movement. In an enclosed valley cold air, being more dense than warm air, sinks to the floor of the valley where it ponds, forming a frost hollow in winter. A German meteorologist was able on one occasion to map the height to which the pool of cold air extended by following the level, along the valley sides, up to which the crops were damaged by frost. This showed as an almost horizontal line round the valley. The concept of a pool or lake of cold air is not, however, really accurate, as will be seen from Figure 3.viii on page 17. The safest positions for delicate crops are found on the middle and upper slopes.

Aspect in a valley is another important consideration for the builder, farmer or orchardist. In order to get maximum warmth, the most favoured locations face towards the equator and slightly to the west. It is obvious that an equator-facing slope will be in least shadow, and the reason for desiring a westerly aspect is that in the morning the westerly slopes have the advantage of a general rise of air temperature in the morning and direct sun in the afternoon, while easterly slopes receive their sun when the air temperature is lowest and has to be raised. The warmest temperatures of the day are experienced after midday when the sun is on the western slopes.

The effects of soil temperatures, slope and aspect may be summarised by the table prepared by Manley which shows the average intervals between damaging frosts in the English Midlands:

Enclosed urban site, large city ...	5 April–5 November	214 days
Favourable hill slope	15 April–1 November	200 days
Normal low-lying ground	15 May–1 October	169 days
Frost hollow	5 June–12 September	99 days
Extensive sandy lowland	20 June–25 August	66 days

Wind has its most spectacular effect on men, buildings and plants when it is in the form of a gale or hurricane, but much less significant winds can greatly affect location of buildings or agricultural practices by damaging crops, removing fine soil, or helping to dry soils or wither plants. The obvious action to avoid damaging winds is to keep to naturally sheltered locations, but where this is not possible, shelter belts of trees or shrubs can be grown, and on a smaller scale garden walls can be built. Trees will always break the force of the wind even if they are in thin plantations. A closely planted wind break is most effective and will shelter an area as much as 30 times as wide as the height of the barrier. It also reduces wind-speed on the exposed side of the wind break, affecting a strip as wide as 8 times the height of the barrier. It is necessary when planting wind breaks on slopes to avoid trapping cold air behind a break and so creating cold spots. Other obstructions such as railway embankments also have this effect.

Unless it is economic to create completely artificial microclimates in glasshouses, the effect of man on temperature and wind is not as great as that on moisture. Man does not alter natural precipitation but, by artificial drainage and irrigation, moisture content of soils can be completely altered. By draining swamps, especially areas of peat, and by putting mole drains and ditches in and around fields, soil moisture is greatly reduced. The capacity of soils to retain moisture is also lowered—often irreparably—through trampling by animals and compaction by machines, so that the pore spaces are closed and it becomes difficult for moisture or roots to penetrate the soil.

The effects of vegetation changes—by grazing, burning, cropping, substitution of species and clearing—on the water entering the soil are also important. By preventing plant growth, transpiration is reduced but runoff is increased. The reduction of humus in the soil and the increase in soil compaction will also increase runoff and therefore the liability to erosion.

Irrigation is seldom as potentially harmful as drainage to the soil and the microclimate of plants, but even this can be harmful if it allows salts to be drawn to the surface by capillary action so sterilising the soil. Irrigation is most useful and effective in regions which have a wet season with excessive precipitation followed by a season of droughts— as in much of monsoon Asia. The saving of the wet season surplus so that it can be used in the dry season has greatly increased production.

A harsh microclimate can be greatly ameliorated by vegetation. Plants absorb and emit less radiation than soil and hence temperatures beneath them are more equable than those above bare ground. During the day, therefore, the air in a forest is cooler than that in the open, but at night it is warmer. The effect can be so great that a slight breeze, analogous to that of a land and sea breeze, with a forest acting as water, can blow between forests and open areas. Vegetation also reduces wind-speed and this, together with the evaporation from plant surfaces, causes humidity to be greater between plants than over bare soil.

11 Weather Forecasting and Meteorological Satellites

All predictions of future weather are statements of the probability of an event; they are not precise statements because atmospheric processes cannot be measured or even observed with that degree of accuracy. Furthermore, a forecast is less likely to be accurate as the time interval between the forecast and the predicted event increases. Short-range forecasts usually indicate the probable weather for 12 or 24 hours ahead, with a more general 'outlook' guide for the following day or two. Extended-range forecasts usually apply to periods of 4–5 days ahead and are much less precise than the short-range type. Long-range forecasts are statements about the probable weather of a coming month, or maybe longer.

Short-range Forecasts

In most countries of the world there are meteorological services which produce daily short-range forecasts based on synoptic charts. These charts are prepared from information received from many parts of the world through an international communications system organised through the World Meteorological Organisation (WMO), an agency of the United Nations. Thousands of observing stations, ships, and aircraft supply coded observations at 0000, 0600, 1200 and 1800 hours Greenwich Mean Time each day. The observations are of present weather; wind direction and speed; amount, form and height of cloud, with visibility; air temperature and dew point; barometric pressure and barometric tendency; weather during the previous 6 hours. This information, together with any additional data, is plotted on to charts using standard symbols arranged in a prescribed order around the map location of the originating station. The analyst can then plot in isobars, contours and thickness lines and put in the fronts. From areas of the world such as the USA or western Europe the data is sufficiently plentiful for accurate plotting, but over the oceans and much of the southern hemisphere, the network of observing stations is thin and the analysis liable to be more inaccurate, and based largely upon the analyst's experience and judgement. The volume of information involved in this international system can be judged by the fact that the Central Forecasting Office for the United Kingdom handles over half a million coded groups of data every day.

When the charting and analysis are completed, the forecaster has to deduce from the charts the likely movements of the pressure systems and fronts and prepare a forecast pressure chart of what he thinks the pattern of isobars and contour lines will be at the time of the next observation. This extrapolation is based partly on judgement and partly on sets of rules derived from studies of earlier sets of charts. It is obvious that such forecasts are subjective in nature and depend to a large degree upon the experience and ability of the forecaster. Most errors in forecasts are the result of incorrect predictions of the timing of events, for speed and direction of fronts and cloud systems are variable. Most forecasts are also for conditions over large areas and are therefore concerned with the general weather of that area, which may not occur at a precise location.

Extended-range Forecasts

This type of forecast depends upon the study of mean circulation patterns and the possible deviations from the mean. The discovery of Rossby waves has had a profound influence on this type of forecast, which, although often inaccurate, is throwing much light on the dynamics of the atmosphere.

Long-range Forecasts

A long-range forecast is a statement of a degree of probability of a certain weather sequence. It is largely based on studies of old climatic records and attempts to see in them patterns of recurring weather sequences. It is usual practice to compare the weather for a particular month with that of the climatic average to see how the particular month

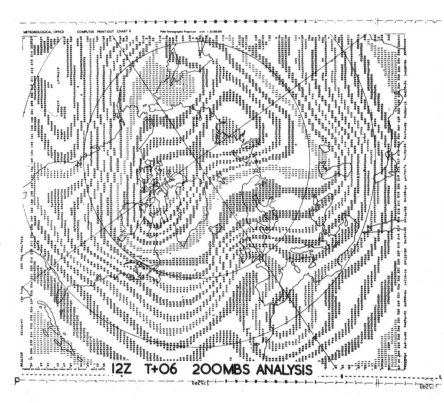

Figure 11.i. A printout from the Comet computer at the Meteorological Office at Bracknell, showing an analysis for the 200 mb surface printed on an outline map of most of the northern hemisphere.
Reproduced by permission of the Controller of Her Majesty's Stationery Office. Crown Copyright reserved

is anomalous. A search can then be made in the records for a year in which a best-fit with that of the present weather can be found. This gives a pattern from which to predict the weather for the next month.

Numerical Forecasting

Meteorology is a branch of fluid mechanics and fluid thermodynamics governed by known basic laws, It might be thought, therefore, that mathematical calculations could be used to predict what would happen in the atmosphere if known changes occurred in it. An Englishman, L. F. Richardson, actually attempted to work out a forecast in 1922 using mathematical methods. His forecast was quite wrong but his methods sound. Unfortunately the variables in the equations are so many that Richardson estimated that it would need 64,000 mathematicians to keep pace with the changes in the weather. The solution to this problem is to use a computer, which can calculate the height of a constant pressure surface at successive times and so predict or forecast the situation at a given time (Figure 11.i).

Figure 11.ii. A key to nephanalysis charts.
After Barrett

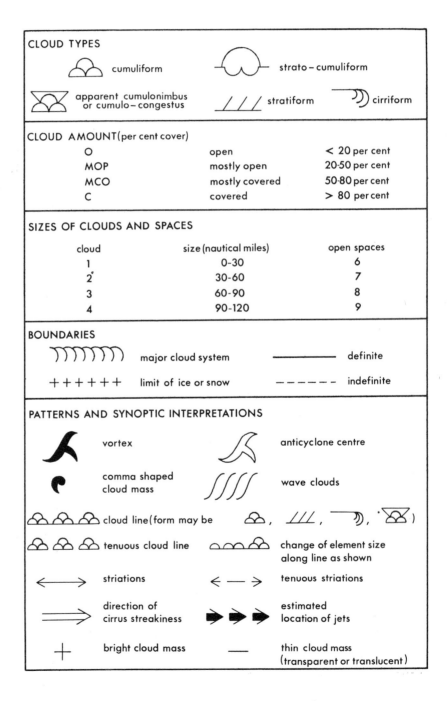

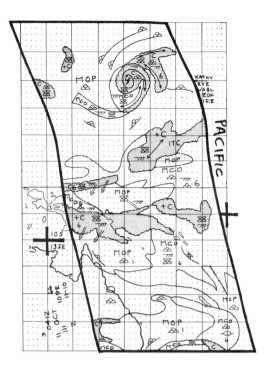

Figure 11.iii. A nephanalysis chart for part of the Pacific to the northeast of Australia, for 11 October, 1966. The ITC is marked by the cloud mass near the top right, and north of it is hurricane Kathy. The stippled areas are significant cloud masses.

Meteorological Satellites

The first TIROS (Television and Infra-red Observation Satellite) weather satellite was launched from Cape Kennedy on 1 April, 1960. The TIROS satellites and the later Nimbus satellites carry television cameras which are now able to send continuous pictures of the earth to ground stations.

In areas with a close network of traditional observation stations, the satellite photographs have so far been regarded as supplementary to other observations, but in areas of poor coverage by ground stations the satellites could greatly increase the accuracy of forecasts, and also our understanding of weather processes.

Most satellites so far have photographed cloud cover and made radiation measurements from which cloud cover at night and the height of the cloud tops can be estimated. Figure 11.ii shows a key to the symbols used. From these photographs a nephanalysis or chart of cloud types, as in Figure 11.iii, can be made and the position of fronts drawn in. Successive photographs of the same area can be used to calculate the rate of development of cloud and pressure systems and also their rate of movement. Nimbus satellites are being developed which will be able to measure air temperature, and in the more distant future, buoys with meteorological instruments can be set to drift on the sea, and balloons to drift through the air. Such buoys will transmit readings to satellites of atmospheric data which the satellites will broadcast to ground receivers. By this means a greater understanding of ocean currents and upper air movements will be obtained.

Plate 12 on page 93 shows an example of a mosaic built up from TIROS photographs which indicates the usefulness of such a system. Similar photographs have also been used for studying ice conditions in polar regions and snow cover in mountains. In the future a new form of climatology may be developed using cloud cover pictures at different seasons.

Value of Weather Forecasts

The World Weather Watch system which commenced operations in 1968 has been approved by over 100 countries. It will co-ordinate an increasingly dense network of surface, upper air and satellite observation stations and be able to provide much more reliable forecasting. It is estimated that the financial savings will exceed £6000 million annually in the fields of agriculture, construction, fuel utilities, forestry and transport. The increased knowledge of the atmosphere should also bring closer the possibility of weather modification by man.

12 Climatic Change and Variation

The earth's climate is known to have varied greatly during the last thousand million years in which geological history can be read from the evidence of rocks and fossils, and to a much smaller extent during the brief period of human history for which records are available. It seems desirable, therefore, to distinguish between the major long-term 'changes' and the minor short-term 'variations', for it is by no means certain that one cause is responsible for both changes and variations.

Climatic Change

There are several kinds of evidence which make it clear that the earth's climate has undergone many changes. The clearest and most dramatic changes are the infrequent but fairly regular ice ages, which left behind the tillites by which we recognise them, and which seem to occur at intervals of about 250–300 million years. We live in an ice age, and it is sometimes forgotten that the more 'normal' climate of the earth would be one in which there were no ice caps and the temperature differences between the equator and the poles were much less than they are now. The evidence of red sandstones, salt deposits and other evaporites suggests that many areas now experiencing humid climates were once deserts; while the presence of coal and oil deposits indicates that Antarctica and the now arid Persian Gulf once had hot humid climates supporting forest vegetation. A reconstruction of past climates is shown in Figure 12.i.

There is little agreement amongst the climatologists, geologists, astrophysicists and others who are concerned with investigating the causes of climatic change, but their more probable theories may be classified into three main groups:

(1) *Terrestrial Causes:*
    (*a*)   changes in the extent, distribution and relief of continents;
    (*b*)   changes in the distribution of oceans and ocean currents;
    (*c*)   changes in the internal heat and volcanicity of the earth;
    (*d*)   changes in the cloudiness and carbon dioxide content of the atmosphere;
    (*e*)   changes in the position of the poles.
(2) *Variations of the Earth's Orbit* (*Planetary Hypothesis*).
(3) *Variations of Solar Radiation* (*Extra-terrestrial Hypothesis*).

Terrestrial Causes

(a) Extent, Distribution and Relief of Continents

An approximate relationship in time between ice ages and orogenies has been the starting point for many theories which ascribe ice ages to the effects of uplifted mountain ranges. Air is forced to rise over mountains and in doing so it expands and cools, and the water vapour within it condenses into clouds. Clouds reflect solar radiation and as a result snow and ice are features of all high mountains, even at the equator. As ice is locked up in glaciers and ice caps, sea level falls and there is a

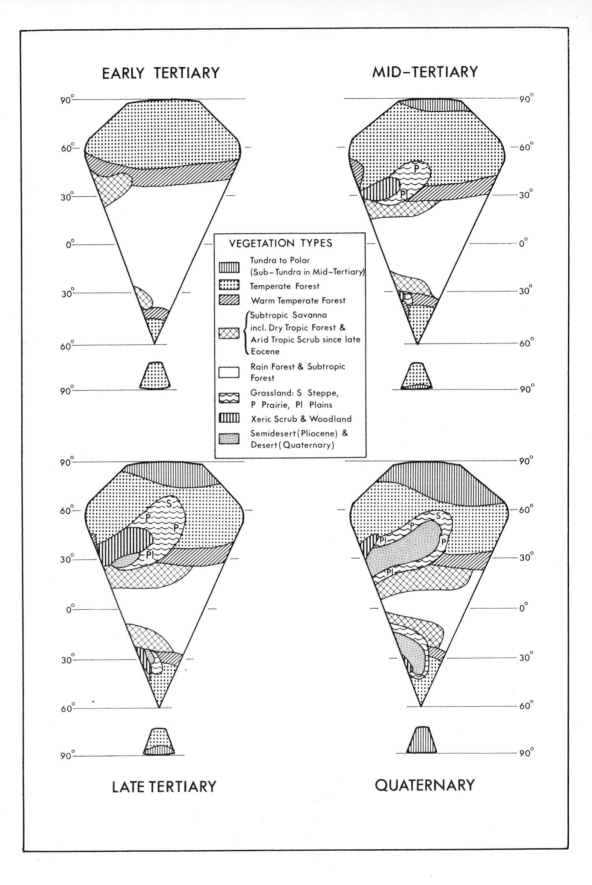

EARLY TERTIARY

MID-TERTIARY

VEGETATION TYPES

Tundra to Polar
(Sub-Tundra in Mid-Tertiary)

Temperate Forest

Warm Temperate Forest

Subtropic Savanna
incl. Dry Tropic Forest &
Arid Tropic Scrub since late
Eocene

Rain Forest & Subtropic
Forest

Grassland: S Steppe,
P Prairie, Pl Plains

Xeric Scrub & Woodland

Semidesert (Pliocene) &
Desert (Quaternary)

LATE TERTIARY

QUATERNARY

◁ Figure 12.i. Axelrod's
reconstruction of Tertiary
and Quaternary vegetation
zones—which are an
indication of climate.

consequent increase in continentality, and a decrease in the circulation
of ocean currents, so that the warming effects of currents moving
polewards are reduced. Finally, the snow-covered surfaces themselves,
with their very high albedo, reflect most of the sun's radiation and thus
increase the cooling effect.

C. E. P. Brooks attempted to work out the effects of self-induced
cooling produced by reflection from snow surfaces, and he concluded
that if the winter temperature at a sea-covered pole, such as the North
Pole, were just above the freezing point of seawater ($-2°C$), and if
there then occurred an initial drop in mean temperature of only $\frac{1}{3}°C$,
the increased area of ice and snow would so cool the atmosphere above
it that an ice cap with a radius of 2400 km and a polar temperature of
$-27°C$, would be produced.

The end of such an ice age would begin when the mountains were so
worn down by erosion that the glaciers and snowfields would diminish
and allow a rise in air temperatures. The attendant rise of sea level
would allow more warm water to pass into the Arctic Ocean through
the Bering Straits, and through the straits between Greenland and
Europe. This would melt the Arctic sea ice and cause a general warming
which could finally dissipate the remaining ice in the world.

The earth's climate after the ice age would become much like that of
the Tertiary era. The polar fronts would vanish and the subtropical
high pressure systems would migrate polewards so that the mid-
latitudes would become rainless in summer, and winter circulation
would also be reduced. The earth as a whole would have warmer and
drier climates. Flohn has estimated that the average equatorial temperature would be $27°C$—as it is now—and that the North Pole would
have an average winter temperature of $1°C$, and an average for the
summer of $8°C$. The mean temperature of the earth would be $19°C$,
compared with $15°C$ at present.

There can be no doubt that the occurrence of mountain ranges
influences the spread of ice conditions but this theory cannot be
regarded as a complete answer to the problem of ice ages because:

    (i) the Quaternary ice age did not develop until after the maximum
       of the Tertiary orogenesis;

   (ii) older ice ages are not all related to orogenies—notably the south-
       ern hemisphere Carbo-Permian glaciation;

  (iii) orogenies do not explain the multiple advances and retreats of
       the Quaternary ice age.

(b) Oceanic Causes

The most notable hypothesis relating ice ages to oceanic events is
that put forward by Donn and Ewing. They suggest that an ice-free
Arctic Ocean would allow sufficient evaporation to increase precipitation, so as to form more snow and thus bigger glaciers on the mountains
of Greenland, Scandinavia and North America. As sea level fell, warm
water would be prevented from reaching the Arctic Ocean and the
Arctic Sea would ice over. With the icing over of the sea, the glaciers
would dwindle because they no longer received sufficient snow, and the
polar ice sheet would melt, The cycle would then begin again.

This hypothesis is not an attempt to find a cause for all ice ages, as it
depends for its validity on the form of the Arctic Ocean basin, which

was certainly not always the same as it is now. Secondly, it does not explain the origin of the first advance of the ice but only the succession of glacials and interglacials. Thirdly, there is some evidence from fossil deposits in northern Canada that the Arctic Ocean was not open and closed at the times required by the hypothesis.

(c) Internal Heat and Volcanicity

One of the earliest suggestions put forward to account for the ice ages was most clearly expounded by Buffon in the eighteenth century. He said that the slow cooling of the once incandescent earth ended with the Quaternary ice age. Buffon did not know of the pre-Quaternary ice ages and Trabert's work shows that the earth's heat has virtually no effect on the atmosphere.

A much more tenable hypothesis is that which ascribes ice ages to the cooling effect produced by volcanic dust in the atmosphere. In its support are the contentions that volcanic dust is released during periods of orogenesis which are known to coincide approximately with ice ages. The effects of a few recent volcanic eruptions have been studied. On 27 August, 1883, the island of Krakatoa in the Dutch East Indies erupted and exploded into the air 53 cubic kilometres of rock. The fine volcanic dust rose more than 32 km into the atmosphere and drifted round the earth. It cut down incoming solar radiation by as much as 40%, and incoming solar energy was reduced for more than 3 years in many parts of the world. The meteorologist W. J. Humphreys made a study of the effects of volcanic dust and concluded that, by reducing incoming radiation but still permitting long heat waves to escape from the atmosphere a volcanic dust pall could reduce ground temperatures by several degrees. Furthermore, the clouds would not cover the earth evenly and so would tend to create or increase temperature differences between one part of the world and another, thus stimulating the circulation. The grains of dust would also act as nuclei of condensation and so increase precipitation and cloudiness.

During the last 80 years, in which reliable radiation and weather records have been kept, there does seem to have been some relationship between cool periods and volcanic activity and between warm periods and lack of activity. To suggest, however, that volcanic activity is a sole cause of ice ages would be to ignore the evidence of pre-Quaternary glaciations, some of which do not appear to have been periods of active volcanism. Even the Quaternary came after a Tertiary era which was active volcanically but also warm. In brief, then, volcanism is probably no more than a subsidiary influence on glaciation and not a prime cause.

(d) Carbon Dioxide Content of the Air and Cloudiness

The varying carbon dioxide content of the air undoubtedly affects incident radiation. When the carbon dioxide content is high, a greater proportion of solar radiation is absorbed (greenhouse effect) and temperatures will be high. A low carbon dioxide content would produce lower temperatures and hence might induce glaciation. The periods of earth history in which carbon dioxide content of the air is assumed to be high are the periods of large scale coal formation and active volcanism, but the relationship in time between these events and glaciation is not regarded as being close enough for them to be a cause of glaciation. An interesting observation is that since the Industrial Revolution, and during the last 100 years, there has been a release of large quantities of carbon dioxide into the atmosphere. This has coincided with a general amelioration of climate which lasted until about 1920, when a period of colder winters and greater storminess began in Europe.

CLOUDINESS

Brooks has calculated that a mean cloud cover over the whole earth results in a $37\frac{1}{2}\%$ loss of solar radiation, and that if this cloud cover were reduced by 1/10 the mean annual temperature would rise by 8°C. Changes in dust in the atmosphere, or an increase in solar radiation to produce more evaporation and hence cloud, are usually suggested as the causes of increased cloudiness. The dust or radiation changes are therefore the prime causes, and cloudiness is an effect; hence cloudiness cannot be regarded as a cause of glaciation.

(e) Position of the Poles

The former tropical climate of Antarctica and the glaciation of South Africa are clear indications of changes in the position of the poles. The early geologists usually assumed that there was a shift of the earth's axis of rotation, but physical considerations weigh against this idea and it is now regarded as more probable that the continents have drifted. Movement of the poles might account for the pre-Quaternary ice ages but cannot account for the Quaternary, as it is established that during the Pleistocene period, and probably in the early Tertiary, the position of the poles was no different from what it is now.

Planetary Hypothesis

In 1920 the Serbian physicist Milutin Milankovitch worked out a 'radiation curve', in other words, a graph showing how much solar radiation is received by the earth. He extended the curves backwards in time through the Quaternary era. He calculated that fluctuations in the earth's orbit and axis of rotation periodically change the pattern of heat received from the sun, so that there are long periods when the summers are cool and the winters mild, alternating with periods of hot summers and cold winters. In a period of cool summers less winter ice will be melted so that there would be an onset of glaciation. Milankovitch calculated that the coolest summers would occur at intervals of 40,000 years. If Milankovitch is correct, we are in an interglacial period of warm summers but can expect a cool summer period and a further glaciation in about 10,000 years' time (Figure 12.ii).

Figure 12.ii. Solar radiation curves after Milankovitch (1938) for latitudes 75°N and 45°N. Glacial phases are according to Köppen.

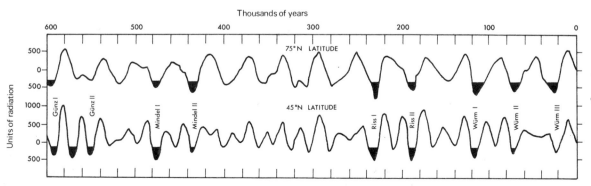

The radiation curves of Milankovitch have not met with general acceptance, although Zeuner and many others have agreed with them, and Emiliani and Wundt have proposed different solutions using radiation curves. The main objections are that it is difficult to understand why there should be breaks of 250 million years or more between periods of cool summers and, perhaps more important, that the curves do not appear to fit the chronology of the glacials and interglacials established by other methods. Furthermore, many workers reject the assumption that high radiation necessarily produces cool summers and mild winters. It would seem that small scale oscillations of radiation, produced by changes in the axis and orbit of the earth, are not great enough to cause glaciation.

Extra-terrestrial Hypothesis

In most of the hypotheses already mentioned the sun has been regarded as a star emitting a constant amount of radiation, but it is reasonable to assume that it has not remained eternally unchanged and that the intensity of radiation may have varied uniformly or periodically, and thus greatly influenced the earth's climate.

Most early theories assumed that decreased solar radiation would be necessary to lower air temperatures and give rise to glaciation, but Sir George Simpson demonstrated in 1937 that increased solar radiation can also cool the atmosphere. The essence of his theory, as modified by Willett, is that increased radiation would increase evaporation and therefore increase cloud cover. This would lower surface temperatures and increase precipitation. The increased snow cover would also reflect more radiation and a glacial advance would occur. Accumulation of ice and snow would cease at the high point of radiation, as temperatures would then be too high, and an interglacial period would occur. Figure 12.iii demonstrates this for the Quaternary glaciations.

More recently (1958) Öpik has suggested that physical changes within the sun would cause periodic variations in radiation which could fit in with the period of glaciations on earth. He suggests that a fall of the sun's heat output by 8% would reduce the global average temperature by the 5°C of the coldest glacials, and that a rise of 9% from the norm would make the global average temperature 22°C. He regards the sun as steadily increasing in temperature so that in about 1000 million years from now—and after three more glaciations—the average air temperature would be 38°C and life as we know it would cease.

Conclusions

There seem to be two main factors which influence earth climates:

(1) Radiation from the sun is almost the only source of energy for atmospheric processes, and it seems reasonable, therefore, for it to have a major influence on climatic change.
(2) Geographical conditions strongly influence climates of individual continents and may therefore be expected to be significant in the development of glacial periods. A feature of glaciations is their periodicity and this also has to be explained.

Our knowledge of the solar system and the earth is not yet great enough for any definitive statements to be made, but it is probable that

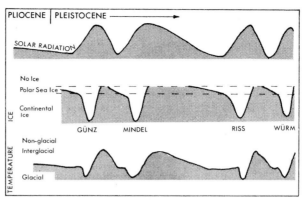

Figure 12.iii. Correspondence of solar radiation, ice accumulation and temperature during the Pleistocene.

After Simpson as modified by Bell

solar radiation variation is the main cause of lowering of earth temperatures, and that the coincidence of this with such terrestrial events as orogenesis with its uplift and volcanism causes glaciations. The terrestrial events may certainly be regarded as controlling the geographical extent of the glaciations.

Climatic Variation

The cave drawings of early man indicate that areas of the Sahara which are now desert were once savanna. More recent records—many of them written—indicate even more clearly that climatic variations of relatively short periods have vitally affected human life. An example is the warm period in Viking times around 1000 AD. The Vikings were able to establish colonies in Iceland and Greenland and to visit North America with little hindrance from ice or severe storms. Between 1000 AD and 1200 AD the Norse settlers in southern Greenland were able to bury their dead in ground that is now permanently frozen. At this time there may well have been little pack ice on the polar seas. From about 1500 AD onwards the climate became more severe, to reach a maximum of cold in about 1800. In this time frost fairs were held on the Thames above London Bridge, the pillars of which were so thick that the flow of the river was sufficiently impeded for continuous ice to form. Below

Figure 12.iv. Variations in Britain's climate in the last 1000 years. After a period of warmth culminating in the thirteenth century there came a 600 year period of harsher weather—the 'Little Ice Age'. The beginning of the twentieth century was warm but since about 1940 there has been a marked cooling.

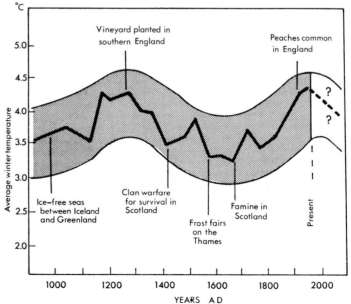

the bridge only ice floes occurred, much as they do on the present-day river Rhine, where the winter weather must be much like that of eighteenth-century London. So cold did it become that the Danish Parliament in 1784 proposed to evacuate the population of Iceland and to resettle it in Europe. Temperatures over the last 1000 years in Britain are shown in Figure 12.iv.

There can be no doubt that short-term climatic variations do occur. H. H. Lamb, of the British Meteorological Office, has attempted to elucidate the patterns of changes in the last 200 years and to construct weather charts for the January and July circulations for each year. Willett in the USA has reconstructed details of the climatic variations in the western plains of the USA for the same period. Willett believes that northern hemisphere weather responds to an 80–90 year sun-spot cycle which has the following features:

(1) The first quarter of the cycle has a low level of solar activity with a low latitudinal pattern of general circulations involving small contrasts between marine-continental and summer-winter conditions.

(2) The second and third quarters of the cycle are periods of increasing solar activity with a poleward shift of the zonal circulation. A warm, wet climate occurs in high latitudes and a warm, dry one in mid-latitudes. This situation favours glacial retreat.

(3) The third quarter of solar maximum activity causes a breakdown of the zonal (i.e. east-west) circulation and the development of blocking anticyclones and of depressions. This pattern produces strong east-west and maritime-continental contrasts and great storminess. The cold winters and warm dry summers do not favour glacial advance except on the western sides of continents in high latitudes.

Willett has also interpreted a 20–24 year cycle, double the length of the well known 11 year sun-spot cycle, which shows minor effects on the weather and which is superimposed on the 80–90 year cycle.

Summary of Evidence of Climatic Changes and Variations

The evidence may be summarised according to the length of time of the change and the type of evidence used:

Variations of 10–100 years: Instrumental recordings, behaviour of glaciers; records of floods, river flows, lake levels; diaries, crop yields, tree rings.

Variations of 250–1000 years: Diaries or historical evidence left by man; fossil tree rings; archaeological evidence; lake levels, varves and lake sediments, oceanic sediment cores; pollen analysis, C^{14} dating, moraines.

Changes of 100,000 years: Flora and fauna of interglacial deposits; pollen analysis; extent of periglacial activity, moraines, loess, old river terraces, old soils.

Changes of millions of years: Deposits, fossils, main dating by radioactivity of rocks.

13 The Classification of Climates and World Climatic Patterns

The basis of all classification is the seeking of characteristics common to many individuals, by which those individuals may be classed or grouped so that simplicity and order can be introduced into what would otherwise be a bewildering collection of disparate units. Classification thereby aids the search for general truths from numerous units of knowledge or data. Most phenomena may be grouped by a method which is either empirical or genetic. The empirical method describes the characteristics of phenomena, and the genetic method classifies according to the origin of phenomena. In climatology, the empirical method (sometimes called the generic method) involves a classification based on the climatic elements, such as temperature and precipitation; the genetic method is based on climate controls, such as air masses.

It follows from the statements above that there are several methods of classifying climates. In order to choose between these methods, their purpose must be considered. The climatologist's use of air masses, which control climates, may not be helpful for geographers, agronomists, biologists or members of some other scientific disciplines. The genetic method of classification has therefore been largely ignored and the many

Figure 13.i. The Köppen-Geiger system of climate classification.

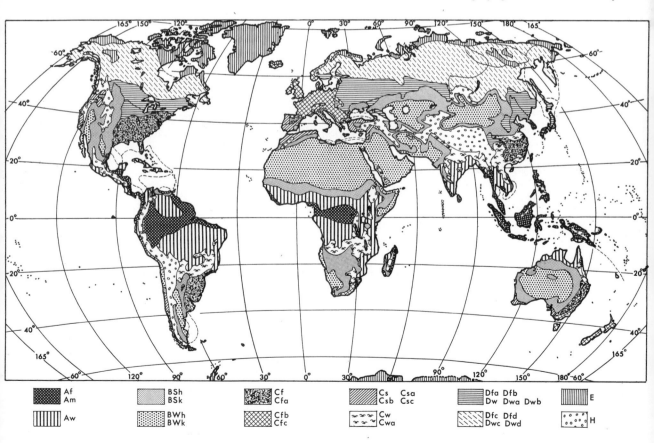

Af Am	BSh BSk	Cf Cfa	Cs Csa Csb Csc	Dfa Dfb Dw Dwa Dwb	E
Aw	BWh BWk	Cfb Cfc	Cw Cwa	Dfc Dfd Dwc Dwd	H

problems of definition which such a method would involve have not been investigated. The most common type of classification is based upon temperature and precipitation. This type of system is most easily applied because it is quantitative, based on easily made measurements and, in areas for which suitable data are lacking, can be extended by the use of obvious vegetation types which grow under known conditions of temperature and precipitation.

The disadvantages of all systems of classification are as follows:

(1) They are all over-simplified and are forced to leave out a multiplicity of other elements or controls.

(2) They all have to draw boundaries round distinctive types of climate, even though it is clearly recognised that this man-made boundary may have no reality in nature and would often be better expressed as a transitional zone.

(3) They are usually only accurate for a core area and become progressively less reliable towards the boundaries of a region.

(4) They are static systems incapable of taking into account short-term variations in climate.

(5) They are often unreliable when using vegetation as a basis for subdivision, for vegetation is often modified by man and may have developed under conditions which no longer exist, so that it does not represent what would develop under the present climate, and in the absence of human interference.

Classification Systems

Wladimir Köppen and C. W. Thornthwaite have produced the best known classifications based on quantitative and empirical data. No really adequate genetic system has been devised, although A. N. Strahler has outlined a possible classification. The problem of the genetic type is that it is not yet amenable to quantitative expression, although potentially it is of more value to the climatologist—and possibly to geographers—than other systems. These three methods of classification will be briefly described here and a more detailed study of the world distribution of climatic types will be found in Chapters 14–18, where a modification of Köppen's classification is used.

Empirical Classifications

Köppen-Geiger Classification

Probably the most widely used system of classification is that of Köppen, who produced his first version in 1900. This was based primarily on vegetation zones and in 1918 it was thoroughly revised to use temperature, rainfall and seasonal variations as a basis. Since 1918, as more information became available, the system has been further refined and revised by Köppen and by Rudolf Geiger. The classification uses a code of letters to designate major climatic groups and their subdivisions. The 5 major groups are named A, B, C, D and E climates.

A. *Tropical climates:* These have an average monthly temperature which always exceeds 18°C. These climates have no winter and a high rainfall which always exceeds the annual evaporation.

B. *Dry climates:* These have no water surplus and the potential evaporation always exceeds the precipitation.

C. *Warm temperate or mesothermal climates:*	These have a coldest month with a temperature under 18°C but over −3°C. At least one month has an average temperature above 10°C. C climates therefore have definite summer and winter seasons.
D. *Snow or microthermal climates:*	These have a coldest month with an average temperature of below −3°C and the average temperature of the warmest month is above 10°C (this isotherm being chosen as the approximate poleward limit of tree growth).
E. *Ice climate:*	These climates have a warmest month with an average temperature below 10°C. They thus have a very cool 'summer'.

The location of these major regions is shown on the map in Figure 13.i. It should be noted that this map distinguishes highland regions (H), with their wide range of climates. The groups A, C, D and E are defined by temperature, but group B by the deficiency of precipitation in relation to evaporation.

Subgroups of the main classification are defined by a second letter:

-f: Moist climates having adequate precipitation throughout the year for vegetation growth; this may be true of A, C and D groups.
-w: The winter is the dry season.
-s: The summer is the dry season.
-m: Monsoon type of precipitation in A climates only, which thus have a short dry season.

Two other letters (always written as capitals) refer only to B climates:

(B)S. *Steppe climate:*	A semiarid type in low latitudes with 38–76 cm of precipitation per year.
(B)W. *Desert climate:*	An arid type usually with less than 25 cm of precipitation annually. Wüste (W) is German for desert.

The following subgroups are identified for E climates: ET—tundra climate; EF—perpetual frost.

A third letter was added to the code by Köppen to distinguish more variations in temperature:

--a: With a hot summer in which the warmest month has an average temperature over 22°C. This applies to C and D climates.
--b: With a warm summer in which the warmest month has an average temperature below 22°C. This applies to C and D climates.
--c: With a cool, short summer with fewer than 4 months having an average temperature over 10°C. This applies to C and D climates.
--d: With a very cold winter in which the coldest month has an average temperature below −38°C. This applies to D climates only.
--h: With dry and hot conditions, in B climates only, in which the mean annual temperature is over 18°C.

--i: With a range of temperature between the warmest and coldest month of less than 5°C.

--k: With dry and cold conditions in B climates in which the mean annual temperature is below 18°C.

The main climatic types of this system are shown in Table 13.1. The Köppen-Geiger system has even more symbols and subdivisions than are mentioned here, but their use is largely for more local conditions than the world patterns with which most users are concerned.

Table 13.1 Köppen-Geiger Classification

Type	Meaning of Major Group	Second or Subgroup	Third Variation
Af	Hot	Rainy in all seasons	
Am	Hot	Monsoonal rain	
Aw	Hot	Dry winter	
As	Hot	Dry summer	
BSh	Dry	Semiarid steppe	Very hot
BSk	Dry	Semiarid steppe	Cold (or cool)
BWh	Dry	Arid	Very hot
BWk	Dry	Arid	Cold (or cool)
Cfa	Mild winter	Moist all seasons	Hot summer
Cfb	Mild winter	Moist all seasons	Warm summer
Cfc	Mild winter	Moist all seasons	Cool, short summer
Cwa	Mild winter	Dry winter	Hot summer
Cwb	Mild winter	Dry winter	Warm summer
Csa	Mild winter	Dry summer	Hot summer
Csb	Mild winter	Dry summer	Warm summer
Dfa	Severe winter	Moist all seasons	Hot summer
Dfb	Severe winter	Moist all seasons	Warm summer
Dfc	Severe winter	Moist all seasons	Short, cool summer
Dfd	Severe winter	Moist all seasons	Very cold winter
Dwa	Severe winter	Dry winter	Hot summer
Dwb	Severe winter	Dry winter	Cool summer
Dwc	Severe winter	Dry winter	Short, cool summer
Dwd	Severe winter	Dry winter	Very cold winter

ET — Polar climate—short summer allows growth of tundra vegetation
EF — Perpetual ice and snow

Thornthwaite's Classification of Climates

Thornthwaite's first classification was published in 1931 and 1933, and the second in 1948. They are fundamentally different although they have some similarities.

THE 1931–33 CLASSIFICATION

The early Thornthwaite classification, like Köppen's, is quantitative and employs a letter code for designating individual climates. Thornthwaite also uses the plant as a 'meteorological instrument', but instead of considering its requirements in terms of temperature and precipitation, he has developed the new and valuable concept of temperature efficiency (T/E) and precipitation effectiveness (P/E). The T/E and P/E cannot be expressed in ordinary climate values and the boundaries are therefore difficult to draw. Furthermore, 32 main climatic regions appear on his world map—about 3 times as many as on Köppen's—and this also makes the system a little more difficult to use.

Precipitation effectiveness for plant growth is determined by dividing total monthly precipitation by total monthly evaporation (called the P/E ratio). The sum of 12 monthly P/E ratios is called the P-E index. The main problem with this method is that evaporation data are not available for most of the earth, so Thornthwaite had to develop a formula from temperature and precipitation relationships as observed at 21 stations in the southwestern USA. The formula was then applied to the rest of the earth. The formula developed is:

$$\text{P-E index} = \sum_{n=1}^{n=12} 115\left(\frac{P}{T-10}\right)^{\frac{10}{9}} \quad \text{P/E ratio} = \frac{P}{E} = 115\left(\frac{P}{T-10}\right)^{\frac{10}{9}}$$

where P and E are measured in inches and T is in degrees Fahrenheit. Based upon this P-E index, Thornthwaite distinguished five humidity provinces, each of which is associated with a characteristic type of vegetation (Table 13.2).

Table 13.2. Thornthwaite's Humidity Provinces

Humidity province	Vegetation	P-E index
A. Wet	Rain forest	128 and above
B. Humid	Forest	64–127
C. Subhumid	Grassland	32–63
D. Semiarid	Steppe	16–31
E. Arid	Desert	under 16

The 5 principal humidity provinces are subdivided into 4 subtypes based upon seasonal precipitation:

r: Abundant rainfall at all seasons.
s: Rainfall deficiency in summer.
w: Rainfall deficiency in winter.
d: Rainfall deficiency in all seasons.

Thornthwaite also developed a formula for measuring thermal efficiency which is determined solely by mean monthly temperatures. This T-E index is the sum of the 12 monthly T/E ratios which are computed by:

$$\text{T/E ratio} = \frac{T-32}{4} \quad \text{T-E index} = \sum_{n=1}^{n=12}\left(\frac{T-32}{4}\right)$$

Based upon the T-E index, 6 temperature provinces are recognised (Table 13.3).

THORNTHWAITE'S 1948 CLASSIFICATION

This method regards vegetation as 'a physical mechanism by means of which water is transported from the soil to the atmosphere; it is the machinery of evaporation as the cloud is the machinery of precipitation'. The combined loss through evaporation from the soil surface and transpiration from plants is called 'evapotranspiration'. Instrumental determination of evapotranspiration is difficult and Thornthwaite was forced to compute it as a function of temperature.

Table 13.3. Thornthwaite's Temperature Provinces

Temperature province		T-E index
A'.	Tropical	128 and above
B'.	Mesothermal	64–127
C'.	Microthermal	32–63
D'.	Taiga	16–31
E'.	Tundra	1–15
F'.	Frost	0

The comparison between precipitation and potential evapotranspiration is a most valuable one and is independent of vegetation. It is most valuable for computing moisture deficiency at given stations for different parts of the year. Because of the lack of data, a world map using this method has not been produced, although potentially it has great value as a rational and practical classification.

Genetic Classification

Air Mass Classification of Climates

Strahler distinguishes 13 types of climate, each of which belongs to one of 3 groups. Group I includes the tropical and equatorial air mass source regions and the equatorial convergence between them. These climates are controlled by the dynamic subtropical anticyclones, which are areas of subsidence and basically dry, and also by the zone of convergence between them. Group II climates are in the zone of intense interaction between contrasting air masses, particularly along the zone of the polar front. In this zone, polar and tropical air masses

Figure 13.ii. A scheme for the classification of world climate based on air mass source regions. Three major climate groups are shown. Land masses are shaded.
After Petterssen and Strahler

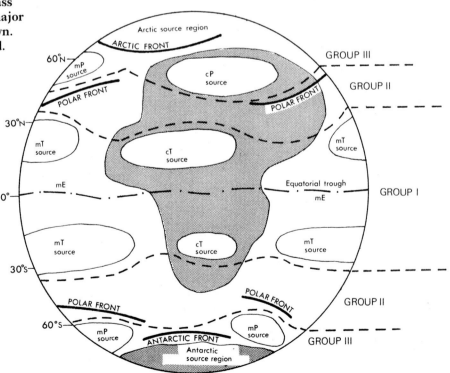

conflict and produce processions of east-moving depressions. The Group III climates are dominated by polar and Arctic air masses which meet along the polar front. The location of these three groups is shown in Figure 13.ii, which uses a schematic continent with a form approximating to that of Eurasia-Africa.

The 13 main climatic types of this classification, together with Strahler's names for them and the Köppen climates which correspond with them, are outlined in Table 13.4.

Table 13.4. Strahler's Climatic Types

Strahler's name	Description	Köppen symbol
Group I: E and T air masses		
(1) Equatorial rain forest	Dominated by mT and mE air masses. Convergence along ITC gives heavy rain and convectional storms.	Af, Am
(2) Trade wind littoral	Trades (i.e. tropical easterlies) bring mT air from moist western sides of oceanic subtropical highs to give heavy but seasonal rainfall.	Af, Am
(3) Tropical desert and steppe	Source regions of cT air over the tropics. Very dry.	BWh, BSh
(4) West coast desert	On west coasts bordering subtropical highs. Subsiding mT air is stable and dry. Foggy on coasts.	BWk, BWh
(5) Tropical savanna	Seasonal alternation of mT and cT air gives wet season at time of high sun, and dry season at time of low sun.	Aw, Cwa
Group II: T and P air masses		
(6) Humid subtropical	Subtropical eastern continental margins with mT air from the western sides of oceanic highs. Heavy rain at time of high sun. Winters are cool with invasions of cP air. Frequent cyclonic storms.	Cfa
(7) Marine west coast	Frequent cyclonic storms with cool mP and warmer mT air bringing variable cloudy weather.	Cfb, Cfc
(8) Mediterranean	mP air dominates in cool, wet winter, and stable dry mT air in hot, dry summer.	Csa, Csb
(9) Mid-latitude desert and steppe	Interior deserts and steppes dominated by hot cT air in summer and cold cP air in winter.	BWk, BSk
(10) Humid continental	Located in centre and east of mid-latitude continents. Polar front variable weather. Largely mT air in summer and cP air in winter.	Dfa, Dfb Dwa, Dwb
Group III: P and A air masses		
(11) Subarctic	Source region of cold cP air. Depressions with mP air bring light precipitation. Extreme temperature ranges.	Dfc, Dfd Dwc, Dwd
(12) Tundra	Arctic front weather with interacting mP, cP and A air. Very cold, but oceanic influences prevent severity of subarctic climate.	ET
(13) Ice cap	Source of Arctic (A) and Antarctic (AA) air. Intensely cold air masses.	EF
Highland climates	Cool to cold and moist climates of considerable local variations. These are not included in the general classification.	

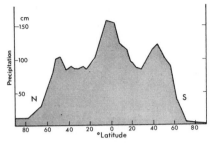

Figure 13.iii. Distribution of annual precipitation arranged by latitude zones. *After Brooks and Hunt*

The Pattern of World Climates

Precipitation, temperature and pressure are the three main controls on the climate of any place. Of these, pressure is the least obvious control, although the movement of pressure systems and the rising and falling of air in them are responsible for the rainfall regime in many areas. Pressure systems and world distribution of pressure have already been discussed.

World Precipitation

World maximum rainfall occurs in the equatorial region of converging air (ITC), although individual record totals do occur outside this zone. Polewards in the subtropical latitudes of diverging and descending air of the subtropical high pressure systems, rainfall is lower, but rises again to secondary maxima at about latitudes 40–50° N and S, where mid-latitude travelling low pressure systems and convergences are responsible for the precipitation. The lowest precipitation zones are in the high latitudes, where low temperatures and subsiding air are responsible. (See Figure 13.iii.)

On a hypothetical continent (Figure 13.iv) the equatorial zone of high precipitation would stretch as a band across the equator. In fact no continent does have such a band because in South America the Andes and in Africa the east African high plateaux impose altitudinal controls (Figure 13.v). The broad, high rainfall zone on the eastern side of the hypothetical continent is produced by the unstable western margin of an oceanic subtropical anticyclone, but on the western side incursions of maritime air are more limited.

Figure 13.iv. World rainfall regions on a hypothetical continent.

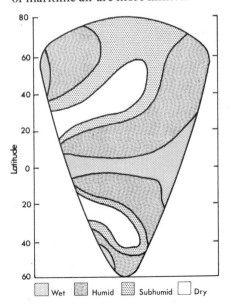

Wet   Humid   Subhumid   Dry

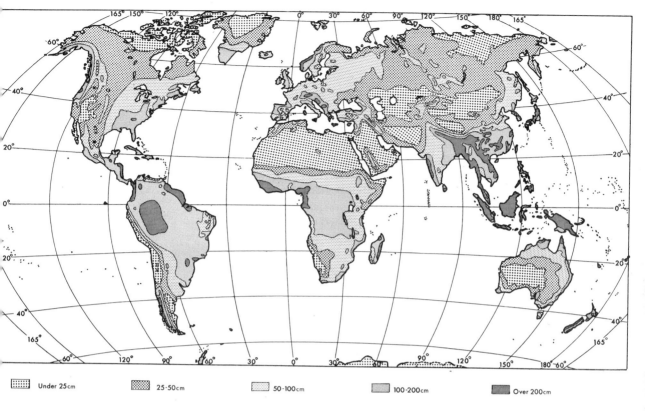

| Under 25cm | 25-50cm | 50-100cm | 100-200cm | Over 200cm |

Figure 13.v. World annual average precipitation.

The dry zones extend from low to middle latitudes. In low latitudes, the western sides of the continents are strongly influenced by the stable eastern ends of oceanic subtropical high pressure systems, and offshore cold water increases the aridity. In middle latitudes, the aridity is caused by a continental location far from the source of maritime air. Marginal locations in middle latitudes benefit from maritime air carried over them by the travelling depressions and so have high rainfall.

Figure 13.vi. Patterns of seasonal precipitation on a hypothetical continent.
After Trewartha

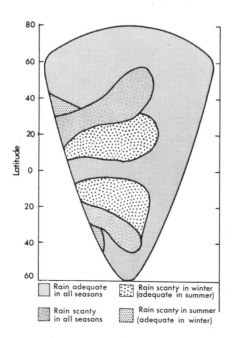

| Rain adequate in all seasons | Rain scanty in winter (adequate in summer) |
| Rain scanty in all seasons | Rain scanty in summer (adequate in winter) |

Figure 13.vii. Schematic cross-section through the atmosphere showing the main zones of horizontal convergence and ascent, and of divergence and descent, together with associated seasonal characteristics of precipitation: (A) during the northern hemisphere summer; (B) during the northern hemisphere winter; (C) zones of seasonal precipitation. Actual climates depart from this scheme because of irregular distribution of land masses.

After Petterssen

The distribution of rainfall is not even throughout the year (Figure 13.vi). In equatorial latitudes, rainfall is not only high but occurs throughout the year because at no time is the convergence (ITC) far away from this zone. In the higher latitude tropics, rainfall is more clearly seasonal because the ITC follows the sun, so that during the summer period of high sun there is a wet season but the winter period of low sun is dry (Figure 13.vii).

In latitudes of about 30–40°, there is a small area, corresponding to the Mediterranean Sea, where summer is dry under the influence of subtropical high pressure systems but where winter depressions bring moist maritime air from the west.

In the humid middle latitudes, travelling depressions bring rain at all times of the year although west coast locations have a maximum precipitation in winter when the low pressure belts and polar fronts are closer to the equator, while in summer these fronts retreat polewards. In the interior of the continents, winter anticyclones and low temperatures keep precipitation low, but in summer convectional rainfall provides the rainfall maximum. The eastern margins of the middle latitude continents have rain throughout the year, but usually with a summer maximum when maritime air is not blocked by the high pressure systems of winter.

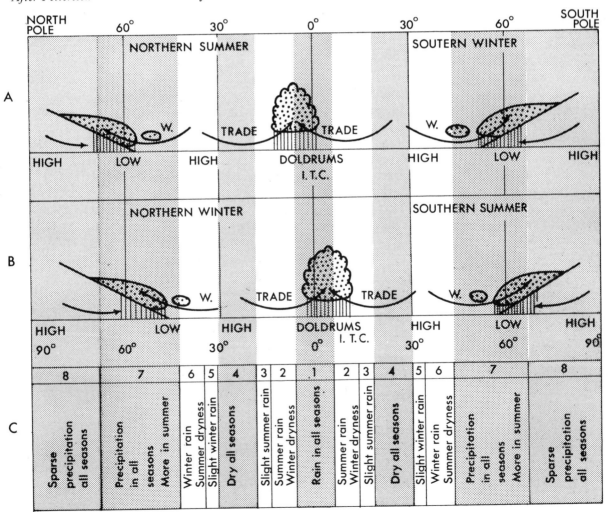

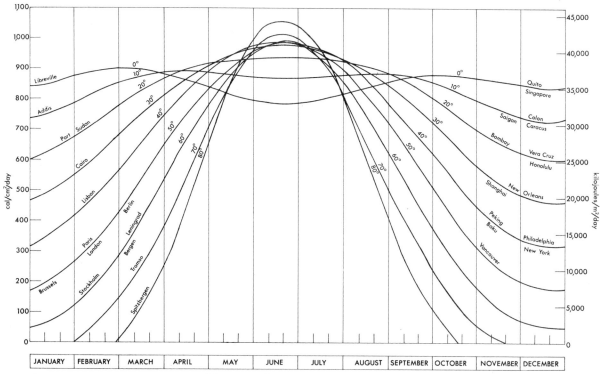

Figure 13.viii shows daily totals of undepleted solar radiation at various latitudes over the course of the year, with latitude lines marked 0° through 80°. The left axis is labelled cal/cm²/day (0 to 1,100) and the right axis kilojoules/m²/day (0 to 45,000). Cities plotted include Libreville, Addis, Port Sudan, Cairo, Lisbon, Paris, London, Berlin, Leningrad, Bergen, Brussels, Stockholm, Tromso, Spitzbergen, Quito, Singapore, Saigon, Colon, Caracas, Bombay, Vera Cruz, Honolulu, Shanghai, New Orleans, Peking, Baku, Philadelphia, New York, Vancouver.

Figure 13.viii. Daily totals of undepleted solar radiation received on a horizontal surface for different latitudes as a function of the time of the year.
After Gates

Plate 12 shows a mosaic of photographs taken by the TIROS IX satellite on 13 February, 1965. Three bands of weather are clearly discernible. In the equatorial regions cloud belts, and in temperate regions cyclonic whirls of the travelling depressions, indicate the causes of the high rainfall of those regions. In the subtropical regions the descending air causes little cloud and explains the predominantly lower precipitation. This mosaic is probably the first clear proof that the general scheme of world climates deduced from widely, and often irregularly, scattered meteorological stations is basically correct.

World Temperature

12. A world mosaic composed of TIROS IX photographs. *NASA*

Figure 13.viii shows that for the warmest few weeks of the year the maximum possible solar radiation which could be received at the earth's surface does not vary greatly for any location, and that the

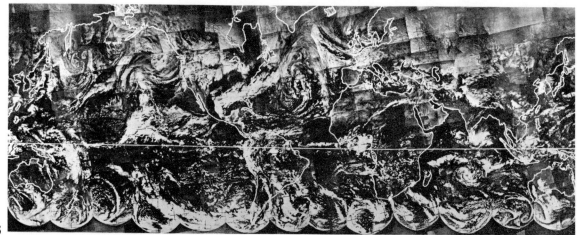

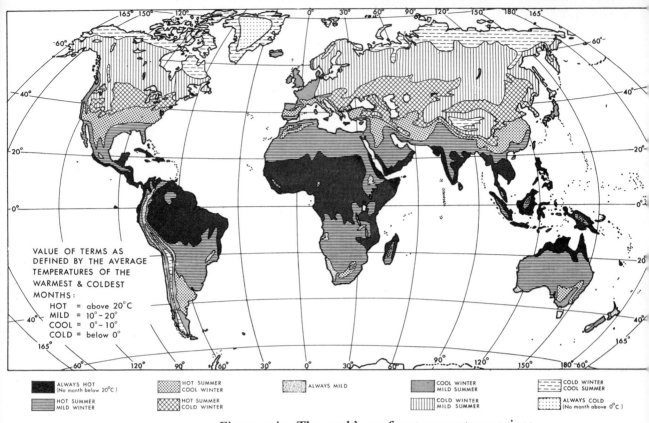

VALUE OF TERMS AS
DEFINED BY THE AVERAGE
TEMPERATURES OF THE
WARMEST & COLDEST
MONTHS:
HOT   = above 20°C
MILD = 10°-20°
COOL = 0°-10°
COLD = below 0°

ALWAYS HOT (No month below 20°C)	HOT SUMMER COOL WINTER	ALWAYS MILD	COOL WINTER MILD SUMMER	COLD WINTER COOL SUMMER
HOT SUMMER MILD WINTER	HOT SUMMER COLD WINTER		COLD WINTER MILD SUMMER	ALWAYS COLD (No month above 0°C)

Figure 13.ix. The earth's surface temperature regions.

Figure 13.x. Mean annual range of temperature in °C.

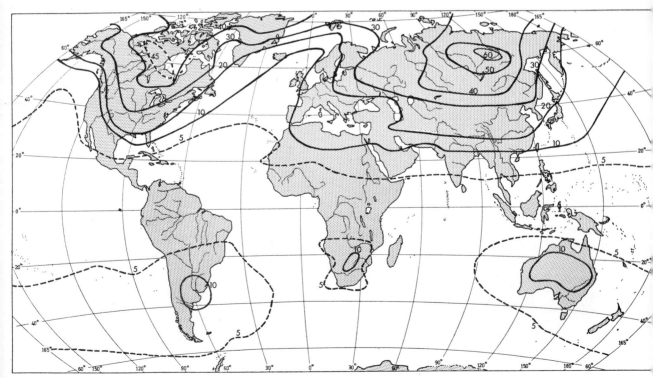

seasonal differences become greatest away from the equator. This simple scheme, however, is greatly modified by cloudiness and other properties of the atmosphere, so that the actual temperature regimes on the surface of the earth are both varied and complicated (Figure 13.ix).

Figures 3.v and 3.vi show mean sea level temperatures for January and July. Such maps are only of value for considering the effects of latitude and continentality on temperatures; they eliminate the effects caused by altitude and as such are not related to reality, but they do make it possible to understand world patterns. Care should always be exercised when using such maps to signify that they are representations of temperatures which would occur at sea level and are not actual temperatures.

As might be expected, the highest mean sea level temperatures occur in the tropics and there is a decline polewards. The temperature belts are generally latitudinal and they shift with the seasons, moving southwards in January and northwards in July. This migration is much greater over the continents than over the oceans; hence, because the northern hemisphere contains much of the land, the greatest changes and extremes occur there and the southern hemisphere is more equable. Both the highest and lowest temperatures occur in the heart of the largest land masses and as a result the greatest ranges are found in northern Asia and northern Canada (Figure 13.x). The effect of warm ocean currents is clearest in the north Atlantic in winter. Cold currents off the coasts of Peru, northern Chile, southern California and Southwest Africa cause the isotherms to bend towards the equator.

14 Climates of the Humid Tropics (A)

Areas within 5° to 10° of latitude (Figure 14.i) of the equator come under the influence of the intertropical convergence and the trade winds. Because this is a belt of shallow depressions, convection and convergence, precipitation is high, often exceeding 200 cm a year.

Equatorial Rainy Climates (Af)

Daily weather in the rainy areas near the equator is remarkably constant throughout the year. The sun is never far from the zenith and there is little variation in the length of day and night. Temperatures seldom fall below 18°C at night and during the day persistent cloud prevents them rising above about 30–35°C. The result is monotonous, humid, hot weather with virtually no seasonal variation.

Skies are usually clear in the early morning and mist clears rapidly, but by mid-morning cumulus cloud builds up. An afternoon thunder-

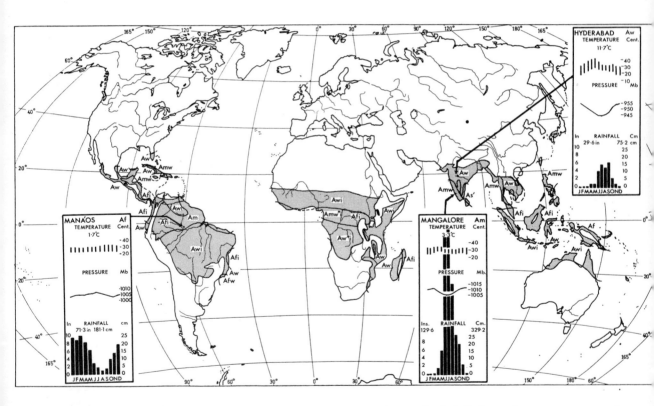

Figure 14.i. Distribution of the A climates according to Köppen, with climate graphs for characteristic stations.

storm is the most common form of rainfall, but where convergence is prolonged or there is a passing wave depression the rain may last for several days.

Rainfall totals tend to be greatest well within the convergence zone, but there is considerable annual variation. Theoretically there should be two rainfall maxima each year just after the time of the overhead sun, but few stations show such a pattern.

Some east coast areas of central and South America and Madagascar have particularly heavy rainfall because of the moist, unstable air brought over them by frequent cyclonic disturbances. Heavy orographic precipitation occurs where hills are present near the coast.

Tropical Monsoon Climates (Am)

The mechanism of the monsoons was discussed in Chapter 8. These climates have high temperatures with a fairly small range, but a great contrast in precipitation between the wet and the dry seasons. During the summer, incursions of maritime air bring heavy rain to the coastal areas shown in Figure 14.i, and this is greatly increased where there are obstructing hill ranges which force the air to rise. The maximum temperatures of the year usually occur in the month before the 'burst' of the monsoon, when there is little cloud. During the monsoon, weather conditions are often similar to those of Af climates, but after the retreat of the monsoon a cooler period of dry weather occurs. During this cooler period, there may be incursions of moist air with travelling depressions which produce overcast skies and light rainfall.

The moist trade winds which cause the heavy rainfall are essentially easterlies, but once in the northern hemisphere they swing round to become southeasterly winds under the influence of the Coriolis Effect. Surges in the trades produce stronger winds and greater discontinuities than usual, and hence shear lines along which convergence and heavy cumulus cloud and showers develop. The monsoon itself is thus not a period of continuous rain, but a wet season within which convectional, orographic and convergence disturbances produce weather variations.

Tropical Wet and Dry Climates (Aw)

Between the wet climates of equatorial regions and the arid zones of the subtropics and tropics are the wet and dry tropics, with a dry period of 2–4 months usually coinciding with the period of low sun and a wet season at the period of high sun. This variation between the seasons is caused by the latitudinal migration of the ITC. The dry season becomes longer with distance from the equator.

Conditions in these areas are similar to those of equatorial regions but there are rather higher and lower temperatures in the wet and dry seasons. The hottest time of the year is in the cloudless months just before the wet season and the coolest just after it, when cloud still persists. In the hottest part of the year daily maxima may exceed 40°C, although 25–30°C is more common. Night temperatures fall to around 15°C. The mean annual range is about 5–10°C.

The rain usually falls from convectional storms in the maritime air and hence is unevenly spread throughout the days of the wet season. In the dry season, convection still occurs but no rain falls from the dry continental air masses of the cool season. The rainfall variability is as much as 25% so that these climates are liable to drought and flood and are difficult for agriculture.

15 Dry Climates (B)

The major feature of all dry climates is that potential water losses from the earth's surface exceed the gains from precipitation, and these climates are therefore deficient in water. It is not possible to set a specific value for the precipitation which forms the boundary of the dry regions because the effectiveness of precipitation is related to the temperature and the season at which the rain falls. Thus 65 cm of annual precipitation may support forests in northwestern Europe, but in the tropics such an amount would barely support the scrub of semiarid lands.

Figure 15.i. Distribution of the B climates according to Köppen, with climate graphs for characteristic stations.

The regions of dry climate are extremely extensive (Figure 15.i) and occupy at least a quarter of the earth's surface, most of it over the continents. The arid lands occur in the tropics and also extend into the subtropical, temperate and boreal zones to form continuous areas of drought.

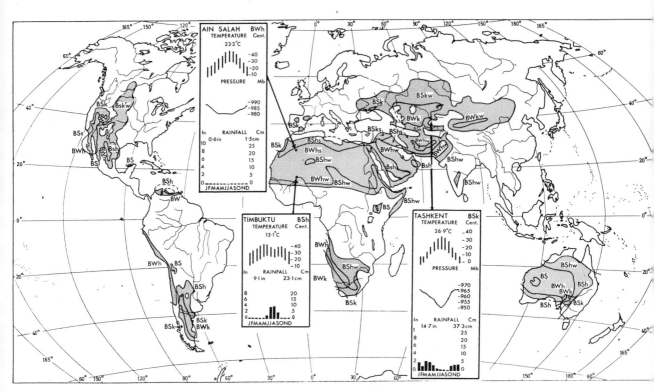

Tropical Semiarid (BSh) and Hot Desert (BWh) Climates

These two types of climate are similar in most respects except that the semiarid zone is transitional to the more humid zones flanking it. Such climates occur between latitudes 10° and 35° with by far the largest area being in the northern hemisphere where the Saharan-Arabian-Indian desert extends about 9655 km from east to west. Australia has the next largest area of tropical dry regions, and smaller areas are shown in Figure 15.i.

The prevailing air masses are the subtropical anticyclones, in which dry descending air is warmed adiabatically to produce dry conditions on the ground. The effects of upper air subsidence inversions extend the arid conditions well into the zones of the trade winds, where the inversion inhibits vertical development of clouds. Where tropical arid climates occur along west coasts of continents, ocean currents modify desert conditions:

off lower California and Sonora the California current produces the Sonoran Desert;

off coastal Peru and Chile the Humboldt or Peruvian current produces the Peru and Atacama Desert;

off northwest Africa the Canaries current produces the Sahara fringe desert;

off Southwest Africa the Benguela current produces the Namib Desert.

The upwelling of cold bottom water renders the air about the coast cool and stable so that, although advection fog and stratus cloud may be produced, precipitation is low.

Within some dry regions, uplands like the Ahaggar and Yemen sometimes cause sufficient lifting for orographic rain to provide less arid conditions.

In the dry air few clouds develop so that there is abundant sunshine and strong heating of the surface every day. During the hottest part of the year, day air temperatures may reach 50–55°C, and 58°C has been recorded in the shade at Azizia, Libya, which counts as an official world record. At night, radiation losses are very great and, although night temperatures may fall no lower than 20–24°C to give daily ranges of 30°C, in winter night ground frosts are common. Winter day temperatures rise to 15–20°C. The yearly range of temperature is thus very high, although some coastal areas have a smaller range because of their cloudiness. The semiarid areas have similar temperatures to the arid zone but with a smaller range.

Precipitation is not only low but also extremely erratic, so that average rainfalls are meaningless as an indication of weather conditions. Arica in northern Chile has an average annual rainfall of 1·6 cm but during one 14 year period there was no rain at all. The truly arid areas do not have a seasonal rainfall but only occasional showers from convective thunderstorms. The semiarid areas, however, do have a more regular regime, with winter rainfall from mid-latitude depressions on the poleward side of the deserts (BShs) and summer rain on the equatorial side when the intertropical convergence brings moist airstreams (BShw).

The absence of vegetation in deserts allows very rapid erosion when rainfall does occur, and the unreliability of rainfall in semiarid districts makes any form of land use difficult; these areas are, therefore, ones of extreme problems for man.

Mid-latitude Arid (BWk) and Semiarid (BSk) Climates

The dry climates of middle latitudes differ from those of the tropics in two main respects: their average temperatures are lower, and they are caused not by descending air masses but by their continental location far from oceanic influences. Mountain barriers are also significant in reducing maritime effects. The continental regions are: the intermontane basins and Great Plains of the northern USA and Canada; the southern USSR; northern China; and an unusual area of southern Argentina which is dry because it is cut off from the moist westerly airstreams by the Andes and suffers from dry föhn winds.

Temperatures are lower than in the tropical deserts, with a greater annual range because of the continental location: Ulan Bator in Asia, for example, has an annual range of 42°C. The lowest temperatures occur in the northern Great Plains and in Siberia, where −45°C has been recorded in winter, while 38°C is not uncommon as a summer maximum along the borders with the tropical semideserts. Diurnal ranges are usually high.

Precipitation is, of course, deficient but in the cooler areas 15–20 cm of rain a year may support steppe grasslands and the precipitation is therefore more effective than in the hotter tropics. In the driest areas there is no regular precipitation regime, but towards the margins of the dry zone precipitation is both higher and more regular. Towards the tropics, the fringes have the winter precipitation caused by incursions of maritime air in travelling depressions. On the poleward edges, the air drawn in by the continental lows may produce some convective rainfall, but in winter the continental anticyclone excludes nearly all moist air, although occasional light snowfalls do occur.

The Patagonian Desert has rather different conditions from the continental deserts, with more maritime influence on the temperatures, which are consequently equable with cool summers and warm winters.

All dry regions are windy and have harsh living conditions. Because of the variableness of rainfall, with spells of wet years followed by extended droughts, settlers tend to occupy the semiarid regions in the wet periods, only to suffer when the dry periods follow. Without adequate irrigation these areas are difficult, if not impossible, for settlement.

16 Climates of Temperate Regions (C)

The climates of the mid-latitudes lack the constant cold of the polar regions and the constant heat of the tropics; they also have a very marked seasonal rhythm with clearly defined winters and summers and a large temperature range. In contrast with the tropics, where the dry season is the dormant period for plant growth, in mid-latitudes the dormant period is winter.

The weather in these regions is produced by the alternating dominance of polar and tropical air masses, with most of the precipitation coming from frontal disturbances within the depressions. The climates are divided into two main groups: the C climates have relatively mild winters and occur on the equatorwards side of the middle latitudes or in maritime locations (Figure 16.i); the D climates have a more severe winter, with snow lasting for at least a month; they cover more polar and continental areas.

Warm Temperate Dry Summer Climate (Cs)

Figure 16.i. Distribution of the C climates according to Köppen, with climate graphs for characteristic stations.

The regions with this climate lie on the western sides of the continents at about 35° latitude. In summer they are under the influence of the subsiding dry air of the eastern part of the subtropical high pressure systems and have hot, sunny dry weather. In winter, polar front depressions and moist maritime air bring rain which can be heavy on exposed areas of high relief. The winter is usually mild and also sunny between the frontal rains.

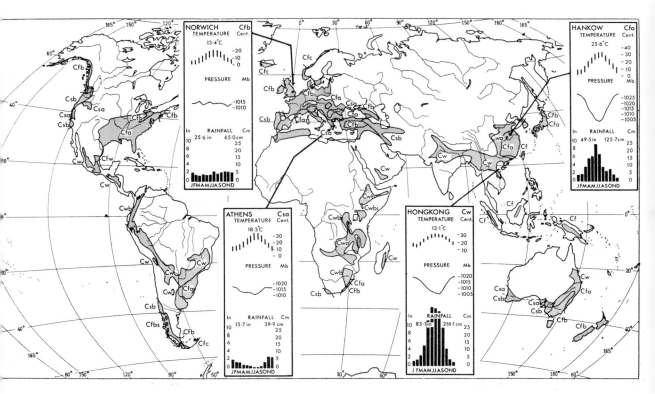

Only in the area of the Mediterranean Sea does this region have a great longitudinal extent. The configuration of the basin allows travelling depressions to penetrate well into the area in winter; a low pressure trough also forms over the warm sea, making it the site of air mass convergence and frontal and cyclogenesis. In both Chile and California, mountains limit the Cs climates to the coast, and in South Africa and Australia they are limited to the southern periphery of the continents. Although this type of climate is often called the 'Mediterranean' type, that region is actually very different from the other regions of similar climate.

Temperatures in midsummer vary with location but are usually high, with mean monthly values around 20–25°C, and a large daily range. In winter, mean monthly values are between 7–13°C, and although daytime temperatures may be pleasantly mild, night-time frosts are common, especially where cold air drains from colder higher elevations, as in the Adriatic and Marseilles areas.

Rainfall varies greatly depending upon locality, but 50–90 cm a year is a characteristic figure. It nearly all falls in winter from fast-moving frontal cloud.

Köppen distinguished between the climates with a warm summer produced by the moderating effect of a maritime location, as in the open Atlantic coast of Portugal and the Pacific coast of California where cold currents produce fog and coastal cloud (Csb), and the hot summer in areas free of such influences (Csa).

Humid Subtropical Climates (Cfa, Cfw)

In similar latitudes to the Cs climates, but on the eastern sides of continents, are the Cf climates. They have more abundant precipitation and it is more evenly spread throughout the year, or else concentrated in the warm season. The differences between the two types of region are caused by the air masses and by contiguous climatic types. Thus while Cs climates are under the eastern limbs of anticyclones in summer, the Cf types are under the western limbs where less stable mT air masses dominate. In Cfa areas in northern India and parts of China, there is a tendency for a monsoon wind system to carry warm, moist, unstable mT air into the interior of the continents in summer and for cool, dry, continental air to dominate in winter. Furthermore, on the east coasts warm ocean currents prevail, in contrast to the cool ones of west coasts. Polewards of the Cf climates are the severe and cold D climates and equatorwards are the moist hot A climates, while the Cs climates grade into warm humid and hot dry climates. The contrasts between the opposite sides of the continents are therefore very great.

Summer temperatures and humidity are high because the unstable maritime air has passed over warm seas before crossing the land. The summers are like those of the humid tropics. Mean monthly temperatures are of the order of 25–30°C. Diurnal ranges are small and the nights are as oppressively humid as the days. In winter, the first wedges of polar air bring a sudden change of weather, with low humidity and a rapid fall in temperatures. The flow of cold air from the interior brings Shanghai's January mean down to 3°C, and even Florida can be plagued with frosts. Since there is no cold land mass polewards of the southern continents, winter temperatures there are higher, with a mean of 11°C in Sydney and 10°C in Buenos Aires for July (the coldest month). These southern areas, and also much of southeastern USA, have mild rather than cold winters.

The precipitation regimes are varied, but annual totals of 75–150 cm are characteristic, with the lowest totals in the areas bordering the semiarid climates. Many areas have a fairly uniform distribution but the eastern Asian monsoon areas have a dry winter as cP air flows out from the continental high pressure zones. Where this air crosses a stretch of sea, as it does to reach southern Japan, then rainfall is high throughout the year. In summer, frontal convergences and convectional thunderstorms give heavy rain, and along the coasts hurricanes and typhoons increase the monthly totals in late summer.

One unusual station is Cherrapunji, in the Khasi Hills of northeastern India, where warm air blowing over the Bay of Bengal and the wet lowlands is forced to rise in the funnel-shaped hills and produce enormous rainfalls. The June average is 290 cm and the annual average 1140 cm.

According to Köppen's scheme the winter-dry interiors of China and northern India are properly classified as Cwa climates.

Cool Temperate Humid Climates (Cfb, Cfc)

This type of climate is characteristic of the western parts of the continents between latitudes 40° and 60° in North America, Europe and South America and between 30° and 50°S in South Africa, Australia and New Zealand. In South Africa and Australia, the location is on the southeastern margin of the continents.

The dominant features of the climate are the warm summers and cool or mild winters, with a rainfall maximum in winter. For the latitudes the climates are remarkably warm in winter because of the offshore warm seas, which are crossed by the dominant depressions bringing mP and modified mT air. There is only an occasional outburst of cP air which brings cold weather in winter, and of cT air which brings hot dry weather in summer. So effective is the warming by the North Atlantic Drift that the shores of Norway are about 15°C warmer than the average for the latitude, and the ports are open to shipping all the year round even within the Arctic Circle. The summer temperatures vary with distance from the sea but July midsummer temperatures of 13–18°C are characteristic.

The most significant characteristic of the weather is its extreme changeability, such that it has been said that Britain has no climate, only weather. Except when the subtropical Azores high pressure system in summer or the continental anticyclone in winter extends on to Britain to give periods of stable conditions, the weather is constantly varying as the passage of fronts and warm sectors of the depressions brings rain and air from both polar and tropical marine source areas.

Precipitation may fall at any time of the year but is greatest in winter when the depressions are dominant, and least in summer when the subtropical highs are more extensive polewards. Actual rainfall amounts depend upon locality, as relief and continentality are most significant. Thus in Britain and New Zealand western coastal upland areas may receive 380 cm or more annually yet 30 km inland in the rain shadow the rainfall may be 65 cm or less. Inland areas where relief favours convection and turbulence, with resulting thunderstorms, may even have a summer maximum rainfall.

In the southern hemisphere, Chile and New Zealand do not experience the stationary stable anticyclones but moving anticyclones with troughs of low pressure and fronts between them. This gives periods of several days of fine weather followed by two or three days of wet conditions. The North Island of New Zealand is occasionally visited by low pressure systems of tropical origin which give very heavy rain.

Among the more unpleasant characteristics of this climate are the days of persistent drizzle and low stratus cloud which result when slow-moving occluded fronts pass. Fog frequency is also high and this, with the frequent gales of spring, winter and autumn, makes the seas off these areas hazardous for shipping.

The Cfc variety of this climate occurs in the extreme south of South America, where the great cloudiness and large number of rain days prevent much summer heating.

17 Cold Climates (D)

The D climates are those whose chief characteristic is a long and cold winter in which a snow mantle is common, but which have a genuine summer in which plant growth can take place. These climates have brief spring and autumn seasons and only a short period of the year is free from frost. Latitudinally, the D climates occur polewards of the C types in the northern hemisphere, and occupy interior and leeward positions on the continents. They are consequently continental in their characteristics and do not occur in the southern hemisphere, where the land masses in latitudes 40–65° are so insignificant. They are excluded from the western margins of the North American and Eurasian continents by the moderating marine influences, but do exist on eastern seaboards, where continental influences are dominant. (See Figure 17.i.)

The temperature range in all D climates is large, with high summer temperatures but extremely low winter ones, so that seasonal as well as daily contrasts are great. During winter the snow cover, because of its high albedo, causes much of the incoming radiation to be reflected and so keeps down temperatures. In the spring it also delays the warming of the air, as much of the solar energy is used in melting the snow and ice. The snow cover is thus partly responsible for the length of winter, but because it protects the underlying soil from deep freezing it enables plants to commence growth soon after the spring thaw. A minor factor in causing seasonal contrasts is the short length of daylight in winter and the long days of summer.

Figure 17.i. Distribution of the D climates according to Köppen, with climate graphs for characteristic stations.

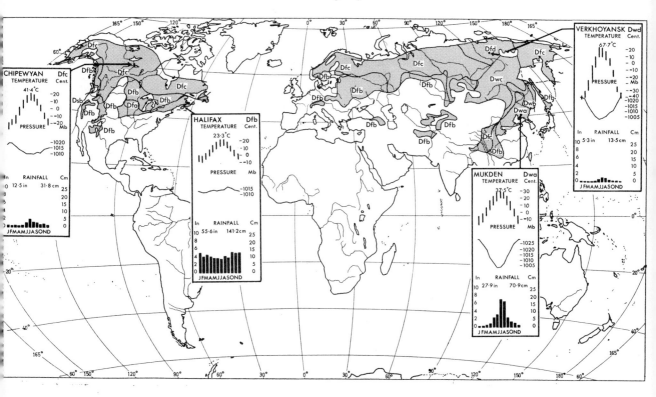

Precipitation occurs mainly in summer, when air temperatures are high enough for considerable quantities of water vapour to be held in the air and unstable air masses and depressions cross the land. During winter the air descends over the cold surface and the continental anti-cyclones with outflowing stable air inhibit precipitation, although light snow showers do occur. Rainfall is largely confined to the summer, when convection and travelling depressions are most common. The summer maximum rainfall thus occurs at the season when it is most beneficial for plant growth.

Humid Continental Cold Climates (Dfa, Dfb)

The latitudinal extent of these two climatic types is very great, being about 35–52°N in North America and 40–60°N in Eurasia, with a rather greater spread southwards in eastern Asia. It is therefore neces-sary to distinguish between areas with warm summers (Dfa) and those with shorter and cooler summers (Dfb).

The cold winter in Eurasia is characterised by a persistent flow of cold air from the anticyclone of central Asia. The dry cP air inhibits precipita-tion, but where it meets mP air which intrudes into western Europe precipitation can be heavy. There is thus a progression from eastern Europe with its heavy winter snowfalls into Asia with decreasing precipitation. In North America there is a constant change of air masses as cP air from the north invades southwards and mT air pushes northwards. The weather at a particular station depends upon the air mass covering it at a given moment. Periods of stable weather are caused by the persistence of an air mass. The lee of the Rocky Moun-tains is a zone of frontal and cyclogenesis and the resultant depressions move eastwards, causing heavy precipitation along the fronts. In winter there is considerable contrast between the prevalent cold and dryness of the northern zone and the warmer, moister conditions to the south. In summer the contrasts between the northern and southern parts of the regions are less pronounced; high temperatures for the latitudes and convectional thunderstorms, especially along fronts, are common features.

In eastern Asia, the outflowing air from the continental high is very cold in winter and precipitation is very low. By contrast, in summer the dominant airflow is into the interior of the continent and heavy rain can result along the fronts and as a result of orographic and convectional instability. These climates of eastern mainland Asia therefore have a kind of monsoonal regime and are distinguished as Dwa and Dwb types because of the very dry winter.

A major feature of the Dwa and Dwb climates is their great vari-ability between northern and southern, and between interior and maritime locations. Precise definition of the boundaries of these climatic zones is particularly difficult and rather pointless.

The Subarctic Climates (Dfc, Dfd, Dwc, Dwd)

The subarctic of the northern hemisphere has the extreme in conti-nental climates, with the world's largest annual ranges of temperature. Nearly all subarctic climates are of the Dfc type, with only small areas of Siberia being so cold in winter that they are classed as Dfd (Verkhoyansk is Dwd) and only eastern Asia having the monsoon effect of the Dwc. These climates occur between latitudes 50° and 60°N.

Polewards they are replaced by the tundra climates and southwards by the Dfb climates, which are commonly regarded as those of the temperate grasslands. The subarctic climates are the zones of the northern coniferous forests. The Eurasian subarctic extends from Sweden and Finland eastwards to the Pacific coast of Siberia and in North America from Alaska to the Atlantic.

Long, bitterly cold winters, very short summers and very brief spring and autumn seasons are characteristic. During the long winters each day has few hours of low intensity solar radiation, so that exceedingly low temperatures of around $-50°C$ are not uncommon. The lowest temperatures have occurred near Verkhoyansk in Siberia where $-68°C$ has been recorded. The cold, dense air masses build up intense thermal high pressures and surges of cold air from the northern interiors of the land masses bring spells of bitter weather to areas near the coasts and further south. The greater size of the Asian land mass and the enclosing effect of the mountain ranges produce lower temperatures than in North America, where mean January temperatures are likely to be closer to $-20°C$ than the more common $-30°C$ in Asia.

Summer temperatures by contrast can be high, for the long hours of daylight produce high mean temperatures, $16°C$ being a common July mean. Midday temperatures may reach $30°C$ in July. All localities in this climatic zone by definition have at least one month with a mean of $10°C$ or over.

The precipitation of all of the subarctic climates is low and of the order 25–63 cm a year, being greater in eastern Canada and north-western Europe than in the extreme interiors. Most precipitation falls in summer as convectional rainfall. The winter precipitation is usually very low but, because the snow accumulates and drifts with the frequent and powerful winds, the spring melt can release large quantities of water.

18 Polar (ET, EF) and Highland Climates

The feature of polar climates is continuous cold and the seasonal rather than daily distribution of daylight and darkness. At the poles the sun is out of sight for about 6 months a year and for the other 6 months is constantly above the horizon. Between the poles and the Arctic and Antarctic Circles, where 24 hours of daylight occur at the summer solstice and 24 hours of darkness at the winter solstice, day and night have lengths of intermediate type. Diurnal variations are therefore of little significance and terrestrial radiation continues throughout the winter, so that the lowest temperatures occur just before the spring equinox.

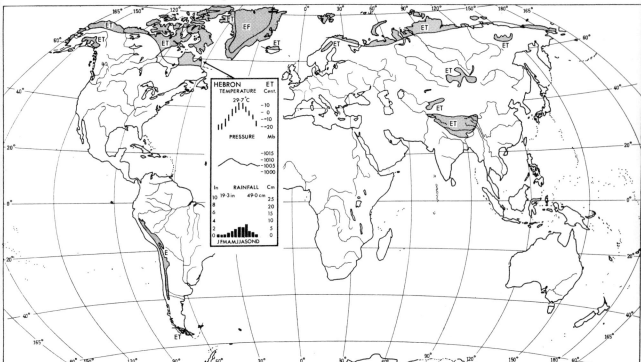

Figure 18.i. Distribution of the E climates according to Köppen, with a climate graph for a humid ET station.

Antarctica is a sea-girt landmass of regular shape and has an ice cap climate (EF). In the northern hemisphere, the fringes of the land masses come under maritime influences and, although the soil is permanently frozen, vegetation can grow during the short summer and this tundra climate (ET) is distinguishable from the ice cap climate of central Greenland. The distribution of land and sea in the Arctic makes the climatic pattern complex. (See Figure 18.i.)

Polar Tundra Climate (ET)

In the cool, short summers of the zone between the northern edge of the coniferous forests and the areas of permanent ice, temperatures are above freezing point for 2–4 months. Many of the days are warm and in the extended daylight plant growth can occur in the shallow soils above the permafrost. The winters are bitterly cold and the snow cover lasts until about May, so that the summer is very short and then the melt-waters, unable to drain through the permafrost, cause much of the tundra to be waterlogged.

Most precipitation falls in the summer months and in the continental areas it seldom exceeds 25 cm a year, but on the coastal areas affected by depressions, especially in Greenland and northern Europe, it can be higher than 75 cm. The winters are very dry as a result of the anti-cyclonic conditions.

The Ice Cap Climates (EF)

On the ice caps, mean monthly temperatures are all below 0°C and there is no vegetation. As yet little is known about the climate of these areas, but the widely held concept of a permanent high pressure system above each of the poles has not been substantiated and there appear to be more incursions of depressions into the polar areas than was formerly thought. During the long winter night no radiation is received and during the summer the snow and ice reflect at least 80% of the incoming radiation so these are the coldest parts of the earth, except for small areas of Siberia.

The high altitude and continental form of the Antarctic produce mean annual temperatures around −50°C near the pole, compared with −30°C in the heart of Greenland and −20°C over the surface of the ice-covered Arctic Ocean. In winter these very low surface temperatures cause temperature inversions to develop, and the cold air flows outwards and downwards, often with such a velocity as to cause local blizzards.

Precipitation is largely from cyclonic storms which invade the ice caps. As a result, the coastal areas of both hemispheres receive more snow than the interior areas. The average snowfall for Antarctica is probably equivalent to about 12–20 cm of rain but little of this falls at the pole.

Highland Climates

The major mountains of the world occupy such a variety of positions in relation to land, sea and latitude that generalisations must be regarded with caution. The outstanding feature of highland regions is the great diversity of actual climates. It is not possible to speak of a distinct highland climate: near the poles, mountains throughout all elevations are cold; those in arid areas like the Ahaggar Mountains of the Sahara have a modified tropical arid climate. The dominant factors which affect highland climates are: altitude, local relief and the mountain barrier effect.

The normal lapse rate of temperature is about 1°C per 100 metres of increased altitude so that, although actual lapse rates may differ from this, temperatures become generally cooler with altitude; however, the clearness of the atmosphere and its decreasing density (and hence opacity to radiation) cause strong temperature variations between areas in shade and full sun, so that freezing temperatures can be experienced in the shade while a very few metres away it is uncomfortably hot. Diurnal temperature ranges therefore tend to be high. Near the equator the diurnal range is often greater than the annual range.

Precipitation normally increases with altitude on a mountain up to a zone of maximum precipitation. Above this zone, the air is drier because of the loss of water vapour in the precipitation at lower levels, and total precipitation declines. On mountainous islands in the oceans, such as Hawaii, which are in the trade wind belt an upper air temperature inversion with subsiding, and therefore dry, air produces an arid climate.

The altitude at which precipitation falls as snow is partly controlled by temperature, but also by the total snowfall, so that although this level generally becomes lower towards the poles, it is also lower on dry slopes than wet ones. Local effects such as föhn winds and the position of the mountain in relation to travelling depressions can also modify the actual climate, as can local relief. Slopes facing towards the equator, and not shaded by local peaks, experience much warmer climates than shaded or poleward facing slopes. Mountain ranges can impede large-scale air circulation so that one side of a range may have wet cloudy climates and the other side arid and almost cloudfree conditions. Particularly good examples of this are the Western Cordilleras of the Americas where, in the zones of westerly air circulation, the west coasts have humid climates yet the eastern sides of the ranges are arid.

PART II   THE GEOGRAPHY OF SOILS

19 Introduction to Soils

To most people the soil is the outer part of the earth's crust in which plants grow. For the pedologist, the soil envelope (pedosphere) requires greater definition. In broad terms, it is that part of the outer crust which extends from the surface down to the limit to which living organisms penetrate. The soil is derived by weathering and biological processes from a parent material which may be the underlying rock, or drift material such as colluvium, alluvium or glacial moraine.

The soil is not a static body but an open dynamic system (Figure 19.i) from which some material is being constantly removed and to which other material is constantly being added. The varied effects of the 5 main factors which influence the soil—(1) parent material; (2) relief of land on which the soil lies; (3) climate; (4) living matter; and (5) time—produce distinctive genetically related horizons within the vertical profile of the soil, and also variations between the soils of different areas by a combination of physical, chemical and biological processes. These variations occur within a continuum, so that the boundaries between different soils are usually broad zones rather than sharp breaks.

The great Russian pedologist V. V. Dokuchaev (1846–1903) formulated the theory that soil can be viewed as a dynamic entity in about 1880. But the real significance of Dokuchaev's work did not become apparent to scientists in the western world until K. D. Glinka's great textbook was published in German and then translated into English by the American C. F. Marbut in 1917.

The essence of Dokuchaev's ideas was his conception of soils as unique bodies, each with its own morphology expressed in its profile, which reflects the combined effects of the 5 factors of soil formation outlined above. Full scientific investigation of the soil should, he said, be able to distinguish the significance and effects of each factor, and so interpret the genesis of a soil.

Properties and Components of the Soil

The profile of a soil as exposed in a roadside cutting or a pit has clearly defined layers or horizons which are distinctive because of their inorganic and organic components. Most horizons are mixtures of sand, silt and clay with some organic matter. Of these constituents, the clay has the greatest significance in chemical reactions because clay particles, being very small (usually defined as less than 0·002 mm diameter), are the most numerous and have the largest surface area in relation to volume. Furthermore, clays are most easily transported through the soil profile and are thus significant in causing differences of composition between the horizons.

The proportions of sand, silt and clay within a soil determine its texture, which in turn determines its permeability to water and root penetration. The soil structure is produced by the aggregation of the sands, silts and clays into clumps which have definable shapes—such as blocky, columnar, prismatic and platy. A soil's texture is changed only with difficulty, but its structure can be easily changed by cultivation—for better or worse.

Soil organic matter and organisms form a complex of the living roots of plants, micro-organisms, the brown and black decomposition product called humus, and numerous intermediate products. Organic matter promotes the development of soil structures and is food for the animals and micro-organisms which enrich the soil and break down its constituents to form plant foods.

For the purpose of study in the field, a three-dimensional body of soil called a pedon is taken to represent a soil type (Figure 19.ii). The area of the pedon varies between about 1 and 10 square metres, depending on the area required for clear representation of the components of

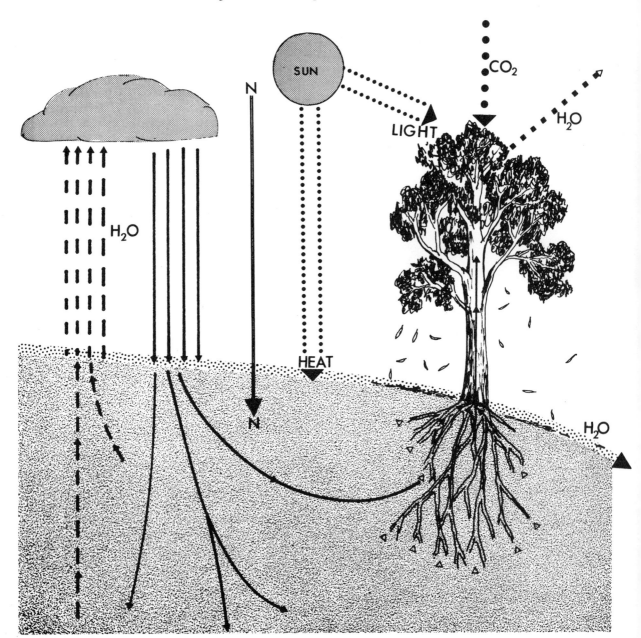

Figure 19.ii. A pedon.

◁ Figure 19.i. Physical, chemical and biological processes operate in the soil. Rain falls and is evaporated or runs off the surface, or circulates in the soil as vapour or capillary movements; it takes part in chemical reactions and is taken up by plants and transpired. Nitrogen is fixed in the soil by micro-organisms and taken up by plants with other minerals. Plants synthesise organic matter and take up CO_2 from the air. Decaying organic matter returns minerals to the soil, where they are available for a new cycle of plant growth.

the soil. The pedon's horizontal dimensions should both be approximately the same. Its vertical depth depends entirely on the soil type under consideration. A pedon is not a mapping unit, only the smallest body of soil which represents a type.

20 The Texture and Structure of Soils

The texture of a soil refers to the size of the mineral fractions in it. Each fraction is defined by its diameter according to an internationally accepted scale which was devised by A. Atterberg in 1912:

Coarse sand	2·0–0·2 mm
Fine sand	0·2–0·02 mm
Silt	0·02–0·002 mm
Clay	Less than 0·002 mm

Sands can be distinguished with the naked eye, but silt can only be clearly seen with a microscope, and clay with an electron microscope. In the laboratory, the fractions in a soil can be separated by mechanical analysis. This involves the removal of stone and gravel and the separation of sand from the broken soil aggregates by sieving. The silt and clay fractions are far too fine to be separated by sieving but, once the humus has been removed from them, they can be dispersed in water and separated by the rate at which they settle from a disperse solution. The coarsest fraction settles first. When the analysis has been finished, the percentages of each fraction can be plotted on a triangular graph-figure which indicates the class of the soil texture.

13. A chart showing the percentages of clay, sand and silt in the soil. The classification is according to that of the US Dept of Agriculture, not the international system.

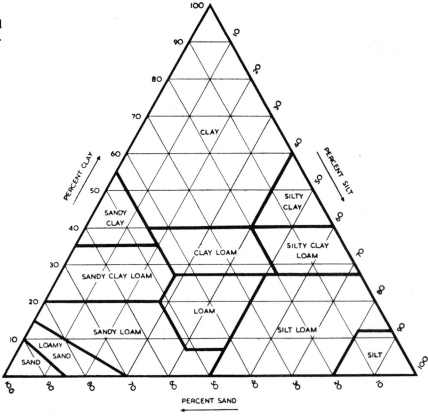

Soil texture can also be estimated in the field by an experienced worker from the feel of a hand-held sample. Clarke gives the following descriptions of soils which do not contain large amounts of humus, non-plastic clay or finely-divided calcium carbonate:

Sands: Soils consisting mostly of coarse and fine sands with little clay. They are thus loose when dry and not at all sticky when wet.

Loamy sands: Consisting mostly of sand but with sufficient clay to give a small degree of plasticity and cohesion when moist.

Sandy loams: Soils in which the sand fraction is still obvious but which mould readily when sufficiently moist without sticking appreciably to the fingers.

Clay loams: The soils are distinctly sticky when sufficiently moist, but with care the presence of sand fractions can still be detected.

Sandy clay loams: These soils contain sufficient clay to be distinctly sticky when moist, but the sand fraction is a more obvious feature than in clay loams.

Sandy clays: These soils are sticky and plastic when wet but still have obvious sand in them. Essentially they consist of sand and clay with little silt.

Clays: Soils which are very sticky when thoroughly moist and very hard when dry. Sand can hardly be detected.

Silts: Soils characterised by a smooth soapy or silky feel.

Silty clays: Soils composed of very fine materials. They are sticky when wet but are modified by the soapy feel of silt.

Silty clay loams: Soils which are less sticky than silty clays and more soapy.

Silt loams: These are plastic and can be moulded and are only slightly sticky but rather soapy.

Non-plastic clays feel loamy and can only be distinguished by the shiny surface which can be obtained by polishing a specimen with the finger nail. Plastic clays are sticky when wet; silts feel soapy or silky; and sands are gritty.

Origin of Soil Particles

The soil particles are derived from the chemical and mechanical breakdown of rocks and larger particles. Mechanical processes—freeze-thaw action, swelling and shrinking with wetting and drying, attrition of boulders and gravels rubbing against each other during transport by streams, ice or mass movements—break down rocks into successively finer particles. Sands and silts are produced entirely in this way, but clay is produced by mechanical action only through the formation of rock flour during glacial erosion. All other clays are the product of chemical action, which may also be responsible for breaking down the largest rocks by volume changes during chemical reactions, and by alteration of minerals.

The minerals of sands and silts are derived from three sources:

(1) crystalline mineral particles from primary rocks;
(2) micro-crystalline aggregates or amorphous deposits formed by weathering or as residues of plant and animal life (e.g. calcium carbonate, iron and aluminium hydroxides or silica);
(3) crystals such as calcite formed in the soil itself.

Silica in the form of quartz is the most common mineral. It is also one of the most resistant of minerals, and the sand and silt fractions of soils commonly have a 90–95% silica content in temperate climates. Chemically it is relatively inert and contributes nothing to plant nutrition but forms the non-reactive skeleton of the soil. In humid tropical conditions, however, it is rapidly leached out of the soil. Other common minerals are feldspars, micas, ferromagnesian minerals, and iron compounds.

Clay minerals are derived largely from micas, feldspars and hornblende, but their structure is complicated and merits a separate study in Chapter 23. Unlike sand and silt, the clay fraction of the soil is composed of secondary minerals formed only after rock is weathered. Clay with humus is the reactive part of the soil.

The particle composition of a soil will depend upon the grain size of the material from which it is derived and the age and environment of the particle, for these factors will determine the degree of weathering to which it has been subjected. Sands will be produced from a parent material of a coarse primary rock or a sandstone, and silt from such materials as loess and volcanic ash which are formed largely of silts blown by the wind. Clays are the final product of weathering of all rocks or they may come directly from mudstones.

14. **Prisms in this soil profile (see above the handle of the knife) are breaking down into blocky structures.**

Soil Structure

15. A very granular clay loam soil.
NZ Soil Bureau

The structure of a soil is determined by the form of the aggregates (which are also called peds) into which the soil components are grouped. Typical soil aggregates are classified and described by these terms: platelike, prismlike, blocklike and spheroidal (Figure 20.i). Aggregates of the first three of these classes are fitting structures so that the side of one ped fits the side of its neighbour. The space between them is usually only open when the soil has dried out, and in a moist state it is frequently only distinguishable as a channel for major plant roots.

Platelike peds are usually horizontal and are found either at the surface or within the soil. The most commonly observed are the skins of clays, of polygonal shape, which form on top of a drying out mudflat or area of clay. Within the soil, they can be formed by the compressing action of heavy machinery or they may be inherited from such parent material as old lake beds or shales; in some cases they may be formed by frost action.

Prismlike structures are of two types—prismatic and columnar. Both types are most common in the middle and lower parts of the soil profile. Prismatic structures are usually the result of slow drying of soils which have a high clay content. The shrinking of the clays and colloids produces cracks perpendicular to the surface of the soil and these cracks are often occupied by roots which help to perpetuate the structure. Prismatic structures are particularly well preserved by tree roots. In cold climates, freezing and thawing can have the same crack-

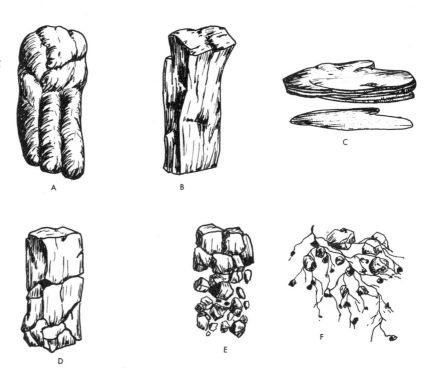

Figure 20.i. Soil structures:
(A) columnar; (B) prismatic;
(C) platy; (D) prism breaking
down to blocky; (E) blocky;
(F) crumbs on grass roots.

forming effect as wetting and drying. Prismatic structures are flat-topped and distinguishable from columnar structures, which have rounded tops. Rounded tops may be caused by changes within the profile but they are most commonly the result of material being washed down cracks and the structure faces, so that subsequent upward swelling forms the dome-shaped top of the column. These structures are particularly common in arid and salt-rich environments.

Blocklike structures may be blocky or cube-like, or rounded and nutlike. They usually occur in the subsoil and are again the result of wetting and drying. They also have fitting faces and this distinguishes them from the spheroidal structures, which are not fitting. Blocks are commonly formed by the breaking down of prisms.

Spheroidal soil structures, or granules, occur mostly in the topsoil. They should be distinguished from the irregular fragments produced by cultivation. Spheroidal structures or granules are formed by pedo-logical processes and are responsible for many of the characteristics of a good agricultural soil (Figure 20.ii). Granules are usually less than 12·7 mm in diameter and porous. They are spoken of as crumbs when their porosity is very high and they look like crumbs of bread. The term 'crumb' is also used very loosely for all small soil aggregates. There is still much discussion of the processes by which granules are formed. The outstanding characteristic of their occurrence is their excellent development under grass and their poor development under forest, where soil organisms are fewer. All grasses have an exceedingly fine and dense root system, which occupies the macropores between the granules. By extracting water from the granule, the roots probably cause it to shrink and so become compact. Roots also provide an environment rich in organic matter which can coat the granules and make them more stable. Some organic matter may act directly as a cement but it may be that the high adsorbtive capacity of humus for water intensifies the effect of wetting and drying, and of temperature changes. Earthworms

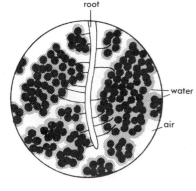

Figure 20.ii. In a freely draining soil the small pores hold the water and the large pores are filled with air. Crumbs that are too large prevent roots reaching water and nutrients inside them. Crumbs that are too small block air channels.

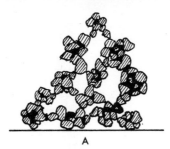
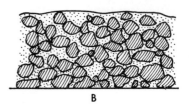

Figure 20.iii. (A) a soil with granular structure with clay-humus binding the coarse grains; (B) a non-aggregated soil with the colloids dispersed.

A

B

and other organisms also promote the formation of granules by passing soil through the gut and combining it with gums and other organic substances.

The chemical constituents of the soil can have a significant influence on soil structure. Many of the chemically active constituents of the soil exist as exceedingly small particles called colloids. They have diameters of between 1 and 0·001 μ. Such particles are far too small to be seen except under an electron microscope. If the soil is rich in sodium—as in some arid climates—the colloids will easily disperse in water and fill the pore spaces between the soil particles. As the soil dries, it will set in a massive state with no fine structures, and hence be agriculturally unworkable. If the soil is rich in calcium, the colloids will flocculate or group together and bind the soil grains into granules leaving air spaces between them and allowing easy cultivation. A microbial product of organic colloids is probably largely responsible for forming ideally small granules of 0·25–2·0 mm diameter (Figure 20.iii).

An agriculturally important characteristic of granules is their stability in water. Those granules which are firmly bound and resist destruction by cultivation or weather are of great value in producing a good agricultural soil and are said to be water-stable; while those which readily break down are unstable. High stability seems to be related to the presence of those organic colloids which are formed only under grass; hence a crop rotation in which grass plays a part improves soil structure.

Figure 20.iv. Structural profiles of three different soil groups. For an explanation of the symbols A to C, denoting soil horizons, see pages 156–9.
After Nikiforoff

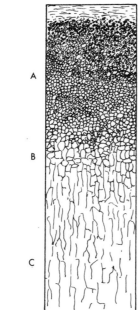

A

B

C

Chernozem

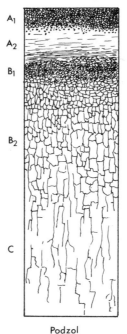

A_1

A_2

B_1

B_2

C

Podzol

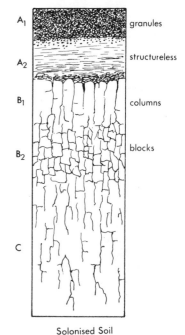

A_1

A_2

B_1

B_2

C

granules

structureless

columns

blocks

Solonised Soil

Not all soils show evidence of structures. Some are said to be massive if they are coherent, and others, like immature sandy soils, may be single grain soils with insufficient fine or colloidal material to bind them. Some soils have a variety of structures in the one profile. Figure 20.iv shows the distribution of soil structures in three different types of soil.

Soil Consistence

This is a term used to denote the physical condition of a soil at known moisture contents. It is described by such terms as: loose, friable, firm, soft, harsh, hard, plastic and sticky. A wet clay soil may be sticky, but as it dries out it will lose its stickiness but still be plastic or easily moulded in the hand, and when it is very dry it may be hard. Consistency is important in that it partly determines the kind of tilth which cultivation will give to the soil. A sticky or hard soil can seldom be cultivated but one which is friable may be in an ideal condition for ploughing or digging. Terms like friable, compact and firm are most useful for describing soils and soft, firm and hard for describing peds.

21 Soil Moisture and Soil Atmosphere

Soil moisture is important for many reasons: firstly, because it supplies the water needed for plant growth; secondly, because it is the solvent for plant nutrients; thirdly, because it largely controls the temperature of the soil and the soil atmosphere; and fourthly, because it largely determines the incidence of soil erosion.

Water Retained in the Soil

Water exists in the soil in the pore spaces between the grains of the soil material. These spaces may be relatively large, as in the space between soil structures or sand grains, or they may be minute and have a diameter little more than that of several molecules, as in some fine clays. The pore space in soils usually varies between about 20% and 50% of the total soil volume. This is illustrated in Figure 21.i, which shows an imaginary soil composed of perfectly spherical grains. The pore space is shown at the two extremes of possible packing of the grains. Table 21.1 shows the pore space available in different types of soil as a percentage of total soil volume.

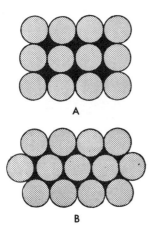

Figure 21.i. The packing of ideally spherical soil particles in (A) gives a pore space (black) of approximately 50%. In (B) it is 25%.

Table 21.1. Pore Spaces in Soils as a Percentage of Volume (After Burges)

Loam	34
Heavy loam	44
Clay loam	47
Clay	52

If water is applied to a soil with a good structure and texture, so that all of the pores are filled down to the permanent water table, the soil is saturated. If the same soil is then allowed to drain freely, the water in the macropores will drain out leaving only a film of moisture on the soil grains, but the micropores will still be occupied by water. A soil in this state will lose little more water by percolation and is said to be at field capacity. Most soils will reach field capacity from saturation in about a day, but in a few cases, as in some of the loess soils of the Great Plains of the USA which are almost entirely composed of silt-size material, field capacity is a rather meaningless term because percolation can continue for some months. Field capacity is an ideal condition for many plants in that they can obtain oxygen from the air-filled macropores and moisture from the micropores. Further loss of water from the soil can result in insufficient moisture being available for plants, which then wilt. This state is said to be the wilting point. The three soil moisture conditions are illustrated in Figure 21.ii.

Water is retained in the micropores, and as a film on the surfaces of grains forming the walls of the macropores, by two forces. One is the attraction of water molecules to surfaces of solids—called adhesion— and the other is the mutual attraction of water molecules—called cohesion. These forces are at their greatest at the face of the solid and become less powerful with distance from it. For this reason, drainage water is easily removed by gravity from large pores but water is firmly retained in micropores.

Moisture Retention and the pF scale

Figure 21.ii. The volumes of solids, water and air in soil at three different moisture contents.

saturation

field capacity

wilting point

Movement of Water in Soils

One way of expressing the tension or suction with which a soil will retain moisture is in terms of the height (measured in centimetres) of a column of water whose weight exerts the same force as the force of tension in the soil. The greater the height of the column, the greater is the tension measured. We may thus express the tenacity with which water is held by soils in centimetres, or that reading may be converted into other forms. One of these forms is in pressure exerted by the water column and hence measured in bars or atmospheres. The negative pressure per unit area exerted by a column of water 10 cm high, under standard conditions of temperature, is about 1/100 of an atmosphere (i.e. 10 mb); that exerted by a 1000 cm column of water is approximately 1 atmosphere of tension. A more convenient form of expression than speaking of 1000 cm columns of water is to use the logarithm of 1000 (i.e. 10^3), which is 3. This is then expressed as pF 3. A tension of 10,000,000 cm is pF 7.

Table 21.2 indicates that the soil moisture can be retained with a tension of about 10,000 atmospheres when the skin of water is only a molecule or so thick and hence held with a maximum force, but by a tension of only 1/3 of an atmosphere when the water film is thick enough for water to move under gravity.

Table 21.2. *Table of Pressure Equivalents for Expressing the Force of Retention of Soil Moisture*

Height of a unit column of water, in centimetres	=	Atmospheric pressure in atmospheres	=	pF
1		1/1000		0—free water
10		1/100		1
100		1/10		2
346		1/3		2·5—gravity flow
1,000	1			3
10,000	10			4
15,849	15			4·2—wilting point
100,000	100			5
1,000,000	1,000			6
10,000,000	10,000			7—maximum force

Three types of water movement within the soil are recognised: unsaturated flow, saturated flow, and vapour equalisation.

Unsaturated flow can be most easily explained by reference to capillary tubes. If two glass tubes are placed vertically with their lower ends in water the water will rise some distance up the tubes. The distance risen will be related to the diameter of the tubes so that the water will rise highest in a narrow tube and least in a wide tube. The rise of the water is the result of an attraction by the glass for the water (adhesion) and at the same time an attraction of the water molecules for each other (cohesion). The column of water will rise in each case

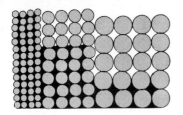

Figure 21.iii. Idealised soil with fine, medium and coarse grains (shaded). Capillary movement of water (black) is greatest in the fine-textured soil.

until the weight of the water just equals the attraction pulling the water upwards. This process is called capillarity and the tubes are called capillary tubes, or just capillaries.

The upward movement of water from the water table is analogous to the movement of water in the glass tubes. In a soil the capillaries are the micropores, but as these are both irregular and discontinuous, and movement of water is often blocked by the closing of pores as clays swell, or by air, the rise of the water must be very irregular. Movement by capillarity can occur in any direction and not just upwards, but in any case it seldom draws water more than 60–90 cm. It is greatest in fine-textured soils and least in coarse-textured ones (Figure 21.iii).

Saturated flow is the percolation or drainage of free water through the macropores under gravity. The water is loosely held and as it is undesirable for plant growth, an essential part of farming practice is to remove it. Unfortunately it also carries away plant nutrients.

Water added to the soil surface will move partly by saturated flow or percolation directly to the water table, and partly by capillarity to areas of higher moisture tension. The extent and speed of the capillary movements will depend upon the texture of the soil, so that in a sandy soil the wetted zone will be narrow and the downward percolation rapid, but in a clay soil the wetted zone will be wide and the movement slow, as in Figure 21.iv.

The movement of water by vapour equalisation is probably of little significance except in some desert areas, where specialised plants may survive because water moves as a vapour from an area of high to low vapour pressure in the soil. Moisture will move from a moist to a dry soil and from a warm to a cool soil although in some cases such movements could cancel each other out. Water which is held at high tensions within the soil by colloids can probably only move as a vapour.

Plants, Soil Moisture and Cultivation

Even in a moist soil some of the water is held at too high a tension for plants to use it, and such water is said to be unavailable. The drier the soil, the more firmly is the water held. Soil texture also influences availability of water for plant use. In a clay soil 20% of the water may be unavailable, whereas in a sandy soil only 1–2% may be unavailable. A clay can retain a higher total moisture per volume than a sand, so they may often yield approximately the same amount of water to a plant. In a drought, however, plants in sandy soils will often die before those in clay soils because water can move more quickly through fine capillaries than through macropores. In dry conditions, plants obtain their water partly by attracting water from the capillaries and partly by growing towards the water.

Cultivation practice is largely aimed at obtaining a correct water balance in the soil for maximum plant use. Excess water is drained away by ditches, tile drains and mole drains so that air can get into the soil. Water deficiency is corrected by irrigation or by methods which are designed to conserve the moisture already in or on the soil. Water loss occurs from evaporation, transpiration from plants, percolation and runoff.

Evaporation from the soil can be partly prevented by using a mulch. In the practice of dry farming, which aims primarily at retaining soil

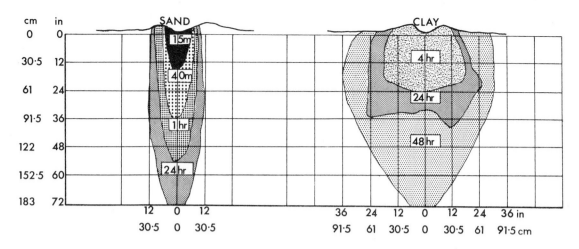

Figure 21.iv. Comparative rates of irrigation water movement into sandy and clay soils. The rate is much greater in the sand, especially downwards. Distances are measured in inches or cm and time in minutes (m) and hours.
After Coony and Pehrson

moisture by having a fallow period and so saving moisture for the next crop, it used to be the custom to leave a fine tilth on the soil surface; it was thought that this prevented loss of moisture brought to the surface by capillarity. In fact, no such effect exists because the water table is usually far too deep, and the fine soil was often blown away by the wind. It is now the practice partly to plough in the stubble from a previous crop or just to cut the roots of the old plants, so that moisture is not lost by further growth and the surface is still protected from the wind. The dead stubble can then act as a mulch and prevent the growth of weeds which lose soil moisture by transpiration.

Surface water is retained and allowed to percolate into the soil rather than run off, either by strip cropping as in North America or by the system of field hedges as in England. The danger of rapid runoff causing erosion by sheet wash or gullying is thereby avoided. In some areas of the Great Plains of the USA or the Canterbury Plains of New Zealand and the Alps, shelter belts of trees are grown to prevent hot, dry winds such as the föhn from depleting the surface soil of its moisture.

In arid climates, the accumulation of salt at the surface of the soil can be a major problem. The salt is derived either from the irrigation water or from the soil, and is moved by capillarity. Irrigation water often concentrates the salt in the soil in areas between irrigation channels, and can cause soil sterility.

The Soil Atmosphere

As much as 50% of the volume of a soil may be occupied by water and gases. The volume of gases present is directly controlled by the volume of water, so there must always exist an intimate relationship between the two. Gases—mainly oxygen—are essential for plant growth and for much of the microbial life in the soil, so the exclusion of gases under waterlogged conditions can control the vegetation and soil processes.

There are two essential characteristics for good soil aeration: (1) ample pore space, and (2) ready access to the pore space for the gases. Ample pore space is a product of good soil structure, or occurs as the result of tillage and good soil drainage. Ready access to the pore space allows an interchange of gases between the soil and atmosphere. This is particularly important in allowing oxygen into the soil and allowing excess carbon dioxide and other waste gases out of the soil. The movement of free air and soil gases can take place readily by the process of diffusion, in which molecules of gases exchange places so that they always tend to produce an equality in the composition of the juxtaposed gases. In this way the free atmosphere receives the waste soil gases and the soil receives oxygen.

Perhaps the most noticeable effect of poor soil aeration is the impoverishment of the higher plants. In the absence of oxygen, root development is hindered, the absorption of nutrients is slowed, and some inorganic compounds toxic to plant life may be formed. In the soil itself, decomposition of organic matter by oxidation is hindered because aerobic bacteria are unable to function. Furthermore, the products of decomposition are completely different. For example, sugars are completely decomposed to carbon dioxide and methane:

$$C_6H_{12}O_6 \longrightarrow 3CO_2 + 3CH_4$$
Sugar   Methane

Less complete decomposition yields toxic acids, and ferric iron is reduced to produce the black, blue and green coloured ferrous compounds typical of waterlogged soils. These contrast markedly with the red colours of soils with active oxidation. For most plants the oxidised states of plant nutrients such as nitrogen are far more valuable and hence poor soil aeration is usually harmful.

Soil temperature is the other important characteristic of the soil atmosphere. Chemical and biological activities require certain temperatures for optimum occurrence. For example, nitrification does not begin until the soil temperature reaches $4·5°C$ and seed germination and plant growth reach their maximum rates at $25–35°C$. The soil temperature depends primarily upon the effective solar radiation it receives. This is influenced by slope, aspect, soil colour and plant cover. The surface temperature of a dry soil may greatly exceed that of the atmosphere—approaching over $55°C$ in a temperate climate—but some centimetres beneath the surface daily fluctuations are hardly noticeable and even seasonal changes lag behind the free atmosphere changes.

As with soil aeration, soil water has a marked influence on soil temperature. Poorly drained soils are usually 'cold' soils, especially in the spring, and at all times of the year the most effective control of soil temperature is by water percolating from the surface. Water regulation is probably the key to what little control man can economically exert on soil temperature.

22 Soil Organisms and Organic Matter

Soil organisms are the animals and plants which exist almost entirely within the soil. A number of animals such as rabbits, or in North America ground squirrels, prairie dogs and kangaroo rats, spend part of their lives in the soil and have some role to play in its development, but they usually live in colonies and are often of local rather than regional importance so they are not included in this discussion. Likewise, only the roots of the higher plants are included.

A broad classification of soil organisms might be set out which distinguishes between animal and plant forms, and between the macro-inhabitants, which are visible to the naked eye, and micro-inhabitants, which are not.

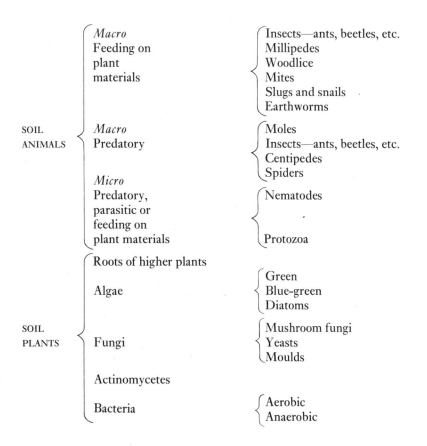

SOIL ANIMALS

Macro Feeding on plant materials
- Insects—ants, beetles, etc.
- Millipedes
- Woodlice
- Mites
- Slugs and snails
- Earthworms

Macro Predatory
- Moles
- Insects—ants, beetles, etc.
- Centipedes
- Spiders

Micro Predatory, parasitic or feeding on plant materials
- Nematodes
- Protozoa

SOIL PLANTS

Roots of higher plants

Algae
- Green
- Blue-green
- Diatoms

Fungi
- Mushroom fungi
- Yeasts
- Moulds

Actinomycetes

Bacteria
- Aerobic
- Anaerobic

Soil Animals

The largest important soil animals are rodents, rabbits and insectivora such as moles. Of these the rodents are of only local significance and so are rabbits, unless they are allowed to increase their numbers to such an extent that they greatly modify the soil and vegetation, as they did in Australia and New Zealand. There they were responsible for largely destroying the vegetation of thousands of hectares and inducing accelerated soil erosion. Burrowing animals as a group are of greatest importance in dry soils like those of the steppes, from which earthworms are almost absent.

Moles are the most effective of the large soil inhabiting animals. Their burrows can form an intricate network extending for many metres laterally beneath the ground surface, and may go as deep as 1·5 metres. Their main food is insects and earthworms so they are usually found in conjunction with the latter. In their burrowing they aerate and drain the soil and intimately mix the surface and subsurface layers, so increasing the depth of the humus-rich layers.

Soil insects—woodlice, spiders, flies, mites, centipedes, millipedes, ants, larvae, etc.—are important because by feeding on animal or plant tissue they break it down and make it susceptible to further attack by bacteria. The bodies of insects also make a significant contribution to the soil organic matter. In tropical areas, termites are particularly important because they bring subsoil to the surface and so mix the soil layers. Termite mounds are ample witness to this activity, and it has

Figure 22.i. Some examples of soil fauna and micro-organisms (at various scales): (A) earthworm; (B) mite; (C) spring-tail; (D) nematodes; (E) millipede; (F) centipede; (G) termite; (H) fungal mycelium; (I) bacterial cells; (J) actinomycetes.

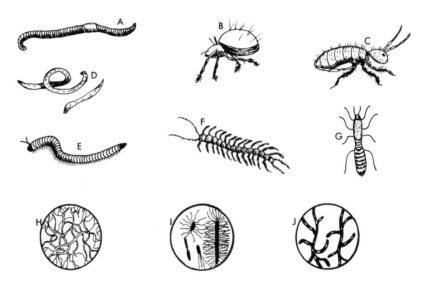

been calculated that termites can bring to the surface 50 metric tonnes of material per hectare in one year.

Earthworms are probably the most important soil animals. By weight they are often more important than all of the other animals together. Bornebusch suggested that on some Danish farms the weight of earthworms equalled the weight of the livestock. They thrive in moist conditions where there is an ample supply of organic matter with a

Figure 22.ii. A larva of the midge *Sciara* excreting fine soil crumbs.
After Kononova

high calcium content. For this reason they are seldom found in acid soils and their absence helps to account for the slow breakdown of peaty material or the raw humus developed under coniferous trees. They are particularly widespread under grass and deciduous tress.

The main function of earthworms is to mix up the soil matter in their intestines, grinding down the coarser mineral fraction and subjecting it

16. A termite mound in Uganda.

to comminution by digestive enzymes. By burrowing through the soil they aerate and drain it, allowing freer development of plant roots. Furthermore, they mix the soil layers and bring their fine excreted material to the surface, so developing the crumb structure of the soil. They also take decaying plant remains down into the soil from the surface and render plant nutrients into a form in which they are concentrated and more readily absorbed by the plant.

Of the soil microfauna *nematodes*, or eelworms, and *protozoa* are the most important. Some nematodes feed on each other but the important ones are those which penetrate plant roots, often doing considerable damage to crops. Protozoa are exceedingly small single-celled organisms like the amoeba. They can cause animal and human disease, but some live on bacteria and may therefore be important in suppressing disease.

Soil Plants

The roots of higher plants may reasonably be classed with soil organisms because in the zone around them, called the rhizosphere, there is a higher availability of organic acids which promotes the breakdown of mineral and organic matter, and so releases an unusually large supply of plant nutrients as well as stimulating the microflora to intense activity.

The importance of the plant root system is very great, because under cultivation and under natural grassland it is the chief source of soil organic residues. Under some conditions, the roots of grass can contribute to the soil as much as 5 metric tonnes of organic matter per hectare in one year, and under cultivation the root systems of such crops as wheat and barley make a similar contribution.

The Microflora

The microscopic size of the soil microflora makes observations of their behaviour and numbers very difficult and so although estimates of their total numbers in a given volume of soil have been made, they may well be wildly inaccurate.

Soil algae are among the least important of the microflora except under certain conditions. Because they can build up their body substance from atmospheric carbon dioxide and nitrogen and require little else, they are often the first colonisers of bare ground such as newly exposed glacial moraines or fresh volcanic debris. They commence the

Figure 22.iii. Part of the root system of a Scots pine showing long roots and short forked mycorrhizas.

biological weathering process by decomposing the primary minerals and synthesising clays. In conjunction with certain fungi they form lichens which are able to trap silt and dust in the air and so build up a very thin soil which higher plants, such as mosses, can colonise. The blue-green type of algae are particularly numerous in the paddy soils of rice growing areas. When such lands are waterlogged and exposed to the sun, the algae can absorb atmospheric nitrogen, which becomes available for plant use when the algae die.

Soil *fungi* vary in size from the microscopic to giant puffballs and mushrooms. They are scavengers that will tackle many types of food, including resistant woody matter, and break it down. The moulds are the most important group of fungi. They have a threadlike or filamentous appearance. The thread, called a hypha, is hollow and increases in length as it grows. A colony of such moulds can produce a web called mycelium, which often forms round the roots of trees. The mycelium secretes enzymes which attack organic matter in the soil and so provide food for the fungus. In some forested areas, especially in coniferous forests, the mycelium forms a mutually satisfactory relationship (that is, a symbiotic relationship) with the tree root, and the fungus is then called a mycorrhiza. The mycelium hyphae penetrate the root and take from it some nutrients, but replace them with those that the tree needs. Some nutrients such as phosphorus are only available to trees through mycorrhizas. So important can this symbiosis be that in newly afforested areas it is sometimes necessary to inoculate the young trees with the appropriate fungus, so that the mycorrhiza can develop. When it dies, the mycelium is attacked by bacteria and its nutrients are made available for plant use.

Actinomycetes, like the moulds, have a filamentous form, but they are smaller than fungal mycelia. They have the important capacity of breaking down resistant humus material and rendering its nutrients available to plants. They can also infest root crops to produce diseases such as potato scab.

Figure 22.iv. On the left a one-year-old pine raised in sterile soil, on the right a similar age pine seedling inoculated with appropriate mycorrhizal fungi.

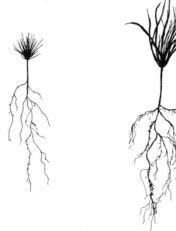

Bacteria have received more study than the other parts of the micro-flora because they have a number of characteristics so important in plant growth that without them plant life could die out. The numbers of bacteria in the soil can multiply rapidly, and it is this capacity to

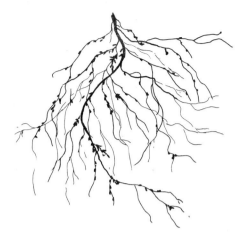

Figure 22.v. Nodules formed on roots of red clover by nodule bacteria.

increase in numbers when conditions are favourable which largely explains their significance.

Each organism consists of a single cell surrounded by a cell wall. They vary in shape from rods to spirals and usually exist in the soil solution or in the colloidal humus of the soil. Most of them feed on the organic matter and help to break it down into humus, but some obtain their energy by oxidising minerals such as sulphur and iron. They obtain the carbon for body-building from carbon dioxide. Aerobic bacteria need free oxygen from the air for survival and so live in well drained soils, where they take part in the oxidation processes which produce nitrates, carbon dioxide, sulphates, phosphates and the red-coloured ferric iron compounds. In water-saturated soils only anaerobic bacteria can survive because they are able to obtain their nutrients from soil compounds. The chemical processes in which they take part reduce soil compounds to form ammonia, methane, sulphides, phosphine and the blue-black ferrous iron. The products of reduction processes are frequently toxic to plants and so saturated soils are usually infertile.

The function of bacteria which makes them essential for plant growth is the fixing of nitrogen in the soil so that plants can use it. Nitrogen is not present in the rocks of the earth, and although a little may be formed during periods of lightning, such small amounts would be rapidly washed away. In the atmosphere it is very inert and can only be made to combine with other substances at very high temperatures or in the presence of catalysts. The occurrence of 3·8–8·5 metric tonnes in every hectare of soil therefore presented a problem to the early soil chemists.

During the nineteenth century it was discovered that two bacteria—*Clostridium* and *Azotobacter*—can fix nitrogen directly by absorbing it from the air and making it available when they die. But these bacteria are of little importance compared with those which work in conjunction with leguminous plants. A leguminous plant, like clover or peas, excretes a substance from its roots which attracts and stimulates the bacteria, which then penetrate nodules on the roots and develop colonies in them. The bacteria feed on the plant and in return supply it with nitrogen which they absorb from the soil air, thus producing a perfect symbiosis.

The bacteria which form the symbiosis with leguminous plants are very selective. The organism which works with clover will not work with lupins, lucerne or soy beans. When a new leguminous crop is introduced, therefore, it is necessary to supply a nitrogenous fertiliser or manure, or preferably to inoculate the plant with the appropriate bacteria. A leguminous crop can fix 280–390 kg of nitrogen per hectare. This becomes available to other plants by the decomposition of organic matter to ammonia, which different bacteria oxidise to nitrite and then nitrate.

Organic Matter and Humus

The organic matter of the soil is derived either directly from plant tissue or indirectly through animal droppings and the remains of their bodies. Plant tissue is composed largely of carbon, oxygen and hydrogen, with very small quantities of essential elements such as nitrogen, sulphur, phosphorus, potassium and calcium. The decomposition or oxidation of this tissue is carried out by the enzyme secreted by the soil flora and fauna, so that the decomposition can be represented thus:

$$\left[C \cdot 4H \right] + O_2 \xrightarrow[\text{oxidation}]{\text{Enzymic}} CO_2 + 2H_2O + ENERGY$$

Carbon and hydrogen compounds in plant tissue

Carbon dioxide is returned to the atmosphere, whence it can again be absorbed by plants, and the other elements are made available as plant nutrients. This decomposition is the first stage in the production of humus; the second stage is the synthesis of complex humic substances. Both stages are the result of enzymatic activity of micro-organisms and of the larger soil animals.

In microflora experiments it was found that fresh litter was successively colonised and processed by:

Mould fungi and spore forming bacteria $\longrightarrow$ Spore forming bacteria $\longrightarrow$ Cellulose myxobacteria $\longrightarrow$ Actinomycetes

each group having a capacity for utilising different parts of the plant and the products of the previous group until, at the end of the humification process, new humic substances were synthesised and some organic substances were mineralised to such end products as water and carbon dioxide.

The rates at which humic substances form under fixed temperature and moisture conditions vary enormously. In general, they form most rapidly in leaves and roots of legumes and other fleshy leaves and most slowly in grass roots and pine needles, where their action is inhibited by the presence of resins. All of the processes are greatly accelerated when the plant remains are comminuted by larger organisms and when they are attacked by the digestive enzymes and intestinal microflora.

Humus is a relatively stable state in the decomposition process in which the organic materials have a black or brown colour and a structureless, amorphous or jelly-like condition. Humus combines with clay to form the clay-humus complex which gives soil many of its properties essential for plant growth and easy cultivation. It can

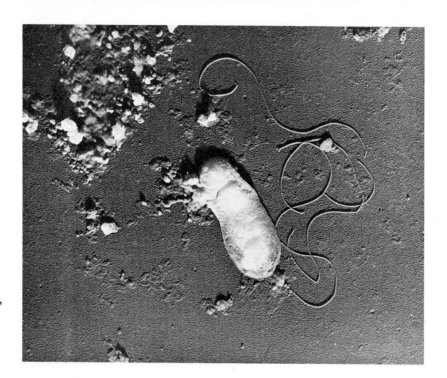

**17. A typical soil bacterium,
×20,000.**
*Electron micrograph by R. D.
Woods, Rothamsted Experi-
mental Station*

assume a flocculated, gel-like form with a high water-holding capacity
which can enable sandy soils to retain moisture. It has very high
swelling and shrinking properties, swelling when wet and shrinking as
it dries, which enable it to break apart massive clay soils and so form
soil aggregates; and it improves structure by binding soil grains with
bacterial slimes. It absorbs nutrient elements into the colloidal material
and makes them available to plants. Lack of humus makes soils powdery
and thus liable to erosion, and this deficiency also reduces their capacity
to hold nutrients.

Humus is present in all soils, although the amount of humus in a
soil may vary from the 80% of some peat soils to 1% or less in sandy
soils. The amount present will depend upon: (1) the rate at which
organic matter is added to the soil in the form of organic residues, and
(2) the intensity of oxidative decomposition and humification. In desert
soils the rate of supply is obviously low, and in well drained sandy soils
oxidation will be very rapid. In saturated soils the virtual absence of
oxidation will allow organic matter to accumulate as peat and humification will be very slow.

The processes of humification are very complex and little understood. Black, amorphous humus forms most readily where microbiological activity is interrupted by periodic droughts or freezing, thus
retarding decomposition and allowing the humus to accumulate and be
incorporated into the soil, which it darkens or melanises. In humid
tropical environments, rapid biochemical alterations lead to the destruction of organic materials and of the black pigment. The humates
then formed are largely colourless so that although a tropical soil lacks
the dark colours usually associated with the humus of cooler climatic
regions, it may have just as high an organic matter content. The
process of production of pale humates is known as leucinisation.

Table 22.1. Humus Percentage in Soils of the USSR (After Kononova)

Soil	Per cent
Tundra soils	1
Strongly podzolic soils	2·5–3·0
Brown forest podzolised soils	4·0–8·0
Deep chernozems	9·0–10·0
Ordinary chernozems	7·0–8·0
Chestnut soils	2·0–2·5
Sierozems	0·8–2·0
Takyr soils	about 1·0
Krasnozems	4·0–6·0

Humus in Soil Groups

Table 22.1 indicates the considerable variation of the humus present in major soil groups. The increasing amount of humus in the soils ranging from the tundra soils to the chernozems is a reflection of climate allowing a great annual increment of plant material to the soil, an increase in the number of soil organisms and, with the exception of the tundra soils, a change from forest to grass vegetation. The chestnut, sierozem and takyr soils have a low exchange capacity and occur in dry climates beneath sparse vegetation. The krasnozems form in hot, humid climates with rapid microbial and chemical action in the soil which counterbalances the rapid supply of plant matter.

Not only does the percentage of humus vary with the soil group but also the composition of the humus and its position in the profile.

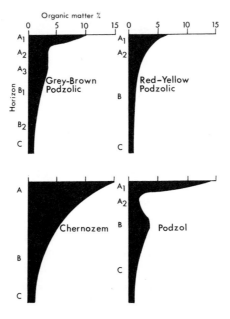

Figure 22.vi. Distribution of organic matter in the profiles of soils from 4 of the great soil groups. The organic matter extends to the greatest depth in the chernozems, where it is derived from decayed grass roots. In the podzol the leached A_2 horizon and the enriched B horizon are obvious.
After Millar, Turk and Foth

Humic acids in strongly podzolic soils, for example, have a high capacity for peptising minerals and are themselves highly dispersed so that they are readily mobile and rapidly leached from the upper soil horizons; furthermore, they do not promote soil aggregation. By contrast, humic acids in chernozems are less mobile (partly because of the presence of lime) and participate actively in soil aggregation. The distribution of humus in four soil groups is shown in Figure 22.vi, in which the contrast between the chernozems and podzols is obvious.

C/N ratio

A measure of the fertility of the soil is the ratio of carbon to nitrogen (C/N ratio) in it. In the process of breaking down organic matter to humus, soil organisms are most effective when the C/N ratio is low. That is why gardeners add nitrogen to compost heaps and why peat and other organic soils are both slow in decomposing, even when drained, and also infertile. The C/N ratio thus characterises the stage of decay and indicates the availability of nitrogen to plants. Podzols have a C/N ratio of between 26 and 18, savanna soils about 14, chernozems about 11 and arid soils about 6. Podzols, then, are seldom fertile, not only because they are leached, but also because organisms can only break down the organic matter slowly to make its nitrogen available.

23 Clays and Colloids

The colloidal state is one which is halfway between a true solution and a suspension. The diameter of the particles in a colloidal state is less than 0·001 mm (i.e. 1 μ). Clay particles have a size less than 2 μ, so many clays, but not all, are colloidal. Colloids may be dispersed through a solution or they may be flocculated into gels or powders.

There are three main groups of clays—the silicate clays, the iron and aluminium hydrous oxide clays, and the amorphous clays such as allophane. The first group are best known, as they are most common in regions with temperate climates; the second group are most common in the tropics, subtropics and cold climate areas.

The Silicate Clays

Clays and clay colloids are laminated and, when viewed under an electron microscope, appear to be made up of a number of plates or flakes. Because of their minute size and laminar form, clays have an enormous surface area for their weight. It has been estimated that 100 grammes of clay might have a surface area of 8·5 hectares. This is a most important characteristic, for it makes most clays chemically highly reactive compared with the other soil particles. Clay has a surface area at least 1000 times that of a similar weight of coarse sand.

A clay crystal, called a micelle, theoretically has a negative electrical charge and hence colloidal particles in a suspension of water repel each other and do not flocculate; but this ideal state seldom occurs in the soil, for the negatively charged micelle attracts positively charged cations in the soil solution. The cations—particularly calcium (Ca^{2+}), hydrogen (H^+), potassium (K^+), magnesium (Mg^{2+}) and sodium (Na^+) —attach themselves to the surface of the micelle, where they are said to be adsorbed. In this way the electrical charges in the soil system are approximately balanced, and whether the colloids remain dispersed or flocculated depends upon the cations available.

The cations are not firmly held by the micelle and are thus spoken of as exchangeable ions, and the replacement of one cation by another is called cation exchange. If, for example, a soil colloid has Ca^{2+}, Mg^{2+}, K^+, Na^+ and H^+ cations adsorbed on its surface, making the micelle neutral in charge, and a fertiliser such as ammonium sulphate were applied to the soil, the ammonium would be adsorbed by the clay, and magnesium, sodium and potassium sulphates and sulphuric acid would appear in the drainage water from the soil, as may be seen from the diagrammatic representation below:

$$
\begin{array}{c}
\text{Ca} \qquad\qquad \text{H} \\
\begin{array}{c}\text{Ca}\\\text{Mg}\\\text{K}\end{array}\boxed{\quad\text{Micelle}\quad}\begin{array}{c}\text{H}\\\text{H}\\\text{H}\end{array} + 3(NH_4)_2SO_4 \\
\text{Na} \qquad\qquad \text{H}
\end{array}
\longrightarrow
\begin{array}{c}
\text{NH}_4 \quad \text{Ca} \ \ \text{Ca} \\
\begin{array}{c}\text{NH}_4\\\text{NH}_4\\\text{NH}_4\end{array}\boxed{\quad\text{Micelle}\quad}\begin{array}{c}\text{H}\\\text{H}\end{array} + MgSO_4 + NaKSO_4 + H_2SO_4 \\
\text{NH}_4 \qquad \text{NH}_4
\end{array}
$$

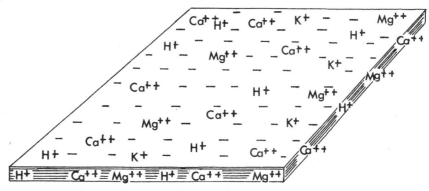

Figure 23.i. Diagrammatic representation of a clay crystal (micelle) showing its sheet structure, many negative charges and adsorbed cations. Water molecules and cations between the sheets are not shown.

By treating a soil with salts or an acid, the cations adsorbed on to the micelle can be controlled. As the cations largely control the physical properties of the clay, the nature of the soil can be affected. Where the soil is composed of micelles with a dominant negative charge, the clays are dispersed and sticky or slimy, and difficult or impossible to manage. Such soils are those with a high proportion of Na^+ cations. They occur in arid regions and have to be treated with gypsum (calcium sulphate) to improve their structure; soils inundated by seawater have the same property. Positively charged micelles, by contrast, are flocculated. Calcium-rich soils have this desirable quality. They are thoroughly flocculated and easily worked to give a fine crumb structure.

The Structure of Clay Particles

It was for long thought that all clays were amorphous but the electron microscope has revealed that silicate clays are crystalline. On the basis of their crystal forms, as revealed by X-rays, silicate clays are divided into three main groups:

(1) kaolinite;
(2) montmorillonite;
(3) illite, which is a representative of the hydrous micas.

Kaolinite is one of the simplest clay minerals. Its flat, platelike crystals are built up into units composed of alumina and silica sheets joined very tightly by shared oxygen atoms (Figure 23.ii). This type of construction is referred to as a 1:1 crystal lattice. The crystal units are firmly bound to each other by oxygen-hydroxyl bonds so tight that few ions or water molecules can penetrate between the crystal units. As a result, the adsorptive surfaces of the crystals are limited to those on the external edges of the lattice structure, for there can be no penetration of the oxygen bonded internal surfaces. Kaolinite's structure therefore determines that it has a low adsorptive capacity for cations,

18. Model of the 1:1 clay mineral kaolinite. The upper surface is all hydroxyls and the bottom surface is all oxygens. A row of hydroxyl and oxygen atoms is in the centre.
Reproduced with permission from: Millar, Turk and Foth, Fundamentals of Soil Science, published by John Wiley and Sons, Inc.

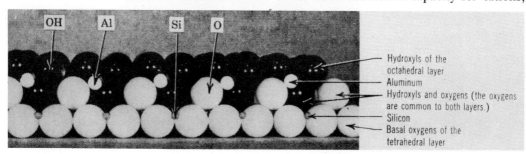

OH   Al   Si   O

— Hydroxyls of the octahedral layer
— Aluminum
— Hydroxyls and oxygens (the oxygens are common to both layers.)
— Silicon
— Basal oxygens of the tetrahedral layer

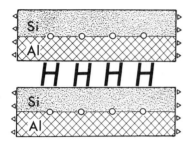

Figure 23.ii. Schematic representation of a cross-section through two kaolinite crystal units. Each crystal unit is made up of one silica and one alumina sheet (1:1 lattice) bonded very tightly by mutually shared oxygen atoms (o). The crystal units are closely bound to each other by oxygen-hydroxyl linkages (H), giving a non-expanding lattice and one which cannot be readily penetrated by water molecules or cations. The adsorbtive surfaces are therefore limited to the edges of the units (small triangles). Many units make up each clay crystal.

19. Kaolinite, ×9,500. *Electron micrograph by R. D. Woods, Rothamsted Experimental Station*

low expansion on wetting and low contraction on drying. Further, it has low cohesion and plasticity.

Montmorillonite has a slightly more complicated structure than kaolinite. In this case the alumina sheet has silica sheets bonded on either side of it by oxygen atoms to form a 2 : 1 lattice (Figure 23.iii). The crystal units are only weakly connected by oxygen atoms so the lattice is weak and can expand and contract rapidly. Cations and water molecules are able to penetrate the weak bonds and the adsorbtive surfaces are very large, so that cation adsorption and swelling and shrinking with wetting and drying are 10–15 times as great as with kaolinite. Montmorillonite is also very plastic and cohesive, and hence very difficult to cultivate.

Illite is one of several hydrous mica clays. It has a similar 2 : 1 lattice structure to montmorillonite, but some of the silicon in the silica sheets has been replaced by aluminium and some of the aluminium in the alumina sheets has been replaced by iron and magnesium. The bonds between the crystal units are formed by potassium atoms, and as these are firm there is relatively little adsorption on the internal surfaces between the crystal units (Figure 23.iv). Illite crystals are therefore less cohesive, plastic and adsorbtive than montmorillonite, and do not swell or shrink so much.

The properties of the three clays are summarised in Table 23.1.

Table 23.1.

Property	Montmorillonite	Illite	Kaolinite
Size (microns)	0·01–1·0	0·1–2·0	0·1–5·0
Shape	Irregular flakes	Irregular flakes	Crystals
External adsorbtive surface	High	Medium	Low
Internal adsorbtive surface	Very high	Medium	None
Cohesion, plasticity	High	Medium	Low
Swelling and shrinking	High	Medium	Low
Cation exchange capacity	Very high	Medium	Low

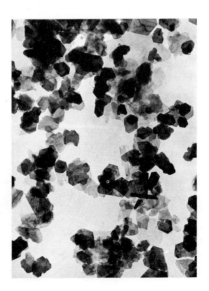

Hydrous Oxide Clays of Iron and Aluminium

Figure 23.iii. A schematic representation of a section through two montmorillonite crystal units. Each crystal unit is made up of two sheets of silica bonded with one sheet of alumina. The crystal units are loosely bound to each other by weak oxygen linkages (o) which allow the distance between the units to vary and also allow penetration and adsorption of water and cations so that montmorillonite has large adsorbtive and swelling and shrinking properties. The small triangles indicate the adsorption surfaces. Many crystal units make up each clay crystal.

Hydrous oxides are oxides containing associated water molecules. Gibbsite, which is aluminium hydrous oxide ($Al_2O_3 . 3H_2O$), geothite ($Fe_2O_3 . H_2O$) and limonite ($Fe_2O_3 . ?H_2O$) (both of the last two being iron-hydrous oxides) are among the most common of these clay particles (note that the water of limonite is a variable quantity and is shown as $?H_2O$).

Relatively little is known about the hydrous oxides of iron and aluminium, but they are known to have crystal lattices similar to those of the silicate clays. They have a smaller number of negative charges per micelle and hence a capacity to adsorb cations which is even lower than that of kaolinite. They are not sticky or plastic, and few swell or

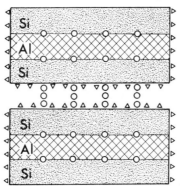

shrink as much as the silicates. Because the silicates are largely formed in temperate climates and the hydrous aluminium and iron oxide clays are formed under tropical climates, there is a marked difference between the agricultural characteristics of clays in the different climatic zones, which can be traced to the structure of the clays.

Amorphous Colloids

Figure 23.iv. Illite has the same general structure as montmorillonite except that strong potassium linkages join the crystal units and prevent adsorption and swelling and shrinking. In the sheets of the crystal units some of the silica is replaced by alumina and in the alumina sheet some iron and magnesium occurs.

The structure of the crystalline silicate clays was revealed by X-rays, but some colloids, which have been called allophanic colloids or allophanes, appear to lack ordered arrangement of atoms within molecules or within solid structures and are therefore said to be amorphous.

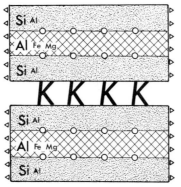

Allophanes are formed at an early stage of weathering of basic silicate minerals from basalt or ultrabasic rock and hence are common in soils formed on basic volcanic rocks or on some volcanic ashes. They can also be produced by extreme mechanical grinding, and hence occur in glacial rock flour and in frost-disturbed soils, because the grinding action reduces minerals to the state of disorder which is the characteristic of allophane.

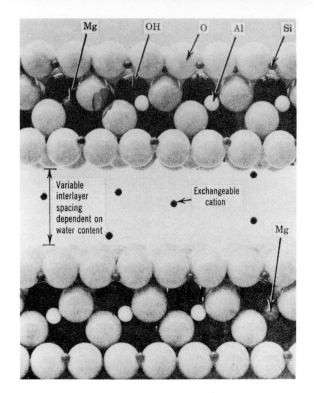

Mg OH O Al Si

Variable
interlayer
spacing
dependent on
water content

Exchangeable
cation

Mg

20. Model of the 2:1 clay
mineral montmorillonite.
*Reproduced with permission
from: Millar, Turk and Foth,
Fundamentals . . .*

Allophane is a gel-like combination of SiO_2 and Al_2O_3, the exact physical properties of which depend upon the water in the colloids. In their hydrous condition these aluminosilicates are held together by random cross-linking due to condensation between hydroxyls attached to silicon and aluminium, and the water is enclosed between the structures (hydrogels). Upon dehydration, the structure shrinks permanently to a more compact form (xerogels). Wet soils contain hydrogels, which give the soil a greasy feel, but dry soils contain xerogels, which make it friable. With long continued weathering, allophanes give way to metahalloysite and ultimately to kaolinite.

Formation of Clay and Colloidal Particles

Clay minerals in the soil are derived either from pre-existing clay minerals in the parent material on which the soil is formed, as in the case of kaolinite derived from a shale, or from the weathering of the primary minerals in parent material. Weathered parent material will

21. Montmorillonite, ×5,000.
*Electron micrograph by R. D.
Woods, Rothamsted Experimental Station*

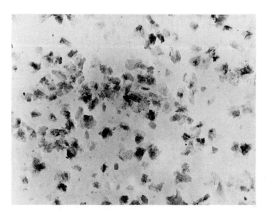

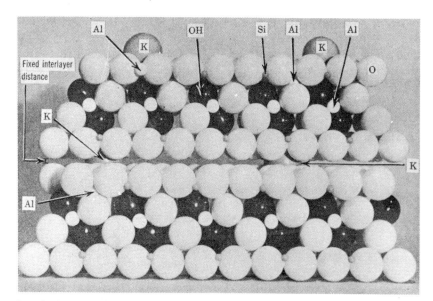

22. **Model of the 2:1 clay mineral illite.**
Reproduced with permission from: Millar, Turk and Foth, Fundamentals . . .

23. **Illite,** × 16,333.
Electron micrograph by R. D. Woods, Rothamsted Experimental Station

break down to form new clay minerals either by minor modifications of the silicates in feldspars, micas and hornblende, or by a more complete breakdown and synthesis of new clay minerals.

Any soil condition which prevents colloidal silica and the various forms of aluminium and iron from reacting with each other interferes with clay formation and induces separation of each element in the soil profile—as happens in podzols and latosols. In podzols, the aluminium and iron become soluble or peptised by the action of organic compounds and are moved by water from the upper A_2 horizon to the lower B horizon. This leaves the silica in an amorphous form in the A_2 horizon, while the iron and aluminium accumulate in the B horizon as haematite and gibbsite. By contrast, in the latosols of the tropics the rapid decomposition of organic matter causes the silica to become soluble in the neutral or slightly alkaline soil solution, and to be leached out of the soil leaving behind the aluminium and iron as gibbsite and haematite.

The kind of clay mineral which is formed in a soil depends upon the prevailing conditions in the soil environment, particularly as influenced by the acidity, the composition of the soil solution and its concentration. For these reasons, different clay minerals may form in each of the soil horizons because of the variation in the degree of leaching and presence of water and organic matter in them. Thus in soils in which calcium carbonate either forms or accumulates, because of the low rainfall or the drainage conditions, montmorillonite predominates. In soils in which calcium carbonate cannot accumulate because of the high rainfall, low temperature or porous soil, vermiculite and kaolin materials predominate. The formation of gibbsite and haematite or other iron minerals is favoured either by a low temperature as in podzols, or by a high one as in latosols. By contrast, illite only forms from parent materials rich in potassium.

Not only do different clay minerals form in the various horizons of a soil but, because of the low rate of clay formation, the growth of any one type of crystal lattice on a surface may vary periodically, as a result of seasonal or other factors which control the soil environment, so that several clay minerals may occur together either as mixtures or as interleaved crystals (Figure 23.v). For example, during active plant growth,

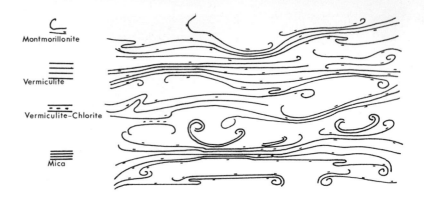

Figure 23.v. Principal features of a dominantly montmorillonite layer silicate clay complex formed by weathering of mica. The spacing of the sheets of the various clay minerals is shown.
After Jackson

when the soil solution is being depleted of its bases and enriched with H^+, the kaolin mineral would tend to form but during a dormant season, when the soil solution becomes enriched with bases and OH^-, the montmorillonite minerals would tend to form.

Because of the variations in each profile, the type of clay minerals occurring in the great soil groups does not present a simple weathering sequence from polar to tropical environments, but in general the tundra and podzol soils of cold climates contain mainly sesquioxides and illite, the chernozems and related soils of temperate grasslands contain montmorillonite as the dominant, and the latosols of the humid tropics kaolinite and sesquioxides. The weathering sequence is always from complex to simple clay minerals, so that kaolinite with its 1 : 1 lattice is a very stable weathering product and only found in very weathered soils. Because of progressive weathering quartz is a component of freshly weathered materials but in time the clay minerals illite, montmorillonite and kaolinite become respectively more important (Figure 23.vi).

This description is something of an over-simplification, for other clays such as chlorite and vermiculite may form at various stages and a stage may be completely missed if weathering is rapid, so that kaolinite can form from illite. This can be shown diagrammatically as in Table 23.2. It will be noted that the weathering sequence is shown as being irreversible and although this is not entirely true, the exceptions are few.

Figure 23.vi. The relative proportion of each of the principal clay components in the fine clay fraction of soils of different ages.
After Jackson and Thompson

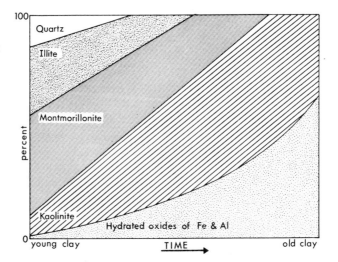

Table 23.2. Weathering Sequences

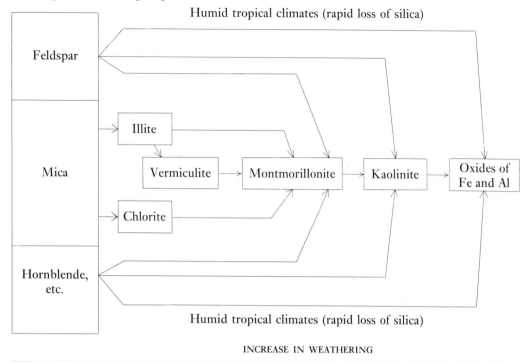

Humid tropical climates (rapid loss of silica)

Feldspar

Mica

Illite

Vermiculite → Montmorillonite → Kaolinite → Oxides of Fe and Al

Chlorite

Hornblende, etc.

Humid tropical climates (rapid loss of silica)

INCREASE IN WEATHERING

Even under the most intense weathering conditions clay formation is very slow, ranging from 0·0001 to 0·002 grammes per 100 grammes of parent material per year. As would be expected, high rates are only found in the humid tropics, where the high temperatures and an abundance of moisture are most common. There is thus a reduction in the rate of clay formation so that it will be found in decreasing order: latosols > prairie soils > brown earth > grey-brown podzolics > podzols—in other words, it is progressively slower in the soils developed under increasingly cool and dry climates.

Organic Soil Colloids

The origin of organic colloids is discussed in Chapter 22. It is necessary to mention them here because they make up a most important proportion of the soil colloids. They are not as stable as clay colloids, being formed and destroyed more rapidly. The humus micelle may not be crystalline but it has many other features in common with clays. It has a very high cation adsorbtive and exchange capacity as seen by the comparison in Table 23.3. Humus also has a very high capacity for holding water and low plasticity and cohesion. It is particularly important in promoting a good soil structure, for it prevents clays cohering and thus promotes granulation.

Table 23.3. Cation Exchange Capacity (measured in milliequivalent/100 g)

Humus	200
Vermiculite	150
Montmorillonite	100
Illite	30
Kaolinite	10

Clays and Man

The physical properties of clays—plasticity, cohesion, swelling and shrinkage—have been mentioned in connection with all clays; it only remains to stress that although the structure of clays is not visible to the naked eye, the effects of this structure can easily be observed. The gardener whose boots become heavy with wet clay notices its cohesion.

24. Cracks formed in drying clay soil.
Photo by T. I. Oliver

The farmer who cannot get a tilth on a soil notices plasticity; the casual observer who sees cracks in a dried out soil sees the effects of shrinkage and he may also notice the large scale mass wasting of the landscape in the form of slumps and flows which are such a prominent feature of some areas.

The farmer in temperate climates will also know that a soil of loamy consistency with only 30% clay will have good qualities because it will be free-draining and at the same time not droughty; it will retain fertilisers long enough for them to be available for plants; it will retain a good tilth; and it will allow good root development. The farmer in the tropics, however, will not have much trouble in preserving a tilth on the hydrous iron and aluminium clays, although he will have problems of rapid leaching of nutrients, irreversible dehydration of exposed soil and the formation of iron crusts.

24 Plant Nutrients and Soil Reaction

Of the factors influencing the growth of plants—(1) light, (2) mechanical support, (3) heat, (4) air, (5) water, (6) nutrients—light is the only one which is not provided in the soil. Plant growth depends upon a favourable combination of these factors, but it is controlled by the least favourable; in other words, the limiting factor is the one which most restricts growth.

The elements essential for plant nutrition are listed below:

	Carbon (C) Hydrogen (H) Oxygen (O)	Available from air and water
Nitrogen (N) Phosphorus (P) Potassium (K)	Calcium (Ca) Magnesium (Mg) Sulphur (S)	Available from the soil solids and needed in large quantities
Iron (Fe) Manganese (Mn) Boron (B) Molybdenum (Mo)	Copper (Cu) Zinc (Zn) Chlorine (Cl)	Available from soil solids and only needed in small amounts

The major nutrients (N, P, K, Ca, Mg, S) are all normally available in a soil and under natural conditions the vegetation preserves a balance, returning to the soil in the humus those nutrients that it has used. Under cultivation, however, crops remove the nutrients and these have to be returned as fertilisers. The minor nutrients, except for iron, are only available in small quantities in a soil and some may not be present at all in certain sands and volcanic ashes, so they often have to be supplied as trace elements in the major fertilisers. Traces of sodium, iodine and cobalt are sometimes also supplied because they are essential for the health of grazing animals although they have no effect on plant growth. Organic matter has been regarded as a critical factor in crop production for a long time. It has the function of improving soil structure and supplying several nutrients, particularly nitrogen and phosphorus.

A total chemical analysis of a soil may frequently suggest that it has a complete supply of nutrients, but in fact these are not always available to the plant. Some nutrients are available because they exist in a soluble form in the soil solution or because, like calcium, they exist as readily exchanged cations on the clay micelles, but many nutrients occur in complex, insoluble forms which are only slowly broken down to the simpler soluble state. This is particularly true of phosphorus, potassium and magnesium, which are not readily available and hence not leached in the drainage water. Calcium, on the other hand, being readily available, is easily leached and so has to be frequently replaced by liming.

Nitrogen is produced from the organic matter by a complex bio-chemical process which forms an ammonium compound which is transformed to nitrite and then nitrate compounds by bacteria. The bacteria will not operate in cold or waterlogged conditions, so a suitable soil condition is essential for the process.

The Soil Solution and pH

The elements do not occur as such in the soil solution, but are present as ions (i.e. an electrically charged atom or group of atoms) such as the negatively charged anions carbonate (CO_3^{2-}), bicarbonate (HCO_3^-), sulphate (SO_4^{2-}), chloride (Cl^-), nitrate (NO_3^-), nitrite (NO_2^-), phosphate (PO_4^{3-}), and hydroxyl (OH^-); or the positively charged cations potassium (K^+), ammonium (NH_4^+), calcium (Ca^{2+}), sodium (Na^+), magnesium (Mg^{2+}), and hydrogen (H^+). The plants absorb the ions from the soil solution through their root hairs.

The relative proportions of hydrogen (H) and hydroxyl (OH) ions have a special significance because they determine the soil reaction, that is, whether a soil is acid or alkaline. The solution with equal amounts of hydrogen and hydroxyl ions is neutral; that with an excess of hydrogen ions is acid; that with an excess of hydroxyl ions is alkaline. The degree of acidity or alkalinity is measured on a scale, called the pH scale, in which:

$$pH = \log \frac{1}{(H^+)}$$

In other words concentrations of the H ions can be written as $1/10^2$, $1/10^3$, $1/10^4$, or 1×10^{-2}, 1×10^{-3}, 1×10^{-4} grammes per litre. For convenience this is expressed as pH 2, 3, 4 respectively. pH 7 is the neutral state; at pH 6 the H ions are ten times as numerous as they were at pH 7 and the OH ions are 1/10 as numerous. The soil solution therefore has 100 times more H ions than OH ions. At pH 8 the OH ions are 100 times more numerous than the H ions (Figure 24.i).

Figure 24.i. In a neutral solution, pH7, the H+ and OH− ions are balanced. In a slightly acid solution, pH6, the H+ ions are 10 times as numerous and the OH− ions are 1/10 of their number at pH7; hence there are 100 times as many H+ ions as OH− ions present. In an alkaline solution, pH8, there are 100 times as many OH− ions as H+ ions.

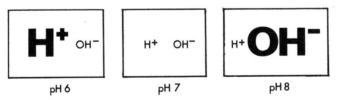

pH 6   pH 7   pH 8

In humid regions most soils have a pH range of about 5–7·5, and in arid regions the range is about 7–9. The extreme acidity of 2·3 has been recorded in an acid peat and pH 11 has been recorded in some arid soils. The importance of the pH is that most plants thrive only when the soil is near neutrality. Although cranberries, strawberries and rhododendrons require an acid soil, few plants will grow at lower pH's than 3·5. Salt-tolerant plants such as salicornia will grow in strongly alkaline soils but few plants can exist at pH's greater than 9.

The second important feature of soil pH is that nutrient availability changes with the soil reaction. Iron and manganese are relatively insoluble on the alkaline side but become more soluble with increasing acidity. Phosphates become more readily available at higher alkalinity. These relationships are illustrated in Figure 24.ii.

pH varies within the soil. It is most easily observed as being different from horizon to horizon, especially in a soil where there is strong

Figure 24.ii. To show the influence of soil pH on the availability of plant nutrients. The thicker the bar the greater the availability of the nutrients.
After Truog

leaching, but it also varies around each colloid, being more acidic closer to the colloid surface. These differences have the effect of forming a variable environment for plant roots and useful bacteria, and so allowing many soil processes to exist in close proximity.

The acidity of a soil solution is an approximate but useful way of measuring the base status of a soil, that is, the actual content of exchangeable mineral ions. In a soil that is saturated with H ions, there will be little 'room' for mineral ions on the clay-humus micelles; hydrogen has a stronger bond to the micelles than most nutrients except calcium. An acid soil has therefore to be treated with readily exchangeable cations before nutrients are added to the soil or they will be washed away rather than exchanged with the hydrogen. Liming is thus an essential part of agricultural practice on clay soils. It has to be carried out at regular intervals because rainwater is, in effect, a dilute solution of carbonic acid (H_2CO_3); consequently the nutrients become detached from the micelles and disappear into the drainage water as carbonates while the hydrogen ions replace them on the micelles. This process of base depletion is known as leaching. Although the pH is an indication of the variations in the fertility of a particular soil, and it may be assumed that if pH falls then the soil has become less fertile, pH cannot be used as a measure of the fertility of two different soils because fertility is also related to soil composition. Soils rich in clay and humus can hold a greater supply of nutrients than sandy soils, and a mere mixture of silt and salt can never be made into a fertile soil. Hence pH values have to be used with care if they are to be reliable guides to fertility.

25 Weathering and the Formation of Parent Material

Soil is an intimate mixture of organic and inorganic material: the organic material is derived from plant and animal tissues and the inorganic material from the rocks of the earth's crust. Before the minerals of the rocks can become incorporated in soil, however, they have to be changed by weathering into a waste mantle, or regolith. In this change, some of the primary minerals have been transformed into secondary mineral particles with sufficient surface area for chemical activity to be significant. They can then mix with the organic debris of the lower nitrogen-fixing plants, which first colonise rock debris and begin the process of soil formation. It is, then, the weathered material of the upper part of the regolith which is the parent material for soil—not the rock itself.

The Rocks of the Crust

These can be classified into the three types: (1) igneous; (2) sedimentary; and (3) metamorphic.

The igneous rocks are derived from magma which, on cooling, forms crystalline rock, with large crystals if the cooling is slow and small crystals if it is rapid. The size of the crystals can have a considerable influence on the rate of weathering of the rock. The composition of igneous rocks varies greatly but quartz, the feldspars, micas, and the ferromagnesian minerals such as hornblende and augite are most common. The rate of weathering of these minerals is approximately in the order in which they are listed, so that quartz is the most resistant and the ferromagnesian minerals are least resistant.

Sedimentary rocks are derived from the weathering products of other rocks, so they are usually composed of primary minerals and also the weathered products of the primary minerals. Their composition is extremely varied, ranging from the almost pure quartz of some sandstones to the assemblage of clay minerals in a shale. A very important component of many sedimentary rocks is calcium carbonate, which acts as a cement in many calcareous rocks and is the dominant material in many limestones. In chalk it can form 95% of the rock. Other rocks such as the marine muds and some loess (10–25% $CaCO_3$) have smaller quantities. Other important cementing materials are iron and siliceous materials.

Metamorphic rocks are the product of alteration of other rocks by heat and/or pressure. Their composition is therefore as varied as their original materials.

The most important minerals found in soils are given in Table 25.1. The enormous importance of silicon (Si), iron (Fe), aluminium (Al) and the plant nutrients calcium (Ca) and potassium (K) should be noted.

Igneous rocks are of importance as providers of parent material in areas of volcanism and on parts of the shield areas of the continents. Much of the surface of the shields of temperate regions is covered by

Table 25.1. Soil Minerals

		Primary minerals		*Secondary minerals*	
	Quartz	SiO_2		Calcite	$CaCO_3$
	Microcline			Dolomite	$CaMg(CO_3)_2$
	Orthoclase	$KAlSi_3O_8$		Gypsum	$CaSO_4 . 2H_2O$
Feldspar	Na-Plagioclase	$NaAlSi_3O_8$		Apatite	$Ca_5(PO_4)_3 . (Cl,F)$
	Ca-Plagioclase	$CaAl_2Si_2O_8$		Limonite	$Fe_2O_3 . 3H_2O$
	Muscovite	$KAl_3Si_3O_{10}(OH)_2$		Haematite	Fe_2O_3
Mica	Biotite	$KAl(Mg . Fe)_3Si_3O_{10}(OH)_2$		Gibbsite	$Al_2O_3 . 3H_2O$
	Hornblende	$Ca_2Al_2Mg_2Fe_3, Si_6O_{22}(OH)_2$		Clay minerals	Al-silicates
	Augite	$Ca_2(Al . Fe)_4(Mg . Fe)_4Si_6O_{24}$			

sediments and drift deposits such as alluvium, loess and glacial moraine or outwash; hence igneous rocks are only of major significance in parts of Africa, Australia, South America and India.

Sedimentary and metamorphic rocks are also frequently masked by organic deposits such as fens and bogs, or by the variety of unconsolidated sediments which may be classified according to their mode of transport (Table 25.2).

Table 25.2. Unconsolidated Sediments

Water	*Ice*	*Wind*	*Mass movement*
Alluvium	Moraine	Loess	Colluvium
Lacustrine sediments	Till	Dune sand	
Marine sediments		Volcanic ash	
		Dust	

Weathering Processes and their Products

Weathering is denudation or rock destruction *in situ* and does not involve transport of the debris. It can be caused by physical and by chemical processes. The main effect of physical weathering is the comminution of rock into smaller particles which thereby makes the particle more liable to chemical weathering. Physical weathering itself does not involve the alteration of the composition of the rock. Alteration is the result of chemical weathering, the end product of which is the formation of clay minerals and the release of soluble products. In general, physical weathering is of greatest importance in climates which are humid and cold; chemical weathering is dominant in humid-warm climates although it does occur in cold and dry climates.

Physical Weathering

The processes of physical (sometimes called mechanical) weathering are: (1) unloading; (2) thermal expansion; (3) crystal growth, including frost action; (4) colloidal plucking; (5) organic activity; and (6) expansion and contraction of clays.

Unloading is of little direct importance in producing a regolith which will provide parent material for a soil, because it takes place at a considerable depth beneath the surface and produces large plates of rock which have to be further broken down. It is caused by the erosion of regolith material from the surface so that the weight of rock on the underlying formations is reduced. The release of pressure allows the rock at depth to expand, and cracks are formed in it parallel to the surface. If the fissured rock is eventually exposed, the plates of rock are revealed. This mechanism, which is often called exfoliation, is thought to be responsible for the development of some rounded rock outcrops of inselberg shape; inselbergs are residual hills left by the removal of weathering products, particularly in tropical climates.

The importance of *thermal expansion*, or the expansion of materials during heating, as a process leading to comminution of rocks has been greatly over-stressed. The theory, which has now been almost completely abandoned, was that because the minerals composing a rock have different coefficients of expansion, their differential expansion and contraction stresses would cause cracks in the rock. It is now thought that the elastic properties of most rocks can accommodate the stresses resulting from expansion, except perhaps during the rapid heating which a rock may experience during a forest fire. The elastic properties of rocks were demonstrated by Tarr (1915), Griggs (1936) and Roth (1965). The last named placed thermistors in holes drilled in a quartz monzonite boulder in the Mohave Desert, USA and found that heating and cooling did cause expansion and contraction of the boulder but that it was uniform and not differential. He also found that the water content of the boulder was considerable, even though the annual precipitation was only 5·1 cm, and that weathering was greatest on the shady or north side of the boulder. The weathering, then, was chemical and not mechanical, and this appears to be the case in even the driest deserts where dew at least is available.

Crystallisation from aqueous solutions and that of water into ice are both important processes by which physical weathering takes place. On freezing, water increases in volume by 9%, and if confined in a crack or a pore space exerts a bursting pressure of about 1380 newtons per square centimetre. In an environment in which freeze-thaw cycles ($-2°$ to $+1°C$) are frequent, rocks can be rapidly broken up and blockfields (felsenmeer) or talus slopes formed. Freezing and thawing are also partly responsible for heaving of ice bodies in soils and the formation of patterned ground in periglacial environments. The formation and growth of salt crystals in rocks in some arid, semiarid and coastal environments may also be an important process aiding rock disintegration. Crystallisation of halite, gypsum, calcite, and other salts from solutions, creates stresses in the materials around them and may produce a minor amount of disruption of mineral grains.

Colloidal plucking occurs when colloidal suspensions dry on the surface of a rock and create a tension which can tear flakes off the surface. This process is most effective when there is frequent wetting and drying and it may be sufficiently powerful to pit some rocks and create depressions which are called taffoni. Some lichens and fungi can have the same effect.

Organic activity is best seen in the action of tree roots growing into a fissured rock and prising it apart as the root grows. The burrowing action of earthworms, termites, ants, etc. also causes physical weathering, but these creatures are probably most important as agents for making channels down which water can percolate and so increase the rate of chemical weathering.

The expansion and contraction of clays with wetting and drying is not normally included in a list of processes of mechanical weathering, but because of the enormous increase of volume which clays such as montmorillonite can undergo in wetting—up to 15 times—they should be considered. Swelling may prise apart rocks but the major effect may be to induce mass movement and so lay bare underlying formations for further weathering.

Because physical weathering is denudation *in situ*, it does not include the comminution of rock during transport by ice, water or wind. Abrasion and attrition during transport, however, do greatly reduce the size of many particles and so make them more suitable for inclusion in the 'fine earth' of the soil—the products of physical comminution which are studied in mechanical analysis. Physical weathering is responsible for the breakdown of all particles except most clays, which are formed by chemical weathering. The rock flour produced by glacial grinding and the fine materials produced when water disrupts crystal lattices are probably the only clay-sized materials formed by physical disintegration.

Chemical Weathering

The production of fine-grained particles by chemical weathering is of major importance because in itself it greatly accelerates chemical actions by forming very large surface areas on which reactions can take place. A small stone with a volume of 1 cm^3 would increase its surface area from 6 cm^2 to 4000 square metres if it were reduced to particles of colloidal dimensions.

Chemical weathering is also important because it causes some minerals to be mobilised and removed from the soil, and other minerals to be replaced, so that new or secondary products can be formed by this alteration. In addition, deposits can be added to the soil or regolith by precipitation of crystals. The processes of chemical weathering are: (1) solution; (2) hydration; (3) hydrolysis; (4) oxidation; (5) reduction; and (6) exchange reactions.

Solution involves the total removal of weathering materials and is only of major significance where the parent material of the soil is a limestone. Calcium carbonate, the main component of limestone, is very

25. Formation of weathering skins at the base of a basalt boulder, and flaking on the exposed face.

soluble compared with the silicate minerals, and is rapidly dissolved when the soil solution contains dissolved carbon dioxide. Solution of limestones is of importance in soil formation because the insoluble non-calcareous residue forms the inorganic soil skeleton. Thus limestone soils have textures varying from clays to sands and have a colour dominated either by the residual minerals or by humus. This accounts for the famed red terra rossa soils of the Mediterranean area which have a high iron content, and the common black humus-rich soils found on many limestones.

Hydration is seldom a purely pedogenic (i.e. soil forming) weathering process, but is significant in altering rocks as in the formation of serpentine from olivine. Essentially, hydration is the adsorbtion of water by a mineral. During the adsorbtion, the H^+ and OH^- ions from the water penetrate between the plates of the crystal lattice, making it porous and subject to further weathering. A good example is the formation of yellow limonite from red haematite:

$$2Fe_2O_3 + 3H_2O \longrightarrow 2Fe_2O_3 . 3H_2O$$
$$\text{Haematite} \longrightarrow \text{Limonite}$$

This process is reversible so dehydration will produce haematite again.

Hydrolysis is the reaction of water with a mineral, and is thus closely associated with hydration. Hydration prepares the way for hydrolysis. In this process water is a reactant and not merely a solvent and the hydrogen ion, H^+, enters into the silicate structure:

$$\text{Feldspar} + \text{Water} \longrightarrow \text{Clay} + \text{Solution}$$

The net effect of hydrolysis is the exchange of an H^+ ion from the water for a cation of the mineral, leading to expansion and contraction of the silicate structure and increasing the pH of the water. An example of hydrolysis is the formation of kaolinite from feldspar:

$$K_2O . Al_2O_3 . 6SiO_2 + 2H_2O \longrightarrow H_2O . Al_2O_3 . 6SiO_2 + 2KOH$$
$$\text{Feldspar} + \text{Water} \longrightarrow \text{Kaolinite} + \text{Potassium hydroxide}$$

In this process the potassium has been replaced in the crystal lattice by hydrogen ions and potassium hydroxide released to the drainage water. The increase in volume which accompanies hydrolysis is the most important physical effect of chemical weathering and greatly exceeds in effect the various mechanisms of physical weathering. Granular decay of coarse-grained rocks such as granite, even in hot desert environments, may be attributed more to hydrolysis of the feldspars than to any other action.

Oxidation is a common and important weathering process carried on both organically and inorganically in a great number of complex reactions. It is the addition of oxygen to a mineral, often by the action of aerobic bacteria. The effects of oxidation are most apparent in rocks containing iron in the sulphide, carbonate and silicate forms. Discoloration by oxides is readily noticeable. In many soils, brown stains indicate the presence of ferric oxide which has been formed by oxidation of ferrous iron:

$$4FeO + O_2 \longrightarrow 2Fe_2O_3$$
$$\text{Ferrous oxide} \qquad \text{Ferric oxide}$$

In oxidations involving compounds of sulphur, sulphuric acid is a possible product and, from the standpoint of further weathering, a most significant one.

Reduction is the release of oxygen from compounds, often by the action of anaerobic bacteria. Reduced conditions occur in those soils or regoliths with slow water movement or waterlogged conditions. Soils in a reduced state liberate large amounts of exchangeable cations—Fe, Mn and Al—and the pH of waterlogged soils becomes very low (i.e. very acid). Ferric iron thus reverts to ferrous iron. Both reduction and oxidation occur in any soil. In wet conditions reduction prevails and the soil quickly uses up any introduced oxygen; in dry conditions, oxidation prevails. The boundary between oxidation and reduction is often narrow, for they are complementary, the balance sometimes changing seasonally, and one cannot then take place without the other.

Exchange reactions have been discussed in some detail in the chapter dealing with clays and colloids. These reactions include the exchange of ions on the micelles and also the modification of the crystal lattices. Many of these reactions involve the organic acids derived from humus.

Factors Affecting Weathering

Few generalisations can be made about the rate of weathering of minerals because of the numerous factors which can influence the process. The three factors of climate, physical and chemical properties of the rock are, however, so important they require special mention.

Climatic conditions determine the temperature and moisture regime in which weathering takes place. Under conditions of low rainfall, mechanical weathering is dominant, and therefore comminution of particles, without alteration of their composition, occurs. With an increase in precipitation, more minerals are dissolved and chemical actions increase, so that chemical weathering of minerals and synthesis of clays becomes more important. In humid temperate climates, silicate clays are formed and altered. Speed of chemical reactions is also greatly increased by a rise of temperature, in fact a rise of 10°C doubles or trebles the reaction rate. Increases in temperature also alter the relative mobility of minerals so that although quartz is highly resistant to weathering in temperate climates, fine-grained quartz particles are more easily weathered in tropical conditions, where iron and aluminium hydrous oxides are more resistant. The iron and aluminium therefore tend to accumulate in tropical soils, which get their red colour from the iron.

Another effect of climate is to control the vegetation and its production of litter. In humid tropical climates the production of organic matter is high—3350–13,450 kg per hectare per year from tropical forests—compared with temperate forests of oak and pine with 900–3150 kg per hectare per year. This means that the supply of organic acids to take part in chemical weathering is high in the tropical forests and low in the temperate ones. The appearance of the soil suggests the reverse because dark humus can accumulate in the cool forests but in the tropical ones it is broken down very rapidly and much of it has a pale colour which makes it difficult to see. The turnover of tropical forest humus is about 1% per day compared with 0·1–0·3% in temperate forests.

The physical characteristics of rocks such as particle size, hardness, permeability and degree of cementation all influence weathering

26. Remains of a sandstone block which has been completely weathered to clay minerals. The concentric weathering skins and lines of old fractures can be seen.

processes. Particle size has already been discussed. Hardness and cementation affect the rate at which weathering can reduce the rock to smaller particles. A siliceous sandstone will obviously be more resistant than a calcareous one. Permeability, however, is probably a more important characteristic, for it will control the rate at which water can seep into the rock and the area of the surface on which it can act. Rocks display a very large range of permeabilities, from low in igneous rocks to high in gravels and sands.

The chemical properties of a particle, particularly its solubility, greatly affect weathering. Highly soluble minerals such as calcite and gypsum are readily leached from a soil if the precipitation is high enough and carbonated or humus-enriched water is present. Under subhumid climates, calcium carbonate is often deposited lower down the profile when the water available is insufficient to remove it from the soil. Many salts are taken into solution and redeposited high in the soil profile under arid conditions.

The ferromagnesian minerals such as hornblende and augite are readily weathered because iron is relatively easily removed from a crystal lattice, and the iron is often visible as a brown stain. The tightness of the lattice is probably important in influencing weathering rates so that biotite (black mica) with a relatively loose lattice is more readily weathered than muscovite (white mica) which has a tighter lattice.

Organic matter characteristics have important effects on the nature and rate of weathering. Colloidal organic matter has a high water retention capacity; in some environments weathering (particularly chemical weathering) is limited by the relative scarcity of water in the weathering mantle. Organic matter also has important effects on pH and ion exchange capacity, which in turn partly control the progress of chemical weathering. Bacteria and other soil organisms play a part in chemical reactions within the weathered mantle. Plants, including fungi and lichens, also contribute to chemical weathering since they abstract elements from rock materials and liberate others in the process. Organic acids, and other products resulting from plant and animal decomposition, greatly increase the solvent power of water so that even ordinarily insoluble substances (e.g. limonite) can be dissolved. The roots of shrubs and trees play an important role in the wedging and prying apart of rocks.

It is not possible to present a list of the relative stability of minerals because it varies with climatic conditions but the following order is probably correct for sand and silt-sized particles in humid temperate conditions: quartz (most resistant) > muscovite, K feldspars > Na and Ca feldspars > biotite, hornblende, augite > olivine > dolomite and calcite > gypsum (least resistant).

Under a humid temperate climate the weathering of an igneous rock might be summarised as:

(1) Mechanical breakdown to finer particles.
(2) At the same time feldspars, micas and ferromagnesian minerals will suffer hydrolysis and hydration. Iron is oxidised and hydrated.
(3) Porosity is increased and calcium, magnesium, sodium and potassium are dissolved and removed.

(4) Secondary hydrated silicate clay minerals are formed and remain with the iron and aluminium silicates. The clay-sized residue will only slowly be altered.

Weathering is a complex procedure. It involves a large number of actions and reactions, many of which are carried out concurrently. In most instances rock weathering cannot be clearly distinguished from pedogenesis (soil formation) and soil horizon differentiation. Within the soil system, weathering, soil formation and horizon differentiation can all be ascribed to additions, removals, transfers and transformation of organic and inorganic material. Most such processes are as common to rock weathering as they are to soil horizon differentiation.

While there is no real end product of weathering or soil formation, development is in the direction of stability and insolubility. Clay mineral formation is most common where extensive periods of time have been available for weathering and soil formation. Erosion, deposition, horizon differentiation, and possibly climatic and vegetational change, are all factors promoting continued weathering and soil development.

26 The Soil Profile and its Development

Figure 26.i. Designations of horizons in a soil profile.

The continuous weathering of parent material and the action of organic matter produce the live thing—soil—which is an intimate body of inorganic mineral particles with living and dead organic matter. Because it is living and not inert it is constantly changing, and in so doing it produces differentiations within its profile so that distinct layers or horizons are formed. The study of horizons and their properties is the chief concern of the pedologist. The significance of genetic soil horizons lies essentially in their functions rather than their morphology, or as Nikiforoff expressed it:

> Individual soil horizons are the working aggregates of a complex thermodynamic system. With respect to their relationship to the whole system, they might be likened to the organs in a living body, each of which is adapted for the performance of some specific functions. Continuous performance of its specific functions imparts to each horizon its individual chemical and morphological character.

A soil horizon may be defined as a layer within a soil which is approximately parallel to the soil surface. It has properties caused by pedogenic processes which distinguish it from the other horizons. These properties are reflected in the characteristics of colour, structure, texture, consistence and presence or absence of carbonates. The properties are usually obvious in the field, although in the early stages of development they may not be distinguishable and occasionally laboratory study is necessary to distinguish horizons.

Horizon Designations

For the sake of convenience, symbols are used to indicate each horizon. There are some slight variations in the system of symbols used but the one which will be outlined here is generally accepted and recognised: O, A, B, C, G are the main letters used for horizons, and subdivisions of each horizon are denoted by numerals: 1, 2, 3. Figure 26.i shows a theoretical profile with all the horizons which are usually distinguished. Any particular soil, however, may have only a few of these.

O Horizon

The O horizon is the surface horizon of organic matter above the surface of the mineral soil. It is subdivided into O_1 and O_2 subhorizons or into L, F and H layers. (In some systems the O_1 is called A_{oo} and the O_2 is called A_o.)

O_1 { *L layer:* Consists of relatively fresh leaves, twigs and other plant debris which do not show signs of decomposition.

O_2 { *F layer:* Consists of partially decomposed litter with recognisable plant remains.
H layer: Consists of virtually structureless decomposed organic matter.

These letters are chosen because O stands for organic, L for litter, F for fermentation, H for humified.

Under forest or grass the constituents of the O horizon may be given one of the names mor, moder or mull (Figure 26.ii).

Mor is a raw humus consisting of little decomposed peaty and fibrous

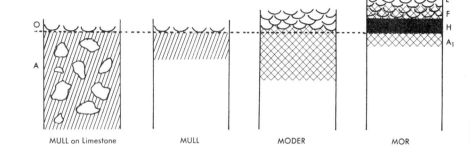

Figure 26.ii. Schematic representation of the different types of humus form.
After Duchaufour

MULL on Limestone   MULL   MODER   MOR

plant remains. It is usually formed under coniferous forests and heaths growing on very well drained soils which are low in available calcium, and are in consequence very acid. It can also occur on badly drained soils, under deciduous forest, and on only mildly acid soils. Mor seems to accumulate on soils as L, F or H layers where earthworms are not present. As earthworms are normally most numerous in calcareous soils and in those sufficiently deep to remain moist throughout the year, mor humus is found on shallow sandy soils and beneath vegetation with a litter poor in bases. Mor has developed on some grass soils which have been rendered acid by repeated application of sulphate of ammonia. Typical pH's of soils with mor humus vary from 3 to 6·5.

Moder is more decomposed than mor, either by fungi, insects or small animals. It is sometimes largely composed of faecal pellets but does not have the crumb structure of mull. A moder is acid in reaction.

Mull humus typically forms under deciduous or mixed forests, although it can occur under coniferous forests especially if they are composed of cedars or certain species of spruce whose litter has a high calcium content. Mulls are usually less acid than mors, with pH's which vary from 4·5 to 8. The organic layer is crumbly and mixed with the mineral particles of the soil, so that there is a transition from the surface mull to the mineral soil. Mulls typically contain many earthworms and other animals which so macerate the litter in their guts that all trace of the original plant structure is lost in the excrement.

Under grass or herbaceous vegetation, most of the organic matter is derived from the decay of the plant roots; it is therefore intimately mixed with the mineral soil and does not accumulate on the surface unless the ground is waterlogged and soil animals and aerobic bacteria are absent. Grassland humus forms are therefore of the mull type unless earthworms are absent, when decomposing grass can accumulate to form a mor type.

A Horizon

The A horizon is the surface horizon of the mineral part of the soil. At the top it is usually a mixture of organic and inorganic matter but at the bottom it has a zone from which materials are lost by leaching. The subhorizons of the A horizon are:

A₁ This is the surface mineral horizon containing a relatively high proportion of organic matter mixed with it. It is usually the zone of maximum biological activity in the mineral soil. Measurements of depth for all horizons are taken, by convention, from the top of the A_1.

A₂ This is a zone of loss of clay, iron or aluminium—in other words, the zone of eluviation (Latin: *eluere*—to wash out). The residual matter left behind is usually quartz, and because of the loss of clay it is frequently structureless. As a result of the losses it is often pale in colour or even bleached. The A_2 is always lower in iron, aluminium and humus than the underlying B horizon.

A₃ This is a transitional zone between the A and B horizons but with characteristics more like those of the A than the B.

B Horizon

This horizon is characterised by accumulation of material washed into it from above, and by chemical deposition. Hence it is an illuvial horizon. It is the horizon of maximum development of the larger structures such as prisms and blocks. It is usually also a horizon of strong colours such as the dark brown resulting from deposition of humus or the red-brown colours of iron.

B₁ This subhorizon is transitional but more like the B than the A horizon.

B₂ This is the subhorizon of maximum development of structure, or of maximum accumulation of clay minerals, or of accumulation of organic matter or the sesquioxides (iron and aluminium oxides), or any combination of these features.

B₃ This subhorizon is transitional to the C horizon but more like the B than the C.

C Horizon and D or R Horizon

C is a mineral horizon which excludes bedrock and which is relatively unaffected by pedogenic processes. It has been modified by weathering and is for the most part outside the zone of biological action. It may be indurated, cemented, gleyed, or it may contain deposits of soluble salts. Historically it has often been called the zone of parent material, but in fact it is not possible to find the parent material from which the A and B horizons have developed because that material has been altered. C material, then, is not necessarily parent material but only assumed to be like it. C_1, C_2, etc. may be used to distinguish between zones within the C horizon but attempts at their precise definition are probably doomed to failure because of the assumptions they would involve.

Underlying rock beneath the C horizon is designated D in European systems and R in the USA.

G Horizon

When a horizon is entirely composed of gleyed materials because of saturation with water for long periods in the presence of organic matter, it usually has a blue, green or grey colour resulting from intense reduction of iron to the ferrous state. If drained, brown colours appear in the soil, especially along root channels and structure faces, because the iron has been oxidised to the ferric state. The horizon may be subdivided into G_1 and G_2, etc. if necessary. Slight gleying is denoted by a lower case letter as in Bg—a gleyed B horizon which looks more like a B than a G. The symbol G has been dropped from the new USA system (7th Approximation).

Additional Symbols Qualifying Horizon Designations

Additional letters have been found useful in denoting special features of horizons and subhorizons. The more important are listed below:

(B) Used to describe a B horizon which is *not* enriched by illuviated clay but has clay formed *in situ*.

(A) Soils lacking a true humus-enriched A but having a marked root development on top of a B.

AC A transitional zone between an A and a C horizon in a soil without a B. It has characteristics of both the A and C. Such horizons occur in skeletal soils and in rendzinas.

B/A A crust of deposited material on an A.

g Gleyed horizon, as in Bg or B_2g.

f Frozen horizon, e.g. Cf in Antarctic soils.

x An indurated zone or fragipan, e.g. Cx.

ca Having an accumulation of calcium carbonate, as in Bca or B_2ca or Cca.

cs Having an accumulation of calcium sulphate, as in Ccs.

h Having an accumulation of illuvial humus, as in Bh or B_2h.

fe Having an accumulation of illuvial iron, as in B_2fe.

c Having concretions. If rich in iron—fec; manganese—mnc; calcium carbonate—cca, etc.

e With strong eluvial features, e.g. Ae.

t Used to denote translocated clay in a B horizon, as in Bt or B_2t.

p Used to refer to a ploughed layer but only in an A horizon—Ap. Ap may also be used to denote other disturbances.

Note: The boundaries between horizons and zones may be clear or indistinct, wavy or straight, indistinct wavy boundaries probably being the more common. The judgement of the observer, therefore, has to be used in describing profiles.

In his classification of soils, Kubiena suggested a general grouping of soils based on similarities of profile and listed the types of profile which are given below. It should be noted that Kubiena and his contemporaries regarded C horizons as modified parent material, but that in the description of C horizons given in this chapter the current American system has been followed in which the C horizon is considered mainly as a zone of oxidative weathering, although fragipans, salt accumulations and gleying can occur in it. Kubiena's profile types are as follows:

A B C soils Those with a leached A above an enriched B over weathered material.

A (B) C An A on top of a B which is not enriched by illuviation but characterised by the formation of clay *in situ*. Weathering occurs in the C, which may be too deep for notable biological activity.

27. AC horizon soil profile, showing the granular A and the C of partly weathered rock fragments amongst A horizon material.
NZ Soil Bureau

B/A B C	A surface enrichment of deposited material.
A C	Soils with a distinct humus horizon but no B horizon. Many soils developed on limestones and many humus-gley soils have this kind of profile.
(A) C	Material with soil life but no clearly discernible humus layer. Some alpine, Arctic and desert soils have such a profile.

Development of the Soil Profile

The following comments are intended to indicate the type of development of a soil which may occur in humid temperate conditions on a freely drained, nearly flat site.

The introduction of biological influences into weathered material residues may be said to be the initial stage in soil formation. The first colonising plants may be lichens and mosses, or marram grasses on sand dunes. Once plants are established, soil flora and fauna increase in numbers and accelerate the accumulation of organic matter, which is constantly increased by the addition of plant litter. The plants derive their oxygen and carbon from the air and the other essential nutrients—particularly N, P, K—and water from the evolving soil in which fungi and bacteria are setting free the minerals. The growth and decay cycles which are constantly repeated in the soil build up the fertility, increasing the rate of chemical weathering and the formation of clay minerals. Under the influence of wetting and drying, flocculation and deflocculation, and in the presence of roots and humus, soil structures develop, and the rooting of higher plants is facilitated.

The differentiation of horizons in soils is the result of the additions, removals, transfers and transformations which are constantly occurring. Water and humus are added frequently; salts, fine particles and humus are removed in the drainage water; humus, clay, iron and aluminium oxides and salts are transferred from one horizon to another; primary minerals are weathered to secondary minerals which may be further modified. The burrowing of animals and interference by man upset the differentiation of horizons but otherwise these processes of differentiation are constantly at work in all soils. The differences between soils are caused by the variations in the balance between the processes.

Development of A Horizons

Accumulation of organic matter is an early step in the development of differentiated horizons in most soils. During the early stages of soil development, the additions of organic matter usually exceed the rate of decay, but after a period of time a balance between additions at the surface and transfers deeper into the soil is achieved. In 150 years soils on glacial drift in Alaska developed to the point at which they were as rich in organic matter as many in the eastern USA, and they then appeared to retain their balance of additions and losses. Environment controls the addition of organic matter, so that it is low in desert soils, high on tall grasslands and in tropical forests, but also losses are low in cold regions and high in the humid tropics.

Once a humus layer is formed soil water, which is already slightly acid because of dissolved CO_2, becomes even more acid because of the humic acids derived from organic matter. This solution has a solvent action on minerals and acts as a catalyst in chemical reactions which cause weathering of mineral matter. From the reactions alkali cations are released, and secondary minerals including clay minerals are formed. The clay minerals retain some of the alkali cations in exchangeable form whilst many are removed by drainage water. Any free calcium carbonate in the A horizon dissolves in the acidic soil solution and calcium then dominates the exchangeable cations. The leaching of the alkalis lowers the pH of the soil, and in the acid reactions iron ions are released from the hydrated oxides and hydroxides as the clay lattices break down. The aluminium ions are then released, so that both the iron and the aluminium are mobilised.

In the presence of basic cations the colloidal organic matter remains coagulated, but in the presence of acids it is peptised, or dispersed, and passes into a sol form in which it can move through the profile. Humic acids also peptise the iron and aluminium sesquioxides, rendering them soluble and so allowing them to be leached. The clay in the A horizon is translocated by mechanical washing into the underlying horizons when sufficient water passes through it.

In summary, the A horizon is one from which material is removed by leaching solutions and by mechanical translocation. It is said to be an eluvial horizon and the process of removal is eluviation.

Formation of B Horizons

Material removed from the A horizon of a soil is not necessarily lost to it, but may be deposited lower down the profile in the B horizon. The B horizon is therefore said to undergo illuviation and to be an illuvial horizon. Because of deposition within them, B horizons usually have stronger colours than the lower parts of A horizons, although the humus in the A horizon can mask this. B horizons are the zone of maximum development of soil structure because the high clay content induces swelling and shrinking. The enrichments of B horizons may be calcareous, ferruginous, of organic matter, sesquioxides or clay.

Calcareous enrichments occur because soluble calcium bicarbonates (calcium carbonate plus carbon dioxide) may be washed down the profile, but in the process the carbon dioxide is lost and insoluble calcium carbonate is deposited. This process is much the same as that which produces stalactites and stalagmites in limestone caves. The depth at which the deposition occurs depends upon the depth to which water penetrates in the wet season. In very dry climates, therefore, calcareous concretions or even a pan of calcium carbonate—called caliche—can form near the surface. With increasing precipitation, the deposition gets deeper until sufficient moisture is available to remove the carbonates completely from the soil. The presence of secondary calcium carbonate in a freely drained profile is the mark of the group of soils called pedocals occurring in subhumid climatic regions in which precipitation is less than the potential evaporation.

In some desert soils a layer of calcium sulphate forms beneath the carbonate as a gypsum deposit. It is more soluble than the carbonates and so carried further in the drainage water. For this reason it is also more likely to be completely removed in the drainage water.

Ferruginous or iron deposits are formed during gleying when iron is

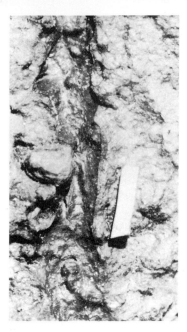

28. A thick clay skin in a soil pore, and thin clay skins on surfaces of peds.
US Soil Conservation Service

reduced to the grey-coloured ferrous state in anaerobic conditions, but in drier periods this may be oxidised to the red-brown ferric oxide. The ferric oxide accumulates preferentially on other ferric oxides, thus encouraging the development of nodules and pans of iron. In extreme cases, bog iron ore is formed in cool climates and 'murram' in the tropics. The alternation of wet and dry conditions in the soil leads to mottling within the profile. The rusty appearance of root channels in grassland soils is a sure sign of seasonal wetness.

The mechanism by which *humus* and the *sesquioxides* are deposited in B horizons is not fully understood. It appears probable that as mobilisation and eluviation from the A horizon occur in acid conditions, deposition occurs in the less acid conditions of the B horizon. The soils in which this process is most marked, called podzols, have a B_1 enriched in dark humus and a B_2 of red-brown colours formed by deposition of the sesquioxides. It is also possible that precipitation of the sesquioxides may be encouraged by restricted percolation caused by mechanically translocated clay. Another possible explanation, applicable to some soils, is that eluviation predominates in a wet winter and that with the drying out of the profile in summer precipitation occurs. The localisation of precipitation in a definite horizon would probably be connected with the nature of the moisture gradient. Tests reveal a higher organic matter content in the B horizons than in the A_2 horizons of many soils. In podzols the humus is dispersed in the B_1 as a coating on the soil particles, but in many other soils it appears as a dark-coloured film or skin on the face of the peds.

Clay translocation is the mechanical eluviation of the A and illuviation of the B. It seems to occur only in soils which have had many years for development and which do not occur in excessively wet climates. It appears that the soil must dry out for at least part of the year. Only very fine clays of $0 \cdot 1 \mu$ or less in diameter are involved.

For translocation to occur, the parent material must contain clay or clay must be formed by weathering. Very fine clays, carrying the same negative charge as the soil matrix, tend to disperse unless salts such as carbonates or free oxides keep them flocculated. Sodium ions and humus induce dispersion. If a soil in which the clay is dispersed is wetted, the clay moves with the percolating water and stops where the percolating water stops. Percolating water may be withdrawn into the soil fabric by capillary action, so that the clay is left on the surface of the peds and on the walls of pores. Wetting and drying of the soil, therefore, appears to be necessary for translocation of clay. The breaking of capillaries by a horizon of coarse texture or the presence of flocculating carbonates seem to encourage clay deposition.

The clays on the ped surfaces and in pores, called clay skins or illuviation cutans, can be recognised in the field with the aid of a 10–20 power hand lens. The clay skin has the smooth, waxy appearance of a coat on the ped. If the ped is broken, there is usually evidence of a somewhat coarser texture within the ped and frequently a different colour. Pores emerging on the lower side of a ped often have irregular lips where the clay protrudes. The surfaces of some coated peds show channels and flow lines formed by running water. Care must be taken to distinguish clay and humus coatings. The humus coatings do not have protruding lips, do not show flow lines, are usually darker in colour and are often patchy.

Pans

In some soils compact or hard layers are formed. Claypans, or zones of clay accumulation, fragipans, or compact horizons which may or may not be cemented, and hardpans, or cemented horizons, are the three main types.

Pans commonly occur in soils on gently sloping or level sites which have excess moisture although they are not completely waterlogged. *Claypans* are accumulations of clay within the profile and may be formed from fine-textured deposits in the parent material or by weathering *in situ* aided by translocation. Soils of the prairies and elsewhere in which they are notable are called planosols. Their low permeability presents many management problems. Seepage along the tops of claypans can sometimes cause mass movement of the slide or flow type on slopes.

Fragipans are horizons that are very slowly permeable, compact or dense, hard when dry and friable when moist. They have silty and fine sandy textures and, although their origin is uncertain, it is possible that some of them are formed by a siliceous cement.

Hardpans are indurated horizons which may be cemented by iron oxide, silica, organic matter or calcium carbonate either singly or in combination. The iron and humus deposits of podzols and the caliche formed from calcium carbonate have already been mentioned. The most widespread type of pan is probably the crusts of sesquioxides (mostly iron) of tropical or former tropical climatic regions called laterite. This pan material occurs as massive layers varying from some centimetres to several metres in thickness. It has a red colour, may be vesicular and may contain entrapped quartz and other minerals. This indurated material is used for building purposes in southeastern Asia. It is thought to have been formed by the hardening of an illuvial horizon after it was exposed by erosion. (See colour plates.)

All pans have the effect of impeding soil drainage and the root development of plants.

C Horizons

C horizons have hardly been mentioned in this discussion because they lie below the reach of the biological forces that operate in the development of A and B horizons. As a result, they tend to have characteristics more clearly dependent on local geology and climate, whereas the A and B horizons show a uniformity of profile morphology when they are developed under similar factors of climate, vegetation and site unless parent material is very significant—as it may be in soils derived from basalt or limestone.

Soil Colour

The colour of soils is one of the most easily observed and useful diagnostic features of a profile. In the field it is usually recorded by comparing the colour of each horizon with a standard Munsell Soil Color Chart which contains a series of charts of colours arranged by hue, value and chroma. Hue is the spectral colour, value is the lightness of the colour and chroma is the purity or strength of the colour. A greyish-brown soil might be recorded as having a colour of 10YR 5/2 when dry

29. A compact silica pan which may be a relict of a former podzol Ae horizon.

and 10YR 4/3 when wet. (It is preferable to record the colour of the wet soil.)

Colour reflects some of the features of soil environment and composition. In New Zealand, for example, the soils of the humid regions are known as yellow-brown earths, those of subhumid regions as yellow-grey earths and those of semiarid regions as brown-grey earths. Colour here seems to reflect climatic conditions.

The soil components which give it colour are primarily iron and humus. Iron gives yellow, red and brown colours, as in the iron-rich soils derived from basaltic parent material or in soils which have been so weathered that iron has been leached from Ae horizons and deposited in B_2 horizons, leaving the Ae bleached white and making the B_2 red-brown in colour. Humus makes most temperate climate soils black or brown, depending largely upon the vegetation type. Grey colours are usually the result of saturation by water causing gleying. Mottling with red and grey colours is caused by impeded drainage with a fluctuating water table.

27 Factors of Soil Formation

The concept of the soil as a body developed as the result of a number of interdependent factors is owed to the Russian Lomonosov (1711–65), but his work did not achieve much publicity and the idea of five major soil forming factors is usually credited to Dokuchaev (1846–1903), who distinguished them as: (1) local climate, especially water, temperature, oxygen and carbon dioxide; (2) parent materials; (3) organisms, both plant and animal; (4) relief and elevation; (5) age of the country.

Jenny expressed these factors as an equation:

$$s = f\,(cl, o, r, p, t, \ldots)$$

in which soil (s) is derived from the factors (f) of climate (cl), organisms (o), relief (r), parent material (p), and time (t). These factors are all parts of the large scale environment and hence this is basically a geographical concept. Not all factors are of equal importance in one soil, and to demonstrate the importance of a particular factor the equation can be variously written:

$s = f\,(cl, o, r, p, t, \ldots)$ climatic control
$s = f\,(o, cl, r, p, t, \ldots)$ biological control
$s = f\,(r, cl, o, p, t, \ldots)$ topographic control
$s = f\,(t, cl, o, r, p, \ldots)$ age control
$s = f\,(p, cl, o, r, t, \ldots)$ parent material control

To complete Jenny's list we should perhaps add gravity which influences all movements in the soil, space in which the soil exists, water and man. Man particularly modifies soil extensively by cutting timber, burning forests and grass, overgrazing the land with his animals, ploughing, manuring, liming, draining, flooding and irrigating the soil.

Time is perhaps a peculiar element in this list because although, like space, it is a condition of existence, and every material object occupies space during the passage of time, neither time nor space cause that object to exist. Every change takes time but the change is not caused by time. Time is therefore not a factor in the same sense as the climate, organisms, relief and parent material are.

In a soil system the factors are invariably interrelated. For example, the parent material contributes minerals, but it also affects infiltration, permeability and drainage and hence the water and gases in the soil. Topography also affects water in the soil by influencing runoff and drainage conditions and hence affects chemical reactions and trans-locations in the soil. The climate determines the effective rainfall through precipitation and evaporation. Organisms also affect the water supply through transpiration and by influencing soil permeability and structure; they also affect runoff. In addition, they contribute carbon dioxide, organic substances, inorganic bases and acids.

A second example is that of temperature and heat supply. The climate is obviously the most important factor in determining the temperature and total heat supply but the parent material influences the temperature at the site through its colour and conductivity, topography through exposure and slope, and organisms through the nature of the cover over the surface, as well as by transpiration, evaporation and air movement. It can be similarly shown that nearly every reaction in the soil is affected by all of the factors of soil formation, and any reaction may consequently result from a number of combinations of the various factors.

One serious objection to the factorial method of studying soil development is the implied assumption that the soils are monogenetic, that is, they have developed under constantly uniform conditions since their initial state. It is possible that in many parts of the world the inert complex of inorganic and organic matter, usually called the soil body, (to distinguish it from the living soil) is thousands, and in parts of the tropics millions of years old, so that climate and hence organisms and topography have changed during the existence of that soil body. The good correlation between certain soil properties and present climatic conditions probably occurs because many of the present-day soils were initiated in older soil bodies which have been little modified by deposition and erosion, or because the relative balance of factors at a site has remained constant. There are some soils of the western USA with a duripan, inherited from the past, which is no longer developing and which represents past climatic conditions.

In the discussion below the factors are treated separately for the sake of clarity although in nature they do not operate separately. The

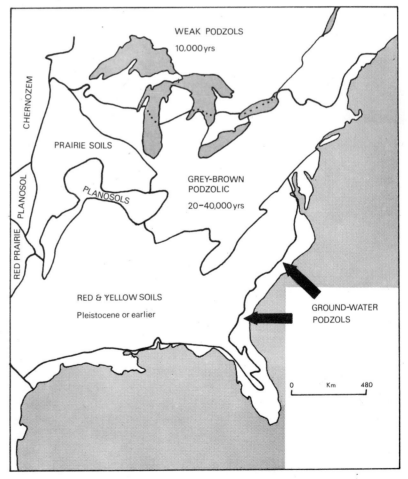

Figure 27.i. The soils map of eastern North America can be read as a map of time available for weathering. In the north, recent glaciation has allowed the podzols less than 10,000 years to develop, the grey-brown podzolics 20,000–40,000 years and the red and yellow podzolics many thousands or even millions of years depending only on the survival of land surfaces. The red prairie soils probably owe their colour, and the planosols their well developed B horizons, to age. The coastal podzols and other coastal soils have varied ages which depend upon sea level changes in the Quaternary. *After Carter and Pendleton*

separation is a reflection of our need to organise the data in order to understand them. Jenny's equation should be solved using quantitative data but these are not as yet available.

Time as a Soil Forming Factor

One problem in discussing the influence of time on soil formation is that of defining the point at which material becomes soil. Alluvium, glacial moraine, volcanic ash and loess may have no horizons but, within days of their deposition, life may colonise them and so produce detectible differences between one part of the deposit and another. Jenny concluded that such materials are soils 'unless they are being deposited under the very eyes of the observer'.

A second problem is that of trying to distinguish the effects of time and soil processes. The differences in the morphology of such major soil groups as podzols, grey-brown podzolics, and red-yellow podzolics in the eastern USA are usually regarded as a function of climate: podzols forming in cool climates, grey-brown podzolics in temperate conditions and red-yellow podzolics in the subtropical climate of the southeast USA. Carter and Pendleton, however, contend that the differences between the soils are more probably caused by time. The podzols have had a weathering time of less than 10,000 years; the grey-brown podzolics a time of 20,000–40,000 years; and the red-yellow podzolics may have had as many as a million or more years to form. The differences in ages are the result of time available since glacially-induced changes of climate (Figure 27.i).

Figure 27.ii. The influence of three different types of parent materials—limestone (L), sandstone (S), and granite (G)—upon soil profile development beneath three different types of vegetation and climate.
After Wilde

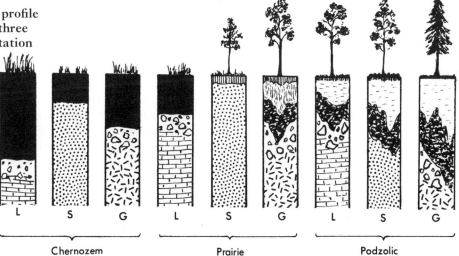

L   S   G   L   S   G   L   S   G

Chernozem   Prairie   Podzolic

A third problem is distinguishing the effects of parent material and of time. Quantitative data are not readily available. In general the softer and more permeable the rock from which parent material is derived, the quicker parent material forms. This may be expressed diagrammatically as in Figure 27.ii. Climatic and vegetation influences can, however, produce variations on this simple progression.

In certain circumstances, however, time for soil formation can be measured by one of several means. The surface on which the soil lies can sometimes be dated. This is particularly true of old erosion surfaces of the type which occur over much of Africa and parts of Australia.

Absolute dating of soil is sometimes possible from historical records, as when the date of deposition of a volcanic ash shower is known, or coastal dune formations can be dated from old maps and records, or different loess and till sheets are dated, or floods leaving alluvium are recorded; radiometric methods can be used to study the age of organic material in the soil. All of these methods have their problems. It is not always known how weathered the original material was and how much clay was present in it.

A few studies may be quoted to give an impression of observations using these methods. After the eruption of Krakatoa in the Sunda Strait (6°S) in August 1883, a mantle of dust and pumice settled on the nearby island of Lang-Eiland; in 45 years a soil profile of 35 cm depth had formed. The mud ejected from Lake Rotomahana (38°S) in the Tarawera eruption of 1886 in New Zealand has weathered more slowly and 4 cm of topsoil has been formed in 80 years. In El Salvador (13°N) the soil was 1 metre in depth after 1993 $\pm$ 360 years; and on the Taupo pumice erupted in 130 AD in New Zealand the soil profile is about 20 cm deep. (See colour plates 1, 2, 3, 4, 18.)

Tamm, working in Jämtland, Sweden (63°N), estimated that a normal podzol with 10 cm of mor humus, 10 cm of bleached sand A_2, and a 25–50 cm B horizon requires 1000–1500 years for development, but that older soils 3000–7000 years old show no signs of greater development. The very slow decomposition of organic matter in cold climates may partly account for this. Tedrow found organic matter 2000 $\pm$ 150 years old at the surface of some Arctic brown soils, and soils of polar-facing slopes usually exhibit less profile development and decomposition than those on equator-facing slopes.

Under some conditions it is possible to see the effects of time on soil development; in some coastal areas, for example, sequences of dunes are formed as a coastline advances and the oldest dune is left inland whilst the youngest is formed at the shore. On these dunes a chronosequence of soils is provided. Flights of river terraces, with the uppermost being the oldest and the current floodplain the youngest, similarly provide a chronosequence. (See colour plates 1, 2, 3, 4.)

In some of the older literature on soils, it is customary to speak of young, mature and old soils. Current opinion is opposed to this because there is no means of measuring the stage at which a soil may be. We do not know what processes have formed it or what changes of climate and vegetation have affected it. If, as Carter and Pendleton suggest, the soils of the eastern USA owe their differences to time and not to climate and vegetation, what is the mature soil of a given place? We may conclude, then, that there is no such thing as 'mature' soil, for soils are dynamic and constantly changing.

Parent Material as a Factor in Soil Formation

The significance of weathered rocks as factors in soil formation has already been discussed in Chapter 25. In general, it may be asserted that this significance diminishes with time as the other factors of climate and vegetation leave their imprint more strongly on the soil, yet the influence of the parent material is often still discernible in soils which have had many thousands of years in which to develop. Some early soil surveyors actually mapped soils on the basis of bedrock, so that it was common to read of limestone soils, sandstone soils and granite soils. The influence of Dokuchaev has encouraged the formation of classifications which are based primarily on climatic or zonal factors, but such classifications still require a category for soils which are strongly influenced by parent materials and these are usually referred to as intrazonal soils.

The weathered rock which provides parent material for a soil may be considered as a passive element in soil formation, unlike climate and vegetation, but similar in this respect to relief and time. Parent material effects soil formation in two ways: (1) through its physical properties, and (2) through its chemical properties.

Texture is probably the most important physical property. In general, acid, igneous and coarse-grained clastic sedimentary rocks give rise to coarse-textured soils. Two extreme cases might be wind-blown dune sands with pale-coloured, coarse-textured, freely-drained soils, and red-brown, fine-textured, more slowly-drained soils formed on basalt. A very fine-textured parent material may give rise to gleying where the water table is high.

Base-rich parent materials, such as basalt, weather to give mont-morillonitic clays; and from base-poor rocks, such as granite, kaolinitic clays are formed. The supply of bases can also affect the type of eluviation in the soil, so that in cool climates a deficiency of bases is associated with strong podzolisation, but this will be weak on base-rich parent materials. Examples of this are the podzols of southern England which

30. The soils shown in these two photographs developed on the same levee and are the same age. The parent material of the first example is a well drained gravel and the profile is deep. The parent material of the second example is a fine silt which is heavily gleyed and soil formation has been slow; the profile is shallow.

occur on acid quartzose sands, but in the cooler climate of Iceland non-podzolised soils form on basic basaltic rocks.

Limestone provides a parent material which is always influential in the development of the soil. Hard, pure limestones tend to give light, shallow soils whilst softer and impure limestones give deeper, heavier soils. Most limestones are either porous or permeable so the soils are usually dry. Many limestone soils—called rendzinas—have much calcium carbonate combined with humus in the profile, which helps to form soil with only A and C horizons, with a strongly developed granular structure and a dark type of humus. By contrast, limestone soils in areas of the Mediterranean develop a red soil known as terra rossa.

The effect of parent material on soil colour is often apparent on well developed soils. Thus the high iron content of basalt gives its soils a red-brown colour in temperate climates. The soils of the old red sandstone and red coal measures of England are also reddish. In the humid tropics, however, black soils with a high clay content form on basalt.

Relief as a Factor of Soil Formation

Relief is a pedogenic factor at both a macro and a micro scale. At the macro scale it has a marked influence on climate and runoff, and therefore on vegetation and erosion. Elevation affects the climate so that there is a general lowering of temperature, with increasing altitude, of about 1°C for every rise of 100 metres (Figure 27.ix.) This is reflected in a shorter growing season, decreased evaporation, greater leaching and, on the windward side of hills and ranges, greater precipitation. Soil formation is therefore slower, and the profile shallower and more likely to be truncated by erosion (Plate 32). Soils at high altitudes are usually stony as well as shallow since the finer materials are easily washed out and deposited lower down the slope, where they help to thicken the soil profile. An idealised toposequence of soils may therefore be described as having thin eroded profiles on upper slopes, a balance in mid-slope with soil formation and erosion near equilibrium, and accumulation at the base of the slope, although in reality mass movement and other

Figure 27.iii. The relationships between soils and topography in an area of uniform parent materials. *After Winters and Simonson*

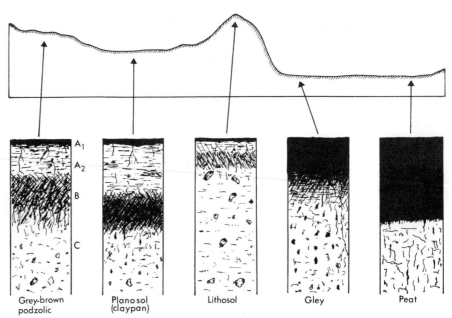

Grey-brown podzolic
Planosol (claypan)
Lithosol
Gley
Peat

A_1
A_2
B
C

processes disturb such a pattern. The thin soils on upper slopes are said to be regressive and those at the base of the slope are accumulative.

Other cases of rapid erosion and accumulation can occur, causing truncation of profiles on slopes so that entire A horizons and even B horizons are removed. This type of accelerated erosion arises under natural conditions, as when periodic violent storms cause large scale mass movement, but it is also frequently the result of man's activity. Rapid accumulation may be the result of man-made erosion but it can also result from such natural phenomena as floods which leave behind alluvium, colluvial deposits left by mass movement, deposits of loess, wind-blown sand or volcanic ash. In these cases the soil is buried and a new soil forms on top of it. The fossil soil, sometimes called a palaeosol, is no longer subject to the same environmental conditions and gradually changes. Buried soils may retain their darkened A horizons and structured B horizons for thousands of years but the processes in them are no longer pedogenic, just geological weathering processes. (See colour plate 24.)

Normal erosion is not necessarily a destructive process, for in exposing fresh rock to weathering it makes available fresh nutrients for incorporation into the soil. Erosion and movement of water in the soil give rise to sequences of soils developed on one type of parent material, so that a particular type of soil occurs at a comparable position on slopes over a large area of uniform climate. G. Milne called this pattern of soils related to geomorphology a *catena*. Other workers have attempted to extend the meaning of catena to include soil sequences related to repeated variations in parent material of the sort found in scarplands, but this type of sequence is best called an association.

Differentiation of the soils in a catena occurs not only because of erosion but because of the lateral and vertical movement of soil water. On flat areas water movement is often vertical, with leaching of bases and perhaps pan formation occurring. In areas where there is a poor supply of bases, kaolinitic clay minerals tend to form. On slopes the soil water movement is mainly lateral; slope soils are constantly enriched by bases and fine materials washed from above so that an alkaline soil with montmorillonitic clay forms. At the foot of slopes both bases and materials accumulate from the A horizons of soils upslope, and where the water table is high, ground water gleys with permanent reducing conditions and peat accumulations are formed. This type of sequence is shown in Figure 27.iii. Where the water table fluctuates pans and mottled gleying may result.

Figure 27.iv. A catena with podzols on the upper slopes under mor-forming vegetation, brown forest soils (BFS) on mid-slopes under vegetation giving a mull humus. On basal slopes soils are gleyed and peat develops in in the valley floor. The arrows indicate the dominant direction of water movement. Leaching is at a maximum on gentle slopes, and valley-side soils are enriched by nutrients leached from higher up the slopes.

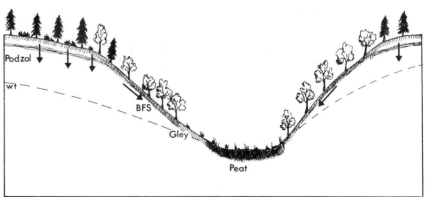

Rimutaka steep land soil skeletal YBE Z

Kaitoke
silt loam YBE

Kaitoke Si L

Ruahine steep land soil YBE Z

Kaitoke hill soils

Heretaunga silt loam YBL L

Waikanae silt
loam Recent

Taita hill soil YBE Z

Gollans silt loam Gley-meadow

Heretaunga stony silt loam YBL I

Waiwhetu silt
loam Recent

Waikanae silt loam Recent

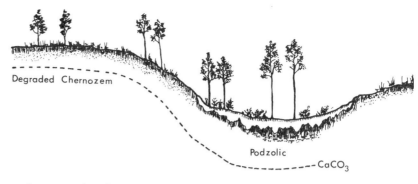

Figure 27.v. Prairie transition forest in southern Russia showing change of soil type with drainage and depth of CaCO₃ accumulation. *After Wilde*

Degraded Chernozem

Podzolic

CaCO₃

31. The location of soil types in relation to topography and parent materials in the Hutt Valley, New Zealand. *Photo by D. W. McKenzie*

An example of a common catena in Britain is shown in Figure 27.iv. Podzols develop on low angle upper slopes under a mor-forming vegetation and with a low water table. On the mid-slopes, brown forest soils with mull humus form where the water table is higher and leaching provides nutrients from the soils further up the slopes. Gleying occurs low down the slopes where the water table is high and peat forms in the swampy valley floor. In some arid regions a similar effect is produced, with well drained non-salty soils on slopes but very salty soils in the basins where the drainage waters accumulate and evaporate.

The variation of soil type with relief and parent material is summed up in Plate 31. The highest hills have skeletal or eroded thin profiles. The lower hills have thicker but still eroded profiles. The floors of the valleys have soils developed upon fan deposits and alluvium, with a chronosequence formed on the river terraces. The area of high water table is shown as having gley soils.

Organisms as a Factor in Soil Formation

The connection between soils and vegetation is so great that the major soil groups of the world can be largely defined according to the types of vegetation they support. Because, however, natural vegetation expresses the summation of the climatic influences under which it grows, it is extremely difficult or impossible to separate vegetation as an independent factor in pedogenesis. This may be illustrated by Shantz's work on the

32. The thickening of the regolith and soil profile downslope as a result of colluviation.

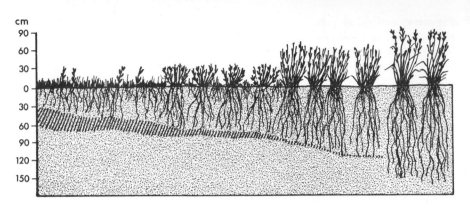

Figure 27.vi. A diagrammatic representation showing the relationship of grass type to the depth of penetration of soil moisture in the Great Plains area. The soil in the root zone is periodically moist, beneath that is a zone of lime accumulation (oblique shading), and beneath that the soil is permanently dry.
After Shantz

soils of the central USA (Figure 27.vi). In a section from the Rocky Mountains in the west to the Missouri river in the east, the shortest grasses with least root development grow in the west and the longest in the east. This is related to the rainfall, which rises from 45 cm per year in the west to 95 cm in the east. The shallow penetration of this water in the west causes the formation of calcium carbonate accumulations at shallow depths. As the precipitation rises, the pan becomes thinner and deeper until near the Missouri river it disappears. Vegetation in this case is dependent on climate and the effects the climate has on accumulation of salts.

An example of vegetation as an independent factor occurs in New Zealand, where the kauri tree (*Agathis australis*) grows in frost-free subtropical rain forest which contains many species. The kauri, however, induces large scale podzolisation. Round the base of each tree accumulates a shallow cone of fallen leaves, bark and twigs which,

33. An egg-cup podzol. The Ae horizon is the bleached zone level with the top of the spade.
NZ Soil Bureau

174

Figure 27.vii. An egg-cup podzol developed beneath a kauri tree.

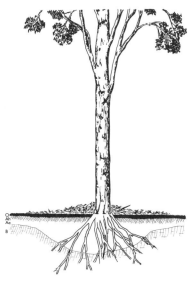

during the long life of the tree (up to 800 years and perhaps more), forms a deep, raw humus. Below this is a strongly leached lens-shaped A_2 horizon which may be 30–60 cm thick near the base of the tree. Because these podzols reflect the shape of the cone of litter, they are called egg-cup podzols (Figure 27.vii). In a few areas the trees are sufficiently close together to make the regional soil a podzol.

A second instance is that described by Tamm from work carried out in southern Sweden. Beech and oak forest is the natural vegetation and beneath this forest is a brown forest soil with a mull humus and high-base status. If, under the influence of man, this vegetation is replaced by heath or coniferous forest, a raw acid humus forms, podzolisation begins, and eventually a podzol soil is formed. In some places where birches have immigrated into the coniferous forest, mull humus forms have developed once again and the acidity has dropped from the pH 4 of the podzol to pH 5·5.

In addition to being influenced by climate, vegetation can also be controlled by relief and soil. Poorly drained hollows will only support bog vegetation and salty soils will only support salt-tolerant plants. Soil fauna must also be considered as factors in pedogenesis, but sufficient has been said of the influence of earthworms, termites and other animals in Chapter 22.

Climate as a Factor in Soil Formation

The principal effects of climate on the soil are through rainfall and temperature. Rainfall determines the supply of moisture to the soil and its organisms, the amount of runoff which can erode the soil, and the character and extent of leaching. Temperature controls the rate of chemical reactions; it has a marked influence on microbiological activity and by its influence on evaporation and transpiration it controls the effectiveness of precipitation. Climate also affects the soil indirectly through the vegetation.

In conditions in which the temperature regime is more or less constant, there is a notable change in the nature of the soil with changes in annual precipitation. A traverse through the USA from west to east would cross alkaline desert soils, then chernozemic soils, and then podzolic soils (Figure 27.viii). The soils of the arid regions have little humus, a low clay content and a high-base status because leaching is insignificant. On the grassland the slightly higher precipitation causes more leaching, but still not enough to remove the calcium carbonate from the soil. The chernozemic soils are rich in humus and clay. As precipitation increases further, the soils become increasingly acid as more leaching occurs. The formation of soil acidity under these conditions of high precipitation and a forest cover may be represented as follows:

$$\boxed{\begin{array}{c}\text{Clay}\\\text{micelle}\end{array}}\begin{array}{c}\text{Ca}\\\text{Mg}\\\text{K}\end{array} + \begin{array}{c}\text{HOH}\\\text{HHCO}_3\end{array} \longrightarrow \boxed{\text{Micelle}}\begin{array}{c}\text{H}\\\text{H}\\\text{H}\\\text{H}\end{array} + \begin{array}{c}\text{KOH}\\\text{Ca(HCO}_3)_2\\\text{Mg(HCO}_3)_2\end{array}$$

Neutral clay $\quad+\quad$ Water and carbonic acid $\longrightarrow$ Acid clay $\quad+\quad$ Removed by leaching

The exchangeable bases of the neutral clay are replaced by the hydrogen ions from the water and carbonic acid, leaving an increasingly acid clay as the salts are leached out of the soil.

The soils crossed in the traverse may be represented diagrammatically to show the increasing acidity (presence of H ions on the micelles) with increasing precipitation:

WEST   EAST

Arid region   Sub-humid region   Humid region

Ca K Ca   Ca Mg Ca   Ca H Ca Mg

| Ca Na K | Micelle | K Mg Ca | | Ca Mg | Micelle | K H H | | H Mg | Micelle | H H H |

Mg Ca Na   Ca Na Ca   Ca H Ca

Desertic   Chernozemic   Podzolic
soils   soils   soils
Desert shrubs   *Grassland*   *Forest*

Figure 27.viii. A SW–NE section across the USA showing the relationship between soil, vegetation and precipitation. Under arid conditions in the southwest, cacti grow on sierozems and CaCO₃ is deposited high in the profile. With increasing humidity, scrub grows on brown soils; short steppe grasses on chestnut soils; long steppe grasses on chernozems; broadleaf deciduous trees and long grasses on the prairie soils; deciduous forests on grey-brown podzolics and needle-leaf evergreen trees on podzols. The CaCO₃ depositional zone gets deeper with increasing rainfall until in podzolised soils CaCO₃ is leached entirely from the profile. *After Thompson*

The effect of precipitation through vegetation is indirect in that it controls the type of plant which will flourish; this in turn affects the soils through the provision of humus, through roots which affect structure, through the presence of soil fauna and the control of runoff. Rainfall in short affects soil directly through leaching and indirectly through the vegetation. A similar sequence of soils can occur on some mountain ranges (Figure 27.ix).

Soil temperature influences soil through vegetation and evaporation, and directly through the rate and type of weathering. Jenny quotes a table by Ramann which, using data from speed of chemical reactions and length of the warm season, suggests that in humid tropical regions weathering is 3 times faster than in temperate regions and 9 times faster than in the Arctic. Such a relationship goes far to explaining the great depth of weathering and soil formation in the humid tropics and the shallow profiles of cold climates. It should always be remembered,

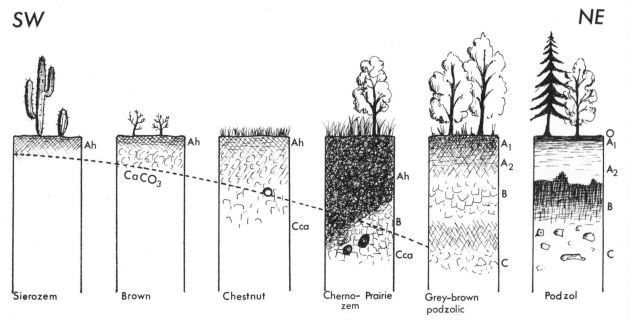

SW   NE

Sierozem   Brown   Chestnut   Cherno- Prairie   Grey-brown   Podzol
zem   podzolic

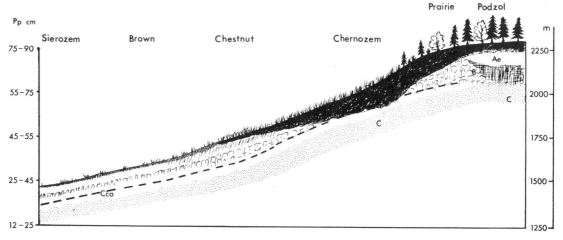

Figure 27.ix. The gradation of soils on the western slopes of the Bighorn Mountains, Wyoming. From the arid climate of the low altitudes there is a gradual change to cool humid conditions at about 2440 metres.
After Thorp

however, that most of the areas of cool climate have suffered marked changes of climate in the Quaternary and that the soils formed after glaciation have not had the long weathering periods of many tropical soils.

The relationship between climate and soils might be summarised diagrammatically in Figure 27.x, in which movements of water, bases and colloids are shown by the arrows and the extent of weathering by the depth of the profile.

Conclusion

The pedogenic factors are not of equal importance. Parent material is a passive factor whose importance diminishes with time; topography is most important as modifying climate and drainage; vegetation is conditioned by climate, topography and the soil itself but it affects the

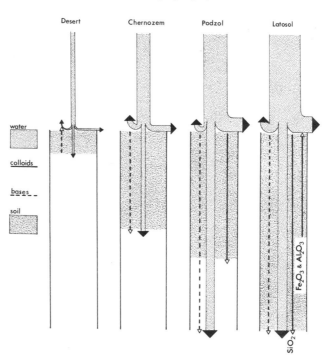

Figure 27.x. Movements of water, colloids and bases within the profiles of four major soil groups. In desert soils, low rainfall and high temperatures cause pre-dominantly upward move-ment of salts. In a chernozem, leaching is limited to the upper 1·2 or 1·5 metres of the profile. In podzols, bases are leached completely and humus and iron colloids deposited in the B horizons. In the latosols, all silica is leached from the profile and iron and alumin-ium oxides are left behind.

soil through roots and humus supply; time has no effect itself but governs the extent in which the other factors operate; climate remains as the one overriding and dominant factor. This relationship was put most succinctly by Robinson, who said:

> The character of the soil is determined by the operation of processes dependent on climate as modified by topography. These processes act both directly and, through vegetation and other biotic factors, indirectly on a given parent material, itself the result of the action of weathering processes on a parent rock. The operation of the weathering and of the soil-forming processes takes place in time which thus affects, although not causally, the final result.

28 The Study of Soils in the Field

The only way to develop an understanding of soils is to study them in the field. This means digging a pit to expose the soil profile, and on the basis of profile studies to map distinctive types of soil. The most rewarding way of doing this is to study catenas, for by doing so the variations of the soil with relief, drainage and vegetation become apparent. The soil surveyor should have the following pieces of equipment: a notebook, a measuring tape, a map and air photo of his area of study, a spade, a hand magnifying lens, a knife, a camera with colour film, an auger, a pH kit, a bottle of dilute hydrochloric acid, a colour chart (see note at the end of this chapter) and two large sheets of polythene or old groundsheets.

When digging soil pits, it is useful to spread out a sheet of polythene on either side of the pit. The turves are cut into uniform shapes to the full depth of the sod and then placed on one of the sheets. The soil removed from the pit can then be placed on the other sheet. When the pit is closed, it is a simple matter to pull the second sheet over the hole and empty it so that when the soil has been tamped down the turves can be put back in place. A good soil surveyor will thus leave no trace of his pit and hence not incur the wrath of landowners.

The pit should be rectangular in plan and large enough for a man to stand in, and it should be deep enough to expose the parent material. The pit should be oriented so that the long sides are illuminated. If it is in the side of a hill, the long sides should run downhill. When the pit has been dug, the illuminated sides should be picked clean with a knife to reveal the profile. It is sometimes necessary to leave the pit to dry before structures become apparent.

The details of the profile are recorded by deciding first the position of the horizon boundaries and then the characteristics of the profile following the scheme set out below. The surveyor should end his work with a description of the profile and a colour photograph for reference, and his map should show the area of each type discovered by pit digging with checks between pits made with an auger.

Profile Description

Site Description

The significant characteristics of the soil site should be carefully recorded not only because they help to define the soil but also because they shed light on soil genesis and on current soil processes.

(1) LOCATION

Location should be given using standard grid references which include map series and sheet number. Grid references only give the location to roughly 100 metres; consequently profile sites should also be precisely located with directions and distances from easily identifiable objects.

(2) TOPOGRAPHY AND SLOPE

The profile site should be described in terms of slope angle, length, direction (aspect) and curvature. An estimate of altitude should be included and evidence of erosion (extent and type) recorded.

(3) DRAINAGE

A general statement should be made indicating that the profile site is:

(*a*)   very poorly drained;
(*b*)   poorly drained;
(*c*)   moderately well drained;
(*d*)   well drained;
(*e*)   excessively drained.

(4) PARENT MATERIAL

The parent material of the soil is the unconsolidated material in which the soil develops (this can be distinguished from the parent rock, which is the rock from which the parent material is derived by weathering.) The parent material may have developed *in situ* from the parent rock, or it may have been transported.

Description should be concentrated on lithology (e.g. thin-bedded siltstones; deep red-weathered sandstones; rhyolitic tuff) rather than on time stratigraphic units (e.g. Oligocene, late Tertiary, Pleistocene).

(5) CLIMATE

General climatic information should be obtained from the nearest climate station. Where possible, details of local and microclimate should be given.

(6) VEGETATION AND/OR LAND USE

Information about vegetation should include kind, composition and status, e.g. well managed ryegrass-clover pasture. The type of agricultural practice (land use) should be noted (e.g. intensive dairy farming, citrus orchard, etc.), together with any other recent modifications to the site made by man.

Measurement of Horizons and Boundaries

The profile description needs to include measurements for each horizon that indicate both its *depth* from the mineral surface (top of the A) and its *thickness*. In many profiles one set of measurements is sufficient to convey a picture of both criteria; in soils where each horizon has a relatively uniform thickness, measurements of depth (e.g. 0–7·5, 7·5–23, 23–50 cm) are generally more convenient. In profiles with very irregular horizons, special care should be taken to express both measurements unambiguously and to record the degree of irregularity. On steep slopes both depth and thickness are measured perpendicular to the ground surface.

Boundaries between horizons vary in distinctness and in shape. The distinctness of boundaries depends partly on contrast and partly on thickness of the transition between the horizons:

Sharp: almost a line.
Distinct: less than 2·5 cm.
Indistinct: 2·5–7·5 cm.
Diffuse: greater than 7·5 cm.

The shape of the boundaries may be described as: *smooth, wavy, irregular* or *discontinuous.*

(Refer to Chapter 26 for details of horizon nomenclature.)

Soil Colour

Colour is the most obvious and easily determined soil characteristic. Although it has little direct influence on the functioning of the soil, a great deal about a soil may be inferred from the colour if it is considered together with the other observable features.

A soil horizon may be uniformly coloured, speckled, or mottled throughout. Since the colour of any given soil material varies with moisture content, its description is best done on the moist soil in the field. The description should begin by noting the principal colour such as brown or grey, appropriately modified (e.g. greyish-brown or brownish-grey), and finally the intensity of the colour added (pale, medium or dark greyish-brown). Wherever possible the Munsell Color Chart should be used, in order to standardise such colour descriptions.

The recording of the presence or absence of mottling forms an important part of the description of colour. The pattern of mottling can be described in terms of the *abundance* of mottles (few, many, abundant, profuse), the *size* of the mottles (fine—0–5 mm, medium—5–15 mm, coarse—more than 15 mm), and the *contrast* between the colour of the mottles and of the ground mass of the soil material (faint, distinct, prominent). Thus the descriptive colour of a specific soil horizon might read: dark greyish-brown with few, fine, distinct reddish-yellow mottles; or using the Munsell Chart, dark grey-brown (10 YR 4/2) with few, fine, distinct reddish-yellow (5 YR 6/8) mottles.

Soil Texture

Soil texture is the particle size distribution of the solid inorganic constituents of the soil, especially those particles less than 2 mm in diameter. The presence of larger particles and organic constituents is recognised by qualifying the textural name (e.g. *stony*—sandy loam, *humic*—silty clay loam).

A soil horizon will hardly ever be shown to consist of material which would fall entirely into one fraction, and therefore a number of soil texture *classes* are defined on the basis of varying proportions of the different sized fractions. There are 11 main classes, which are described in Chapter 20. Textural classes with the term sand in the name can be modified for very fine, fine and coarse sand, e.g. very fine loamy sand (see Chapter 20 for details).

ASSESSMENT OF SOIL TEXTURE

Texture largely determines the total surface area of the soil mass, and hence influences such handling properties as stickiness and cohesion.

When assessing texture, a small sample of soil is moistened to about the sticky point and thoroughly worked between the fingers to break down structural units. With experience, the texture can then be assessed fairly accurately on the feel of the soil material.

The main particle size fractions are characterised by the following properties:

Sand: Individual grains can be clearly seen or felt. Soil mass feels gritty and is usually loose and in a single-grain condition. Not sticky or plastic (see *Soil Consistence*).

Silt: Individual particles too small to be seen or felt. Mass when moist can be moulded but will not smear to give a thin flexible 'ribbon' of soil material. Does not feel noticeably sticky but has a marked soapy or slippery feel.

Clay: Individual particles are even finer in diameter. When dry, usually forms very hard lumps. Silicate clays are quite plastic when set and will form a long thin flexible 'ribbon' when pressed between thumb and forefinger. Sticky and cohesive.

The presence of unduly high amounts of intimate humus gives a greasy feel to the soil mass. Sandy soils feel heavier and clay soils feel lighter than soils of average content.

Soil Consistence

Soil consistence refers to the general handling properties of the soil material as a whole. This handling characteristic results from the combined effect of structure, texture, organic matter, porosity, moisture content, type of clay mineral and bases present on the clay mineral. It is a handling characteristic of the soil experienced when the soil is handled, dug or augered. Consistence of a soil varies with moisture content and any descriptions of consistence should give the state of moisture, whether wet, moist or dry.

The following are definitions of some terms used in describing soil consistence:

Sticky: Characteristic of soil material which, when pressed between thumb and forefinger, adheres to both and tends to stretch and pull apart rather than pulling free from either digit. Tenacious.

Plastic: Characteristic of material which will readily form a thin roll when rolled between thumb and forefinger.

Friable: Under gentle pressure between thumb and forefinger, soil material crushes easily to small structural units and will cohere again when pressed together.

Soft: Soil mass very weakly coherent and fragile; breaks to powder or individual grains under very slight pressure and will not cohere again when pressed together.

Loose: Non-coherent; soil falls off auger and runs through fingers.

Firm: Soil material crushes under moderate to strong pressure between thumb and forefinger.

Compact: Combines firm consistence with close packing of particles. Digs cleanly and gives a good 'bite' with auger.

<table>
<tbody>
<tr><td>*Hard:*</td><td>Moderately resistant to pressure; can be broken in the hands without difficulty but is barely breakable between thumb and forefinger.</td></tr>
<tr><td>*Extremely hard:*</td><td>Extremely resistant to pressure; cannot be broken in the hands.</td></tr>
</tbody>
</table>

The terms plastic and sticky usually apply to soil material in the wet condition; loose, friable, firm and compact to material in the moist condition; and loose, soft, hard and extremely hard to material in the dry condition. A soil horizon may possess a compound consistence. Such a case is covered by describing the consistence of the soil mass as a whole and then the consistence of its constituent units, for example a loose mass of hard granules.

Cementation of Soil Material

The term cementation is used to describe a brittle, hard consistence brought about by some substance other than clay minerals, such as calcium carbonate or oxide of iron.

<table>
<tbody>
<tr><td>*Weakly cemented:*</td><td>Mass is brittle and hard but can be broken in the hands.</td></tr>
<tr><td>*Strongly cemented:*</td><td>Cannot be broken in the hands but easily broken by a blow with a hammer.</td></tr>
<tr><td>*Indurated:*</td><td>Very strong cemented; requires a very sharp blow with the hammer to break the soil material.</td></tr>
</tbody>
</table>

Soil Structure

Soil structure refers to the massing of primary soil particles, whether mineral or organic, into aggregates which are separated by surfaces of weakness. Thus soil structure units should not be confused with artificial masses of soil material caused by ploughing or digging (*clods*), or with local segregations of various compounds which have irreversibly cemented soil particles together (*concretions*).

A soil may be said to be structureless when there is no observable aggregation. Structureless soils may be of two types:

<table>
<tbody>
<tr><td>(1) *Single-grain:*</td><td>Non-coherent mass of primary particles—usually a coarse-textured soil.</td></tr>
<tr><td>(2) *Massive:*</td><td>Coherent material with uniform cohesion of particles throughout the mass with no planes of weakness—usually a soil of clay texture.</td></tr>
</tbody>
</table>

Structure is a physical condition between the two extreme forms of the massive and single-grain structureless condition. The morphological classification of structure is into platelike, prismlike, blocklike and spheroidal (see Chapter 20).

Soil pH

Using a BDH Universal Soil Testing Indicator, place a sample of soil (untouched by hand, for sweat may cause an acid reaction) in the larger end of the small elongated container provided with the Indicator. Add enough Indicator to immerse the sample completely; then after about a

Table 28.1. Effects of Carbonate Test

% Carbonates ($CaCO_3$) present	Audible effects of reaction	Visible effects of reaction
0·1 →	None ————————→	None
0·5 →	{ Faintly audible increasing to slightly } ——→	None
1·0 →	{ Faintly audible increasing to moderate } ——→	Slight effervescence
2·0 →	{ Moderate to distinct, heard away from ear } ——→	Slightly more general effervescence
5·0 →	Easily audible ————————→	{ Moderate effervescence Bubbles to 3 mm easily visible }
10·0 →	Easily audible ————————→	{ General strong effervescence Bubbles to 7 mm easily visible }

minute, when the reaction has taken place, tilt the container slightly to separate the solution from the sample. Evaluate the pH by comparing the colour of the solution with the colour chart provided.

Soil Carbonates

The application of a little dilute hydrochloric acid will indicate the percentage of carbonates in a sample. If the acid is splashed on to the profile, the pH test must be carried out first. It is safer to do the carbonate test by putting the soil sample in a small dish. Effects of testing are given in Table 28.1.

A typical brief soil profile description is given below:

0–7·5 cm   Pale brownish-grey (2·5 Y 6/2) silt loam with abundant small, strong brown mottles and stains along root channels; breaks out in large, friable clods (up to 10 cm in diameter) which crush to weakly developed granules; few small stones; abundant roots; few earthworms; distinct boundary to the horizon.

7·5–18 cm   Pale grey (5 Y 7/2) silt clay loam with few strong brown mottles; breaks down to large, friable clods (up to 10 cm in diameter) which crush to weakly developed medium granules; few small stones; abundant roots; few earthworms; horizon moist becoming wet at base; distinct boundary.

18–30 cm   Pale grey to pale olive (5 Y 6/2–6/3) silty clay loam; weakly developed coarse prismatic structures; strong brown to reddish-brown (7·5 YR 5/8–6/8) mottling inside prisms, and pale grey (5 Y 7/2) on faces of prisms; moist; porous only through fissures; a few roots and earthworm tracks decreasing with depth.

Note: Munsell Soil Color Charts are manufactured by:
Munsell Color Co. Inc.,
2441 N. Calvert St,
Baltimore 18,
Maryland, USA.

As these charts are extremely expensive, many people prefer to use the Japanese version in which the colour chips are of adequate quality, but the legend is partly Japanese. They are obtainable from:

Fujihira Industry Co. Ltd,
No. 11–6, 6– chome, Hongo, Bunkyo-ku,
Tokyo, Japan

The BDH Universal Soil Testing Indicator is manufactured by British Drug Houses Ltd and is available from most chemists.

29 The Classification of Soils

The purpose of classifying phenomena is to assemble our ideas and understanding of phenomena in an ordered form. A good classification of soils should also help us to remember the significant characteristics of each soil; to synthesise our knowledge of soils; to see relationships between soils, and between soils and their environment; and finally to develop predictions of their behaviour and responses to use and manipulation by man. A problem which cannot be avoided is that soils change their position in a classification because of changes in external factors, so that a soil formed on alluvium, for example, will change towards the type of soil generally found in its region (the zonal soil) if deposition ceases and drainage improves, or a brown forest soil will become podzolised if its broadleaf forest is replaced with a heath and coniferous forest vegetation.

There are two main types of classification of soils: genetic and morphological. Genetic classifications are based upon a knowledge, or assumed knowledge, of the processes which formed the soil. Thus genetic classifications may be based on such processes as podzolisation, laterisation and calcification. This type of approach is being abandoned: partly because it now seems that processes like podzolisation are not distinct but part of a set of additions, removals, transfers and transformations of the soil constituents which give rise to horizon differentiation in all soils; and partly because as our knowledge of soil processes advances, it becomes necessary to change the classification system. The morphological approach is based upon the features of the soil as it occurs in the field and requires an understanding of the factors of soil formation—climate, organisms, parent materials, topography and time. In practice most classifications use both genetic and morphological criteria.

Classifications

The great Russian pedologist *Dokuchaev* produced his first classification in 1886 and his final one in 1900. His system may be regarded as the first attempt to develop a classification which embraced the soils of the world. Dokuchaev was particularly impressed by the clear zonation of soils in Russia, which is related to climate and vegetation. The European and Russian plains are covered with glacial and loess deposits which vary very little over large areas, so that in the absence of marked relief the soil-climate-vegetation relationship seemed obvious. Furthermore,

Figure 29.i. Horizontal and vertical zonation of soils in European Russia.
After Zakharov

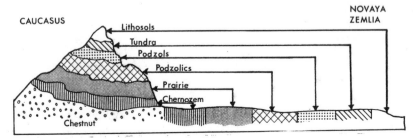

186

Tundra · Podzol · Grey-Brown Podzolic · Chernozem · Chestnut · Sierozem

CaCO₃   CaSO₄

Figure 29.ii. A N–S section across European USSR showing the relationship between vegetation and soil type.
After Rode et al.

the zonal pattern was repeated in the Caucasus Mountains (Figures 29.i and 29.ii). The genesis of the soil through processes related to climate and vegetation was seen by Dokuchaev to be reflected in the morphology of the profile, so that the soil could be defined by its thickness, humus content, colour, composition and structure of the horizons, and because the morphology was related to the genesis the history of the soil could be read in the profile. Dokuchaev's collaborator Sibirtsev introduced the terms zonal, azonal and intrazonal, and their work laid the foundations for the classifications of the next sixty years.

Marbut in the USA was greatly influenced by Russian pedology, especially as presented in Glinka's textbook. Marbut produced a number of classifications, of which the 1936 one is the most detailed. He laid great emphasis on the fully developed or 'mature' soil on a 'normal' or freely drained site on gently undulating topography. The great advance Marbut made was in changing the concept of soil as weathered rock, held widely in USA, to soil as an independent natural body.

Baldwin, Kellogg and Thorp made the final revision of Marbut's system in 1949 (Table 29.1). Although widely adopted, the US system still had a number of defects. It was still largely based on the environment rather than on the soil itself; it was concerned mainly with the virgin soil rather than being able to take in disturbances caused by man; its nomenclature was based on soil colour rather than on the outstanding properties of the soil; its three highest categories of zonal, intrazonal and azonal were too few. The zonal soil is the mature soil on a normal site reflecting the factors of climate and vegetation; the intrazonal soils are those reflecting poor drainage, salt accumulation or parent material; the azonal soils are those with poorly developed horizons because of erosion or youth. The whole concept of a zonal soil, fully developed at equilibrium with its environment, raises considerable problems and these will be briefly discussed at the end of this chapter.

New US Classification System (7th Approximation)

A completely new classification system was introduced in the USA in 1960. It was subsequently revised and then in 1965 the seventh revision or approximation was published and became the official USA system. The 1965 system is completely different from any other classification, using a newly developed series of terms, a method of defining terms and

Table 29.1. US 1949 Soil Classification in the Higher Categories

Order	Suborder	Great soil groups
Zonal soils	(1) Soils of the cold zone	Tundra soils
	(2) Light-coloured soils of arid regions	Desert soils Red desert soils Sierozems Brown soils Reddish-brown soils
	(3) Dark-coloured soils of semiarid, subhumid, and humid grasslands	Chestnut soils Reddish chestnut soils Chernozem soils Prairie soils Reddish prairie soils
	(4) Soils of the forest-grassland transition	Degraded chernozem Non-calcic brown or shantung brown soils
	(5) Light-coloured podzolised soils of the timbered regions	Podzol soils Grey wooded, or grey podzolic soils* Brown podzolic soils Grey-brown podzolic soils* Red-yellow podzolic soils
	(6) Lateritic soils of forested warm temperate and tropical regions	Reddish-brown lateritic soils* Yellowish-brown lateritic soils Laterite soils*
Intrazonal soils	(1) Halomorphic (saline and alkali) soils of imperfectly drained arid regions and littoral deposits	Solonchak, or saline soils Solonetz soils Solod soils
	(2) Hydromorphic soils of marshes, swamps, seep areas and flats	Humic gley soils* (includes wiesenboden) Alpine meadow soils Bog soils Half-bog soils Low humic gley soils* Planosols Ground-water podzol soils Ground-water laterite soils
	(3) Calcimorphic soils	Brown forest soils (braunerde) Rendzina soils
Azonal soils		Lithosols Regosols (includes dry sands) Alluvial soils

* New or recently modified great soil groups

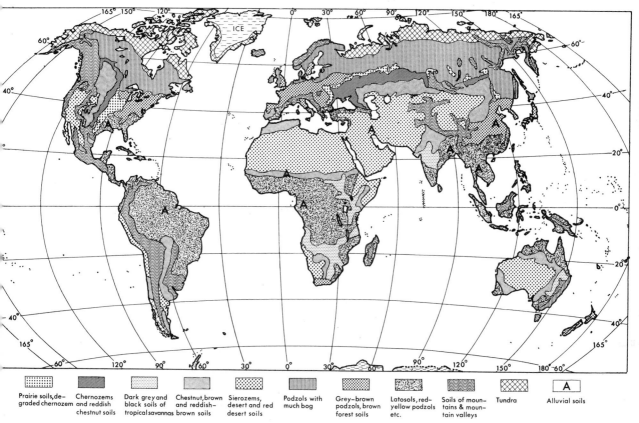

Prairie soils,degraded chernozem   Chernozems and reddish chestnut soils   Dark grey and black soils of tropical savannas   Chestnut,brown and reddish-brown soils   Sierozems, desert and red desert soils   Podzols with much bog   Grey-brown podzols, brown forest soils   Latosols,red-yellow podzols etc.   Soils of mountains & mountain valleys   Tundra   Alluvial soils

Figure 29.iii. Zonal soils of the world. *After Kellogg*

horizons by measurement and chemical analysis, and an organisation of soil orders based upon features of the soils rather than features of the environment. The names of the ten soil orders and their equivalents in the 1949 system are shown in Table 29.2.

Table 29.2.
7th Approximation Soil Orders and Approximate Equivalents in Revised Classification of Baldwin et al.

Present order	Approximate equivalents
(1) Entisols	Azonal soils and some low humic gley soils
(2) Vertisols	Grumusols
(3) Inceptisols	Ando, sol brun acide, some brown forest, low humic gley and humic gley soils
(4) Aridisols	Desert, reddish desert, sierozem, solonchak, some brown and reddish-brown soils, and associated solonetz
(5) Mollisols	Chestnut, chernozem, brunizem (prairie), rendzinas, some brown, brown forest, and associated solonetz and humic gley soils
(6) Spodosols	Podzols, brown podzolic soils and ground-water podzols
(7) Alfisols	Grey-brown podzolic, grey wooded soils, non-calcic brown soils, degraded chernozem, and associated planosols and some half-bog soils
(8) Ultisols	Red-yellow podzolic soils, reddish-brown lateritic soils of the US, and associated planosols and half-bog soils
(9) Oxisols	Laterite soils, latosols
(10) Histosols	Bog soils

Table 29.3. Main Orders and Suborders in the 7th Approximation (cont. on facing page)

Order	Suborder	Distinguishing characteristics
Entisols		*Soils without natural horizons or only the beginnings of horizons*
	Aquent	Saturated with water at some season
	Psamment	Textures coarser than loamy fine sand to 50 cm or more depth
	Ustent	Characteristically associated with arid to semiarid regions
	Udent	Usually moist, textures finer than loamy fine sand and less than 50 cm deep
Vertisols		*Soils with large amounts of expanding lattice clays in climatic areas with pronounced dry seasons*
	Aquert	Chroma number less than 1·5 in upper 30 cm; may have mottles
	Ustert	Chroma number more than 1·5 in upper 30 cm, with no mottles
Inceptisols		*Soils that form rather quickly and have little evidence of significant illuviation*
	Aquept	Saturated with water at some season; may have fragipan or duripan
	Andept	High content of allophane (claylike minerals) and/or clay
	Umbrept	No more than minor amounts of allophane or volcanic ash
	Ochrept	A thin A_1 or Ap resting on a changed or altered horizon (cambic)
Aridisols		*Soils that are usually dry with a calcic horizon immediately underlying a light-coloured surface horizon (ochric epipedon)*
	Orthid	An ochric epipedon and a cambic horizon, a duripan, or an eluvial calcic, gypsic, or salic horizon
	Argid	An ochric epipedon and an argillic or natric (columnar structural argillic) horizon
Mollisols		*Soils that have a dark-coloured surface (mollic epipedon) but excluding those with a fraction of clay dominated by allophane, or a silt and sand fraction dominated by volcanic ash throughout*
	Rendoll	Parent materials with more than 40% $CaCO_3$ equivalent
	Alboll	An ochric horizon immediately below the mollic epipedon
	Aquoll	No light-coloured subsurface horizon and is strongly hydromorphic
	Altoll	Chiefly soils that have been called chernozems
	Udoll	Lighter coloured than aquolls, parent materials contain less than 40% $CaCO_3$
	Ustoll	A mollic epipedon more than 50 cm thick and common to abundant worm holes or worm casts (krotovinas)

The *advantages* of the 7th Approximation are not yet all obvious as it has not been in use for long and has only been seriously tested in the USA and its territories. It is clear, however, that it is:

(1) Not dependent upon assumptions about genesis of soils and will not have to be modified as our knowledge of soil genesis advances.
(2) Based upon soil properties and characteristics which can be measured, the categories being based upon measurements of diagnostic horizons.
(3) Designed to include soils which have been affected by man.
(4) Of use for correlating other systems because of its precision.
(5) Formed with terms which are closely defined and new, with no ambiguities.

Order	Suborder	Distinguishing characteristics
Spodosols		*Soils that have a bleached A_2 (ashen) horizon (spodic) that is demonstrable after ploughing and cultivation*
	Aquod	Saturated with water at some season and have characteristics associated with wetness (mottles, coatings, concretions, etc.)
	Humod	No characteristics associated with wetness and a spodic horizon enriched with humus and aluminium
	Orthod	Spodic horizon enriched with both iron and humus
	Ferrod	Spodic horizon enriched by iron without comparable increases in humus and aluminium
Alfisols		*Soils that have no mollic epipedon, oxic or spodic horizon*
	Aqualf	Saturated with water at some season
	Altalf	Pseudo-mollic epipedon (umbric) but no argillic horizon
	Udalf	Lack characteristics associated with wetness definitive of aqualfs
	Ustalf	Some part of the solum dry for 3 months or more
Ultisols		*Soils that have no oxic or natric horizon, but do have an argillic horizon*
	Aquult	Saturated with water at some season
	Ochrult	Have higher chromas than aquults
	Umbrult	Have an umbric or mollic epipedon, and the argillic horizon diagnostic of ultisols
Oxisols		*Soils that have an oxic horizon: an altered subsurface horizon consisting of a mixture of hydrated oxides of iron or aluminium, together with variable amounts of non-expanding lattice clays and accessory diluents, like quartz, that are highly insoluble*
	Aquox	Saturated with water at some season or have soft plinthite within 30 cm of the surface
	Humox	Very rich in humus, base saturation less than 35%
	Ustox	A subsurface horizon dry for 60 consecutive days in most years; base saturation more than 35% in oxic horizon
	Idox	Dry for more than 6 months; base saturation more than 50% in all horizons
	Orthox	All remaining oxisols
Histosols		*Soils that are characterised throughout by a high organic content*
	Saprist	Highly decomposed organic materials to depths of more than 60 cm
	Hemist	Partly decomposed organic materials more than 60–90 cm thick lying over any other diagnostic master horizon
	Fibrist	Relatively undecomposed organic materials more than 60–90 cm thick lying over any other diagnostic master horizon
	Leptist	Too thin for other suborders, or lacks diagnostic master horizons

The *defects* of the system are probably not all appreciated and as they become apparent many of them will be corrected by further additions and revisions. Some defects are as follows:

(1) It requires laboratory data for fitting some soils into the system.

(2) It does not appear to be related to the applied sciences such as forestry and agriculture.

(3) It is not possible to predict the boundaries of soil types by using field experience or air photos because the position of those boundaries is controlled by measurements of horizons.

(4) It is a very complicated system for use in underdeveloped countries, where it is perhaps only necessary to classify soils by their ability to grow certain crops.

(5) It divides soils which are part of a continuum into different categories, and places unlike soils in the same order.
(6) It makes use of the features of diagnostic horizons rather than those of the whole soil.
(7) It separates soils as much on the degree of their development as on the kind of development.

In addition to these defects, the system has disadvantages which make it unsuitable for use on a world basis at the time of writing (February 1968), because no world or even continental map has yet been published. Because of the complexity of the criteria used, it might well be that no accurate world map of soils can be produced for many years. Another weakness at present is that the definitions of the histosols and oxisols are not yet complete.

Table 29.3 lists the main orders and suborders with their features. The 7th Approximation may not yet be suitable for world use, but because it will inevitably become more important with time an aquaintance with its terminology is necessary. It is, in spite of its defects, probably the best classification yet devised, if only because its terms are clear and precise.

Equilibrium Concepts

Because of the unsuitability of other available systems for use on a world basis, this book uses the US 1949 classification. Inherent in that is the idea of a zonal soil, developed on a 'normal' site and in a 'mature' state in which it is no longer developing but in a state of equilibrium with its environment. Such a concept raises problems, for unless the soil is regarded as being 'fossilised' and devoid of life, changes will occur in it because the environment is constantly changing. It is obvious that soils are not 'fossilised' and it is also obvious that if the environment changes then there will be a time lag before the soil becomes adapted to the change. The equilibrium has then to be a dynamic one, maintained by the circulation of materials in the soil-vegetation system, in which down-wearing of the landscape and removal of eroded soil from the surface is no faster than the sinking of the soil into fresh parent material. The equilibrium so occurring is that of an open system through which there is a continuous flow of energy and matter. The evidence for such a dynamic equilibrium is sometimes preserved in the soil as an indication of polygenetic origins, and on a wider scale because there are a number of distinct soil groups occupying broad geographical areas in which there is a clear relationship between soils, climate and vegetation.

The disadvantage of a soil classification system based upon a concept of a mature equilibrium soil is that climatic changes have occurred in the recent past, and the ecological status of many types of vegetation—particularly grassland vegetation—is doubtful. Furthermore, Carter and Pendleton have suggested that several of the zonal soils of the USA are in the process of developing to other types. The grey-brown podzolics, for example, may be developing into red-yellow podzolics. For these reasons the 7th Approximation system of quantitative criteria is far more reliable than the criteria of its predecessors.

1   2   3

1–4. A soil chronosequence. These soils have formed on a series of coastal dunes. 1 and 2 have formed since AD 1886 and 3 and 4 are older than that. They are arranged in order of age and show profile development.

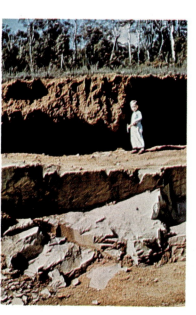

4

5. Laterite on colluvial material over granite, Beechina, Australia. *Soil Survey of England and Wales*

6. Latosol, western Nigeria. *Soil Survey of England and Wales*

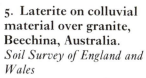

7. **Red-yellow podzolic soil, southeastern USA.**
Photo by J. D. McCraw

8. **Red desert soil, Hawaii.**
Photo by J. D. McCraw

9. **Grey-brown desert soil, California.**
Soil Survey of England and Wales

10. **Solonchak in an old meander belt of the Euphrates, Iraq.**
Soil Survey of England and Wales

11. **Chernozem, Germany.**
Photo by J. D. McCraw

12. **Degraded chernozem, Germany.**
Photo by J. D. McCraw

13. **Grey-brown podzolic soil, USA.**
Photo by J. D. McCraw

14. **Peaty gleyed podzol, Scotland.**
Photo by J. D. McCraw

15. **Arctic brown soil on volcanic ash, Iceland.**
Photo by J. D. McCraw

16. **Rendzina, New Zealand.**

17. **Ground-water gley soil on estuarine alluvium, Somerset, England.**
Soil Survey of England and Wales

18. **Volcanic ashes. 1 metre of white Rotomahana mud over 5 cm of black Tarawera ash, both erupted in 1886. At the base of the column is Taupo pumice erupted in 130 AD. Waimangu, New Zealand.** *Photo by J. D. McCraw*

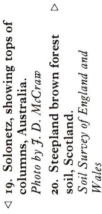

▽

19. Solonetz, showing tops of columns, Australia.
Photo by J. D. McCraw

20. Steepland brown forest soil, Scotland.
Soil Survey of England and Wales

▽

21. Tundra peat, Norway.
Photo by J. D. McCraw

22. Antarctic soil showing lag gravel and surface pan.
Photo by J. D. McCraw

▽

23. Peat formed of rushes, Rukuhia, New Zealand.
Photo by J. D. McCraw

24. Palaeosol in volcanic ash, Raglan, New Zealand. The fossil A horizon is shown above a prismatic structured B horizon.

30 Latosolic and Related Soils

The soils of the humid tropics have developed in an environment which is significantly different from that of other major types of soil, not only in respect of climate but also in respect of the great length of time available for weathering. The climate is important because constantly high temperatures and high humidity are responsible for a soil climate which is conducive to rapid chemical weathering. All chemical reactions increase in rate of reaction with a rise in temperature—a rise of 10°C causing a doubling of velocity—so that in the soils of the humid tropics weathering processes are approximately three times as rapid as those of cold climates. In addition to this, the landscapes of the humid tropics are often old—as on the plateaux of Africa, India and Brazil—so that weathering and soil formation have reached an advanced stage. By contrast, areas of colder climate have been subjected to the changes of the Quaternary era which have spread glacial moraines and loess widely, allowing contemporary soil processes as little as 10,000 years in which to operate. The humid tropics were little affected by the Quaternary climatic changes and the soil bodies contain material weathered during several millions of years under a uniform vegetation cover. There is no doubt that soils similar to those found in the humid tropics once covered large areas of what are now temperate regions, but their remains are only relicts of a past climatic situation.

Weathering

Under typical climatic conditions on gently sloping surfaces covered by forests or scrub, weathering in the humid tropics involves hydrolysis of the rock forming minerals in which the silicates are mobilised and, like the bases, are leached away leaving iron and aluminium oxides. This weathered material, rich in sesquioxides, forms the parent material for the soils. Where the rocks are low in bases and the soil solution has a low pH, colloidal silica and alumina combine to form the 1 : 1 lattice clay mineral kaolin. On basic rocks, however, the solution has a high pH and the silica remains in solution to leave iron oxides and gibbsite. The iron sesquioxides may be in the form of limonite, which is yellow in colour, or red haematite. In essence, then, there are two main types of weathering product—the kaolinitic formed on acid rocks, and the ferrallitic formed on basic (ferromagnesian) rocks.

The individual particles are all very small, mainly in the silt and clay fractions, and they are all secondary minerals. Some of the particles are cemented together by iron oxides, and occasionally by aluminium and silicon hydroxides, to produce a pseudo-sand—so called because of the size of the particles. The weathered materials and hence the soils consequently do not have the usual properties of clays, even though they are composed of clay minerals. They are not plastic, do not swell on wetting, are permeable, have a very low base exchange capacity and are easy to cultivate. When the material is red, the soil is often called a tropical red earth or a latosol; and when the amount of kaolinite present is low, it has been called a laterite, although for reasons discussed at the end of this chapter the more usual modern term is ferrallite—indicating the iron and aluminium oxide composition. Ferrallites are only formed from the weathering of basic igneous rocks on well drained sites on ancient surfaces. They are not formed if the weathering mantle is subjected to strong ground-water influences, for ground-water nearly always contains colloidal silica which reacts with the aluminium hydroxide to form kaolinitic clays. In practice, a complete range of weathering products and soils can be found between the extremes of the kaolinitic and ferrallitic type.

Organic Matter

The soils of the humid tropics are different from those of other regions not only in the extent and depth of weathering, but also in their organic matter. In the forested humid tropics the turnover of organic matter is rapid. The amount of organic matter lying on the soil surface is therefore approximately the same throughout the year but it seldom exceeds a thickness of 15 mm. The litter is rapidly incorporated into the soil by the numerous soil organisms, particularly earthworms and termites. The soil climate favours very rapid breakdown and the formation of a nearly colourless humus, so that most tropical soils appear to be poorer in humus than they are. In areas with a dry season the litter may be on the soil during the driest period and is only humified during the wet season—unless, as in the savannas, it is consumed in the annual grass and scrub fires.

The Soil Profile

34. Latosolic soil profile. The tape is scaled in feet.
Photo by Roy W. Simonson, US Soil Conservation Service.

Very little is known about the development of soil profiles in the humid tropics and it is often extremely difficult to decide at what point the soil gives way to the non-soil of the underlying weathered material. In areas of ancient land surfaces there is little opportunity to see chronosequences, but in Indonesia the young volcanic landscapes offer an opportunity of studying soil development. This is shown diagrammatically in Figure 30.i, from the work of Mohr and Van Baren on volcanic ash.

Weathering begins as soon as rain falls on a fresh layer of volcanic ash. Bases are leached, sesquioxides form, and at some time alumina combines with mobilised silica to form kaolinite or, if the environment is rich in bases and drainage impeded, montmorillonite. As vegetation begins to grow and more organic materials are formed and released, further bases are leached. At some depth calcium and magnesium begin to accumulate and in this environment silica becomes less soluble and is precipitated so that it forms a cemented layer or siliceous hardpan.

The continual leaching of bases causes the soil to become more acid and the sesquioxides accumulate, their yellow and brown colours becoming more noticeable and redder with age. At depth kaolin accumulates as a white clay and above it iron oxides form, especially where alternate wetting and drying occurs. The iron oxides form a pan which may be in the form of nodules or pellets.

The formation of the white kaolin layer over the cemented siliceous pan reduces percolation and is the cause of the formation of the ironpan. This becomes progressively thicker as it encroaches on the permeable red earth and alumina is constantly washed out. The continual loss of nutrients and the increasing impermeability cause a decline in the vegetation until it is so impoverished that the remaining red earth is

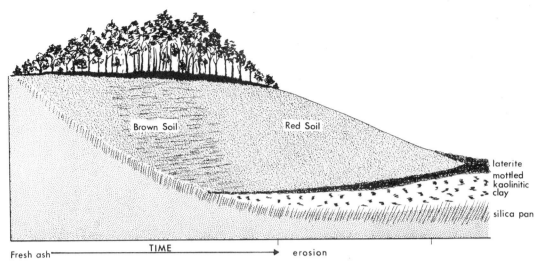

Labels in figure:
Brown Soil
Red Soil
laterite
mottled kaolinitic clay
silica pan
Fresh ash — TIME → erosion

Figure 30.i. The development of a weathering and soil profile with time from a volcanic ash in the humid tropics. It shows the growth of vegetation, and then the results of erosion when vegetation is cleared and a lateritic crust forms the surface.
After Mohr and Van Baren

eroded away and laterite, (i.e. the ironpan deposit) is left forming the surface.

No indication is given of how long this process will take, and not all pedologists agree with Mohr and Van Baren. Their scheme has the advantage that it brings together many features of tropical soils which are often left as unexplained fragments of knowledge. Further research is needed before this hypothesis can be confirmed.

The profiles of most tropical soils have a very thin A horizon overlying weathered material in which there are few pedological structures and little indication of B and C horizons. A scheme of horizon nomenclature developed by the late C. F. Charter has been found applicable to the soils of west Africa, although it has not been much used elsewhere. Charter found that much of the soil material, even on gentle slopes, is not derived from the underlying weathered rock but from colluvial deposits which are moved downslope by creep and wash. Commonly between the sedentary weathering mantle and the colluvium there is a stone line of angular fragments of quartz derived from veins in the igneous and metamorphic rocks. The activity of termites and clay eluviation produce a differentiation of the mineral fractions so that the lower part of the colluvial creep layer is more gravelly than the upper part. The finest material is often carried by earthworms to the surface and left as casts. The result is a profile with the following features:

CrW: Horizon formed from worm casts.
CrT: „ formed by termites.
CrG: „ of gravel accumulation.
S_1: „ with little decomposing rock visible.
S_2: „ with many decomposing rock fragments.
S_3: „ of altered rock which still retains the structure of fresh rock.

The CrW horizon usually contains more silt and clay than the CrT, and the greatest concentration of clay is in the S_1 or S_2.

Catenas

Because of the effects of topographic position on erosion and soil water movement, soil types vary over short distances so that the soil pattern is a mosaic of types. Figure 30.ii shows a catenary sequence in which the

Figure 30.ii. Soil profile variations down a slope in the humid tropics. The symbols used are Charter's; (R) indicates rock.

iron concretions

gravel

iron pan

clay

CrW

CrT

CrG

S

R

Figure 30.iii. A frequently occurring catena in the humid tropics. A lateritic crust (LC) on the plateau, red soils (RS) on the slopes and black soils (BS) in the valleys.

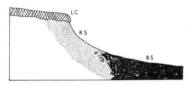

LC

RS

BS

soils of the upper slopes are thin and contain much decomposed rock. The soils thicken downslope as the result of colluviation. In many areas on old land surfaces the upper slopes are covered with iron crusts; these are the relicts of old laterites derived from the soils formerly covering that landscape which is now only preserved on summits. Below this cap the soils are of the red variety, but in basins where calcium, organic matter and soil water accumulate the soils are often black or grey. This sequence gives a catena of the type shown in Figure 30.iii.

Tropical Black Soils

If drainage in a humid tropical area is impeded, then gleys and ironpans or ground-water laterites are formed near acid rocks. Near basic or newly weathered rocks, leached colloidal silica and bases accumulate in

35. Gilgai microrelief in India. The scale of the relief is greater than average.
US Soil Conservation Service

depressions to form black tropical soils—sometimes called margalitic soils—which get their black colour from the black humus which is stable in the presence of calcium carbonate.

In the presence of calcium and frequent water saturation the clay formed is montmorillonite. This has strong swelling and shrinking properties, so that in the seasonally dry tropics deep cracks develop in the soil. The higher the clay content of the soil, the deeper the cracks. Swelling and shrinking also results in the breaking down of the soil into small aggregates which may be blown into the cracks during the dry

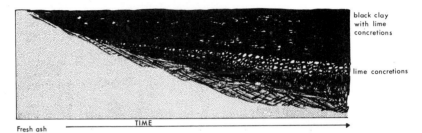

Figure 30.iv. Development of the profile of a black tropical soil with time.

season. In this way considerable quantities of surface soil are transferred into the subsoil. With the coming of the wet season the entire soil swells, the cracks close and the fine grains are mixed in with the subsoil. This process of natural tillage to a depth of 2 metres or more obscures differentiation in the profile. Calcium carbonate concretions form within the profile but mostly as a layer just beneath the depth of cracking. The black soil passes abruptly into the underlying rock. In Figure 30.iv this is shown as volcanic tuff. The swelling and shrinking of the montmorillonite produces a microrelief of enclosed basins and knolls or ridges. The ridges vary from several centimetres to half a metre

Figure 30.v. Gilgai.

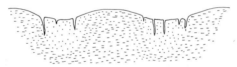

high and rarely to perhaps 2 metres. This type of microrelief is known as gilgai (Figure 30.v).

Black or grey tropical soils rich in calcium occur in many parts of the world and are given a number of local names, e.g. regur in India where they cover 520,000 square kilometres of the Deccan basalts, tirs in north Africa and black earths in Australia. It should be emphasised that they are not chernozemic for they have a different structure, clay content and humus from the temperate grassland soils.

Classification of Tropical Soils

The terminology and classification of tropical soils is very confused. The 1949 US classification uses the terms: reddish-brown lateritic soils; yellowish-brown lateritic soils; laterite soils. The term lateritic was used because it describes the process of soil formation in which silica and bases are leached and sesquioxides and kaolinite left as the soil minerals. Modern scholars are opposed to using terms referring to distinct soil forming processes, and prefer to regard soil formation as the resultant of a variety of chemical, physical and biological reactions, all of which are potential contributors to the development of every soil.

197

36. Typical laterite quarries close to Angadipuram, south India, near Buchanan's type exposure.
Photo by C. S. Fox. Reproduced with permission from: Prescott and Pendleton, Laterite and Lateritic Soils, *published by C.A.B.*

Laterite also has much confusion of meaning attached to it (see below). To overcome these difficulties, US soil scientists proposed the term latosol to refer to the typical soils of the humid tropics with a soil mass of sesquioxides, 1 : 1 lattice silicate clay minerals, low cation exchange capacities and red colour. The term has been widely used as an *omnium gatherum* in preference to lateritic. The 1960 US system used oxisol as its general term and many European systems use ferrallitic as a general term for soils with iron and aluminium sesquioxides, and ferritic for those with iron alone. Kaolisols are those tropical soils rich in kaolin. (See colour plate 6.)

Agricultural Soils

Most mature tropical soils are friable and easily worked because of the absence of swelling clays. They are, however, deficient in plant nutrients, particularly phosphorus, which is usually 'fixed'. Nitrogen is also deficient, but is not usually the limiting factor because the other nutrients are so deficient. The soils are usually acid, but liming can be harmful because the calcium can dominate the exchange complex and the other nutrients like potassium and magnesium are released too rapidly into the drainage water. For these reasons humus is by far the best fertiliser, but its nutrients are broken down and leached very rapidly. It is therefore necessary to treat tropical soils with great care, avoiding excessive exposure to the sun which accelerates breakdown

Figure 30.vi. Idealised profile through a lateritic crust (LC), over a mottled zone (MZ) and pallid zone (PZ).

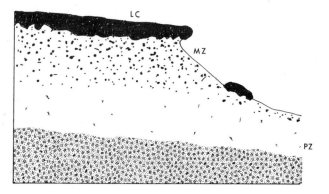

of organic matter. Much research will be necessary before these problems can be overcome. In Chapter 35 there is a discussion of shifting cultivation which is a traditional response to the problem.

Laterite

The word laterite was originally suggested by Buchanan in 1807 as a name for the highly ferruginous deposit he first observed in Malabar, India. He called it laterite because it was frequently dug out of the ground while it was moist and soft, and left in the sun to dry and harden into bricks. The Latin word *later*—a brick—relates to this building material. Unfortunately later workers often assumed the word laterite referred to its red colour. A second source of confusion arose when the term was used to refer to any red-coloured soil material of the humid tropics. Such is the confusion that the 1960 US classification has abandoned the term altogether and now uses plinthite to refer to the sesquioxide-rich clay which hardens on exposure. Many geologists, however, still use the term laterite in Buchanan's original sense. (See colour plate 5.)

A typical profile through an exposed laterite (Figure 30.vi) would show below the ironstone crust, which may be almost any thickness from 2–30 metres, a zone of white kaolinitic clay containing red mottles of iron. Downwards the mottles disappear and a white pallid zone occurs. Beneath this is the parent rock. The pallid zone is not always present. Ironstone beds occur over very large areas of the humid tropics and frequently form the resistant caps on hills. Beneath them residual blocks litter the slope, and they are sometimes cemented as a layer in the colluvial soil by the iron oxides in percolating ground-water (Figure 30.vii). Small ironpans are continuously forming in valleys or at breaks of slope where iron-rich water concentrates (Figure 30.viii).

It has already been suggested in Figure 30.i that laterite will be formed at depth as part of the normal soil forming processes and will be exposed as the topsoil is removed by erosion. There are probably several combinations of processes which can give rise to the same result. Hardening on exposure does not always occur and is probably

Figure 30.vii. Landscape features with which laterite is commonly associated: (1) plateau cap of laterite; (2) fragments of laterite fallen or slipped from above; (3) a stone-line of laterite fragments cemented by iron-rich percolating water; (4) modern sediment on an erosion surface.
After Sivarajasingham et al.

37. Use of laterite as a building material at Ranjol, Hyderabad, India.
Photo reproduced with permission from: Prescott and Pendleton, Laterite . . .

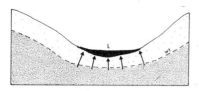

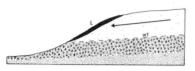

Figure 30.viii. Lateritic crusts (L) formed by percolating water. The arrows indicate the direction of movement of iron-rich solutions. The water table is shown (wt).

the result of a high degree of crystallinity in the iron. It is probable that alternate wetting and drying is necessary for hardening to occur.

Laterite seems to develop most commonly and extensively on flat surfaces where the soil is waterlogged for long periods, and which have a marked dry season during which iron is precipitated at the upper boundary of a fluctuating water table. The underlying pallid and mottled zones lose their iron either by lateral seepage or to a limited extent by capillary rise of iron-rich solutions.

The exposure of hardened laterite as a result of natural and induced erosion has rendered large areas sterile. The process is reversible, as the iron may be rendered soluble by organic compounds. Where vegetation can be re-established, the laterite can be broken down, but on old, hard, impermeable exposed laterites vegetation will not develop. The formation of laterite indicates the great care necessary in the intensive use of tropical soils and the dangers of soil destruction and land sterility caused by erosion and excessive exposure to the sun.

38. Residual hills in the Australian desert capped by resistant laterite.
Photo by R. J. Blong

Soils of the Tropical Savannas

Polewards of the humid tropics the dry season becomes longer and hence the period of the year in which chemical weathering can occur becomes shorter. The vegetation becomes sparser and the supply of organic matter to the surface of the soil is also reduced. As a result of these changes the weathering and soil profiles become thinner. Lateritic crusts are common in the savannas but are almost certainly relicts from a time when the tropical forests were more extensive. Further away from the equator, the iron accumulations in the soil decrease until they are represented only by nodular concretions in the soil and then under more arid conditions by coatings on soil particles. In conditions of increasing aridity, the free alumina in the soil declines and silica becomes less mobile and so forms a more important part of the soil skeleton. The organic matter content of the soil becomes progressively lower and clay illuviation also declines. In general, then, chemical weathering rates decline, leaching is reduced and soil profiles become shallower.

Local names have been given to the soils of the savannas but because investigation of these soils has only just begun, many classification systems have been content to distinguish the soils on the basis of colour and of iron content. Thus the wet savanna soils are red, and give way to reddish-brown and brown soils with increasing aridity.

Mediterranean Soils

The soils of the seasonally dry subtropics, areas of a Mediterranean climate, have only recently been described as a distinct great soil group. De Villar recognised them in Spain in 1937; Prasolov described them also in 1937 and called them cinnamon soils; and in 1933 Prescott called such soils in Australia grey-brown soils. Although the term cinnamon soils is still used, the terms red Mediterranean and brown Mediterranean soils are more common. The differences between the latter two soil groups are more in morphology than process.

In the hot and rainless summer the upper soil horizon dries out and weathering is very slow. In the lower horizons there is sufficient moisture available for weathering to continue throughout the year, and clay formation is therefore greatest in the B horizon. During the summer capillary water rises in the profile, carrying dissolved substances which are precipitated when the water evaporates. Calcium carbonate is the most important precipitate in the middle and lower parts of the soil profile and takes the form of a very fine white powder. The calcium carbonate is leached downwards in the winter by water saturated with carbon dioxide derived from humus, but the periodic rise of soil solutions results in a permanently neutral reaction in the profile. The saturation of the clay-humus complex with calcium ensures stability of the humus.

Red Mediterranean soils have an ABC profile with an A horizon of humus accumulation 25–50 cm thick, red-brown in colour and having a granular structure. The B horizon has a high clay content and also shows evidence of strong illuviation of clay, for the peds have coats of oriented clay and there are clay linings on cavity walls. The B horizon is red or red-brown in colour and it has a strong blocky or prismatic structure. Concretions or streaks of calcium carbonate are common lower in the profile.

Brown Mediterranean soils have profiles very similar to those of red Mediterranean soils except that the colours are browner and a hardpan may occur low in the profile. Brown soils are often associated with cork oak in humid areas or with maquis and moist sites and siliceous rocks.

Figure 30.ix. Typical occurrences of terra rossa. (1) Limestone; (2) terra rossa under garrigue shrub vegetation; (3) brown soil under remnant evergreen oak forest; (4) eroded terra rossa on slope; (5) colluvial terra rossa in basin.
After Ganssen and Hädrich

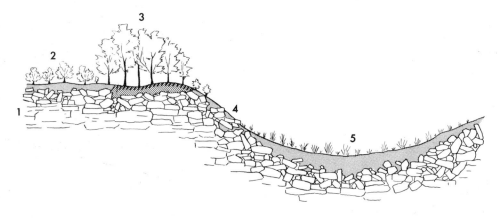

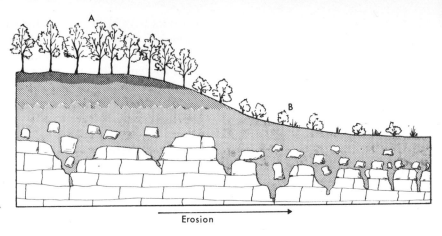

Figure 30.x. Degradation of a brown soil formed under evergreen oak forest (A), by erosion and reddening (rubification) of the surface horizons as the subsoil is converted into a terra rossa beneath garrigue shrubs (B).

Erosion

In areas of limestone, which are common in many parts of the Mediterranean Basin, rendzinas may form but there are also areas of terra rossa. These red-coloured soils are now often only remnants, for in many areas extensive soil erosion has resulted from deforestation. Terra rossa has an unknown origin. In some areas it may be a colluvial deposit, derived from eroded red and brown soils, which has accumulated in basins where it is now cultivated (Figure 30.ix), and it may be a new soil forming on a former B horizon of a brown soil which has been eroded (Figure 30.x). It may also be the result of forest clearance from brown soils and establishment of garrigue vegetation. It is also possible that the parent material of terra rossa may be derived from an interglacial period of moist subtropical weathering. It is clear, however, that terra rossa soils are in harmony with present climatic conditions even if they only form on older soil bodies.

(*Note:* Terra rossa should not be confused with terra roxa, which is a base-rich, low-humic clayey latosol of the São Paulo area of Brazil.)

Soils of the Humid Subtropics

In the humid subtropics the absence of a cold or dry season and of excessive heat allows soil processes typical of the cold and hot humid climates to occur but with less intensity. Forms of laterisation and podzolisation may therefore be found in one soil.

Red and Yellow Podzolic Soils

In the eastern Black Sea area of Russia, in southeast China and the southeast USA the combination of podzolisation and laterisation produces red and yellow podzolic soils under broadleaf forests. Colloidal silica is leached into the drainage water (as in the hot humid tropics) but the sesquioxides released in the acid organic solutions (podzolisation) are kept near the soil surface by the action of the vegetation and are not therefore leached. Both types of soil are deeply weathered, acid and devoid of free lime. The A_1 horizons are dark brown with incorporated humus and a crumb structure. The A_2 is leached and lighter coloured; in the yellow podzolic it is thick and yellow, in the red podzolic it is browner and thinner. The B horizons are rich in illuvial iron, aluminium and clay, yet because this is predominantly kaolinitic they are friable, permeable and strongly structured. The B horizons

39. Profile of a red podzolic soil. The scale is in feet.
Photo by Roy W. Simonson, US Soil Conservation Service.

give the colour name to the soils. In general the yellow podzolics form on more acid parent materials on poorly drained sites where hydrated iron gives the yellow colour, and the red podzolics are on better drained sites where the iron of the B horizon is not hydrated and is red. (See colour plate 7.)

The rapid cycling of humus nutrients allows these soils to be fertile as long as they are under forest but once exposed they must be well supplied with N, P, K and lime. Over large areas they have been severely eroded, especially in the USA where the exposed B horizons are now the agricultural soils.

31 Desertic, Saline and Alkaline Soils

Recent research has suggested that chemical weathering, using the little water available from rain, seepage and dew, is the dominant process in deserts. Such weathering tends to produce large quantities of coarse particles by the solution of the materials binding rock grains in sedimentary rocks and by decomposing the matrix materials holding the crystals in igneous and metamorphic rocks. When fine materials are blown away (see Plate 45), some of these coarse deposits remain *in situ* as a lag gravel coated with iron oxides (desert varnish), others are moved by the occasional flood. The result is that desert surfaces may be sheets of gravel, bouldery, bare rock or more rarely sandy wastes. In such areas the formation of true soils is almost impossible but where water can accumulate, as in depressions or at the foot of fans, weathering is more rapid, plants can grow and soil solutions can move and produce profile differentiation.

A feature of many desert soils is the crust formed on the surface by dessication leaving the calcium carbonate, gypsum, or more rarely ferruginous compounds to cement the soil particles. These crusts are distinct from pans, which consist of accumulations of clay and fine material derived from hydrolytic decomposition, and which occur at some depth in the soil profile. (See colour plates 8, 9, 10, 19.)

Sierozems

In the extremely arid areas profile differentiation is seldom discernible, but in the more humid fringe areas true soils can form. Because of the greater vegetation cover grey colours predominate and in the USA the grey desert-margin soils are called grey earths or sierozems (also spelt serozems). These soils have calcium carbonate and calcium sulphate in the profile, but with the greatest concentration within 30 cm or so of the surface. The surface of the soil is often covered by a layer of wind-scoured pebbles which protects the underlying soil from wind erosion. The soil texture may be uniform, but clay formation in the moister B horizon often produces heavier textured subsoils. The lack of organic matter, water and organic or inorganic colloids renders the sierozems infertile. To improve them is costly because the salts have to be leached out by washing with irrigation water, deep drains have to be provided to prevent re-accumulation of salts, and nitrogenous fertilisers have to be added.

Alluvial Soils

The most productive soils of many deserts are those formed on alluvium deposited by flooding rivers. These are not zonal soils as they are not able to develop distinctive profiles because of the continuing addition of sediment to the surface. It is calculated that the sediment from irrigation water adds a thickness of 5 mm a year to some of the areas of the Indus. The stratification of alluvial soils is therefore depositional rather than pedological in origin.

Like all soil in areas where potential evapotranspiration exceeds precipitation, alluvial soils have accumulations of salts. The salts are derived from ground water and in coastal areas from salt carried inland by the wind as salt spray. The salts of sodium are most common, with calcium, and to a lesser extent magnesium, carbonates. Calcium is also present in the slightly soluble hydrated sulphate gypsum ($CaSO_4.2H_2O$). Desert soils are not the only ones in which calcium carbonate accumulates, for it forms concretions in the chernozemic soils of the subhumid grasslands.

Saline and Alkaline Soils

The soils described in this section occur most commonly, but not exclusively, under dry climates. They owe their distinctiveness to the presence of an excess of sodium salts or to the preponderance of sodium among the exchangeable bases. There are three commonly occurring groups, each with a name derived from the Russian.

Solonchak (sometimes spelled solontchak or solontshak) is also known as saline soil or white alkali soil. The capillary rise of salt-rich ground water and its evaporation at the ground surface cause precipitation of salts on the soil surface. The salt crystals fill the soil pores and push apart the soil particles as the crystals grow, so that the A horizon is loose and puffy and may have a salt crust on its surface. The salts are the carbonates, sulphates, chlorides and nitrates of calcium, magnesium and sodium. In those solonchaks in which calcium carbonate is the dominant salt the organic matter is stabilised and accumulates and the A horizon has a crumb structure, although as in all solonchaks the B horizon is virtually structureless. The $CaCO_3$-rich soils are easily developed for cultivation but where other salts dominate the soils have to be treated with acid fertilisers before they can be used. Solonchaks occur mainly in depressions or near the base of slopes.

Solonetz, or black alkali soil, has a dark colour caused by finely dispersed deflocculated organic colloids. Solonetz develops as an improvement in drainage causes leaching of the soluble salts from a solonchak. Two conditions are necessary for the solonetz process:

(1) the presence of about 15% or more exchangeable sodium in the soil adsorption complex;
(2) the possibility for the soil solutions to shift downwards.

As the solonchak is converted to solonetz by improved drainage, the soluble salts are removed but the exchangeable sodium on the clay micelles causes the colloids to disperse (peptise). The downward percolating colloids are flocculated at depth when they reach a horizon containing quantities of calcium carbonate or gypsum; the flocculation takes place chiefly because of partial exchange of calcium and sodium in the adsorption complex.

As a result of peptisation and shifting of colloids, the upper part of the profile has an eluvial solonised A_2 horizon with an unstable laminar or foliate structure. This horizon is white or pale grey in colour because of the loss of iron oxides and humus and the relative accumulation of silica (a process which is similar to podzolisation). The illuvial B horizon of a solonetz is always very compact. When dry it is coherent, and when wet strongly swelling, viscous and impermeable. The sharp

40. Profile of a desert soil. The scale is in inches.
Photo by Roy W. Simonson, US Soil Conservation Service.

volume changes with wetting and drying of the particles saturated with sodium result in vertical fissuring and conspicuous columnar structure (Figure 20.iv). The B horizon is coloured brown by the abundant iron colloids. Beneath the B horizon, layers of salts form in order of solubility with the least soluble $CaCO_3$ at the top, gypsum ($CaSO_4$) below and NaCl and Na_2SO_4 at the bottom.

Solonetz are unfavourable to plants because of their extreme alkalinity (pH 8·5–11) in the presence of sodium and the impermeability of the B horizon. Improvement of these soils involves the use of acid fertilisers, such as those containing sulphur which oxidises in the soil to sulphuric acid, and the adsorbed sodium is replaced by calcium by applying gypsum.

Solod (also spelled soloth, soloti and solodi) is produced, according to Russian pedologists, when sodium is leached from a solonetz to leave a structureless acid soil. As the impermeability of a solonetz B horizon increases, less and less atmospheric moisture penetrates the soil and periodic stagnation of surface water occurs. In this stagnant water an abundant flora of diatomaceous algae develops. The diatoms build their skeletons from silica and use that in the solonetz horizons, gradually destroying them. The silica of the diatom skeletons is redeposited in the soil but in rearranged forms so that the B horizon becomes permeable and the colloids are leached downwards. The columnar B horizon of the solonetz is replaced by a solodised whitish horizon rich in silica. At the surface a humic A horizon forms. The resulting soil is a solod.

41. The rounded tops of columns of a solonetz. The scale is in inches.
Photo by Roy W. Simonson, US Soil Conservation Service.

Soil Sequences

Solonchak, solonetz and solod represent an evolutionary sequence, although solod can form directly from solonchak. The three soils occur in association as complexes related by relief and hydrological conditions. The low-lying areas may be solonchak, those on moderate relief solonetz and those on higher ground solod. The difference in level need only be a very few metres to have a strong influence on the type of soil which forms (Figure 31.i). Solonetz and solod may be formed by human activity or by natural causes, such as a climatic change or lowering of the water table which allows more leaching of a solonchak.

Figure 31.i. Solonchak soils with a salt crust on the floors of basins, solonetz on midslopes and solod on higher slopes in a semiarid area.

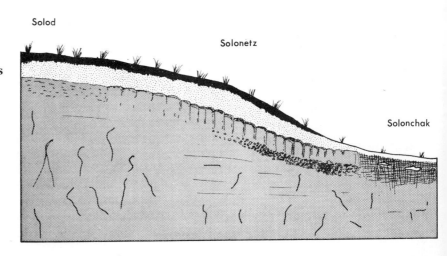

Solod

Solonetz

Solonchak

Reclamation of Saline Soils

Where saline soils have sufficient calcium available to prevent loss of structure, washing the soils with water may be all that is necessary to make them fertile. It was found, for example, that a 2 metre column of water was effective in reclaiming a 1·3 metre column of soil in the upper Colorado river basin, and that an additional application of water every fourth year would keep the salt content down to harmless quantities. Well controlled irrigation schemes prevent the accumulation of salt by deep drainage but when the water table is allowed to rise—as has happened in many areas—the salts accumulate near the surface and destroy valuable agricultural soils. When excess sodium salts are present in the soil, loss of structure and toxic alkalinity can be prevented by dressing the soil with lime or gypsum ($CaSO_4 . 2H_2O$).

Many arid regions have a temperature regime favourable to crop production so that if irrigation can be successfully carried out they can be very productive. Already about 13% of the arable lands of the world are irrigated and about 25% of the world's food is produced on them.

32 Chernozemic Soils

The chernozems or black earth soils of the subhumid temperate grasslands occupy an important place in pedology, partly because through studying them Dokuchaev laid the foundations of many modern ideas about soils, and partly because they occupy an intermediate position in many soil classifications as they are the boundary between the leached and non-leached soils. Marbut divided zonal soils into the two orders of pedalfers and pedocals. The pedocals, or non-leached soils, are those in which deposition of salts occurs—particularly calcium compounds—because the potential evapotranspiration exceeds the precipitation. The pedalfers, or leached soils, occur in humid climates where leaching is a major soil process. The terms pedocal and pedalfer became obsolete when the new US classification system was introduced.

Calcification

Deposition of calcium salts, or calcification, occurs in the range from chernozems to desertic soils. The reason why calcium rather than some other base ion accumulates in the soil is that calcium, like iron, is a common constituent in the rocks of the earth's crust, and when a soil solution is concentrated by evaporation calcium salts, particularly calcium carbonate, are among the first to be precipitated. The calcium salts in the soil solution may be drawn upwards by plant roots, or to a limited extent by capillary action, or they may be released by the breakdown of organic matter and then washed down the profile. Around the boundary between the pedocals and pedalfers the calcium carbonate deposition zone is 0·9–1·2 metres deep, coming closer to the surface in increasingly arid areas, and going deeper or disappearing in more humid conditions. The other important calcium salt is gypsum ($CaSO_4$. $2H_2O$), which is precipitated slightly less readily than the carbonate and is deposited in a layer deeper than that of the latter. The calcium carbonate in the profile may appear as whitish-grey streaks which look like fungal mycelium, as greyish soft nodules or as hard, whitish pans. Soils with high lime contents are sometimes called calcimorphic.

Chernozems

Chernozems (also spelt tshernosems) have the major characteristic of being black. In the Ukraine the black A horizon is 1·3–2 metres deep, but elsewhere and in North America it is much less and 50 cm would be a more common figure. The litter layer is usually thin and formed of a mat of dry and dead grass. The A horizon is generally granular, with the granulation increasing with depth until at the bottom of the A the granules are like peas.

The transition to the B horizon is indefinite but is noticeable by the increase in the size of the aggregates and the browner colour. The calcium carbonate nodules and streaks occur in the B, and in Russia crotovinas, or old rodent holes, filled with carbonates or humic material are a distinct feature. Crotovinas are not a feature of the American

**42. Profile of a chernozem.
The very dark A₁ horizon,
and the white calcium
carbonate nodules in the C
horizon are very clear. The
scale is in feet.**
*Photo by Roy W. Simonson, US
Soil Conservation Service*

Prairie Soils

chernozems. The depth of the B horizon is usually determined by the lower limit at which accumulation of lime ceases.

The black colour and granular structure of chernozems is caused by the grass vegetation and the presence of calcium. The organic matter is derived not from the surface litter but from the close network of grass roots which largely die off each summer during the drought season. The dead roots are humified during the next spring when water again becomes available. Good grassland may contain over 12·5 metric tonnes per hectare of roots compared with 2·5–5 metric tonnes of above-ground material, and over 2·5 metric tonnes of dry matter per hectare may be added to the soil each year as roots. The black colour which persists through the granules gives the impression that chernozems are higher in organic matter than they actually are. The range of organic matter is between 5 and 20% by weight, 8–10% being most common. It is the flocculation of the organic colloids and their stability in the presence of calcium which prevents the organic matter being broken down. Humus mineralisation is also slow during the dry summers and cold winters. The pH of a chernozem is approximately neutral all the way down the profile and there is little translocation of silt or clay, or leaching of sesquioxides.

The chernozems are particularly well developed on loess parent materials in Russia and on glacial deposits and other rocks in North America. They are the soils of the tall grass prairies, and in Russia, North America and parts of South America have been turned into the great wheat growing soils of the world.

With increased leaching under a more humid climate the chernozems give way to prairie soils and with increasing aridity to chestnut soils. (See colour plates 11, 12.)

In the more humid areas of the grasslands which are found to the east of the chernozems in North America and to the northwest in the USSR, leaching becomes sufficiently effective to remove all of the calcium carbonate from the profile, but it does not cause movement of sesquioxides, which remain stable. Because of the leaching of lime, the prairie soil is slightly acid throughout the profile and some clay translocation takes place so that a weak claypan may be formed. The organic matter content decreases with depth in the profile, and the A horizon is generally thinner than that of a chernozem. The B horizon is often a reddish-brown.

Because the prairie soils are close to the transitional zone between the grasslands and forests, they are often replaced by brunizems, grey forest soils or degraded chernozems. The brunizem is a forest soil in origin, from which the trees have been cleared by burning or browsing so that grass has taken over the regraded the soil—that is, the weak podzolising processes have been halted. A degraded chernozem is produced when forest invades the prairie edge and weak podzolisation of the chernozem or prairie soil occurs. Some authorities regard all prairie soils as degraded chernozems, interpreting the prairies as a vegetation relict from a drier period, that persists where forest development has been hindered by fire or animal browsing.

It is obvious that the humid edges of the grasslands will have a mosaic of soils which are related not only to the present vegetation and

Figure 32.i. Shallow soils developed on similar calcareous material but only 61 metres apart. The prairie soil is under grass and the forest soil under species of oak.
After Wilde

Prairie Soil   Forest Soil

climate but also to past events. In such conditions sharply different zonal soils can exist within a very few metres of each other (Figure 32.i).

The soils of the prairie-grassland transition are all fertile and they form the productive basis for the agriculture of the 'corn belt'. It is to this type of soil that the farmer of western Europe tries to convert his soil by the use of leys and farmyard manure.

Chestnut Soils

With increasing aridity the chernozems give way to the chestnut and brown soils of the steppes. The chestnut soils derive their name from the colour of the edible chestnut (not the horse-chestnut). Because of the sparser vegetation of the steppes, the A horizon is lighter in colour than that of the chernozems and tends to become greyer towards the desert margin. The granular aggregates of the chernozems are replaced by platy or laminar structures. The pH of the A horizon is mildly alkaline and more strongly so in the B horizon, where calcium carbonate and gypsum accumulate. The chestnut soils gradually grade into the brown soils and then the sierozems as the arid climates are approached.

33 Brown Forest, Podzolised, Arctic and Antarctic Soils

The humid, cool and cold forested regions of the world mostly occur in the northern hemisphere. Their natural vegetation is deciduous, coniferous and mixed forest, and although this has been extensively modified the soils still owe most of their morphological characteristics to the forest vegetation. It is possible arbitrarily to divide the soils into two groups—those with a mull humus formed under deciduous forest and those with a mor humus mainly formed under coniferous forest and heath. The mull humus forms in a brown soil which is well supplied with bases, has a moderately high nutrient status and considerable biological activity within its profile. A mor humus is acid and lacking in biological activity and releases leaching solutions into the underlying soil which is thereby podzolised.

There is every possible grade of soil between the calcimorphic brown soil and the extreme podzol, but even when these intergrades are well defined not all pedologists agree on the names given to them. Because the criteria used to distinguish the soils are not consistent from one country to another, and because the importance of each soil is also very variable many of the names overlap. This type of confusion has made necessary a precise classification like the new US system but this has not yet been applied to Europe, where the traditional terms are often used. (See colour plates 13, 14, 15, 20, 21, 22.)

Brown Soils with a Mull Humus

Most broadleaf deciduous trees like oaks and beeches are more demanding of nutrients than the needle-leafed coniferous trees of the boreal forest. The lime content of oak leaves is 3%, for example, of birch 1·5% and of pine 1%. These deciduous trees thus have a base-rich litter compared with the base-poor litter formed under the needle-leafed trees. The activity of earthworms and other organisms is greater under the broadleaf trees, and a mild mull humus is formed which is readily incorporated into the soil. In the neutral or slightly acid topsoil the clay, humus and iron oxides remain flocculated and stable. They colour the soil brown and the soil type may be called a brown forest soil.

If rainfall is particularly high or the litter is moderately acid, a moder may form or a crypto-mull (i.e. a very thin scarcely visible mull humus). The increased acidity will deflocculate the clay and allow it to be mechanically translocated into the B horizon where it will be deposited as clay skins on the peds; there is no single English term for this process so the French term *lessivage* may be used. Soils in which *lessivage* is the major process may be called leached brown soils. (Leached is not an accurate translation, for leaching is a chemical rather than a mechanical process).

Under a mor humus, acid soil solutions not only mobilise the clay but release the humus and sesquioxides of the A_2 horizon in a chemical process. The clay, humus and sesquioxides are then deposited in the B horizon. This chemical process is podzolisation.

Brown Forest Soils

These soils develop in warm temperate subhumid climates such as those of western and central Europe, where precipitation is low and has a summer maximum. The mull is often deep and well mixed, and the soil has a high base status. The low permeability of the soil caused by the clay and low precipitation inhibits leaching and translocation, so only soluble salts and lime are leached and these are usually returned to the soil in the litter. The profile shows a deep litter over an A_1 horizon of brown-black mull with 5–6% organic matter and a fine crumb structure. The A_2 horizon is slightly acid, and has only 1% organic matter: it is brown. The B or (B) horizon has a clayey texture, small blocky structures and only slight enrichment in clay and iron oxides.

Brown Calcimorphic and Acid Brown Soils

On limestones or other rocks with high calcium carbonate such as loess and chalky boulder clay, a brown calcimorphic soil is formed with a pH of 6–7. On siliceous rocks such as the granites and schists of the Hercynian mountains of Europe, an acid brown soil with pH 5–6 is formed. The two types of soil mentioned here may be grouped together with brown forest soils as brown earths by some classifications.

Leached Brown Soil

Under more oceanic climates than those of central Europe, as in western Europe and in cool areas of the eastern USA, there is a more intense translocation (*lessivage*) of clay and iron and marked deposition of clay as skins on the blocky peds of the B horizon and along root channels. The A_1 horizon is loamy and an A_2 horizon of grey-brown fine blocks overlies the yellow-brown Bt. This type of soil corresponds to the grey-brown podzolic of North America. In both Europe and America these soils are the basis of traditional mixed farming. They respond well to fertilisers and are not readily eroded. Manuring and leys have often turned them into a soil type close to that of the prairie soils.

In some brown soils, clearance of the deciduous forest and replacement with heath or other acid vegetation has caused a degradation so that a sequence develops:

Mull: Brown forest soil ⟶ Leached brown soil

Mor: Podzolised soil ⟶ Podzol

43. The profile of a grey-brown podzolic soil.
Photo by Roy W. Simonson, US Soil Conservation Service

Podzolised Soils

The podzols are often regarded as the zonal soils of the boreal forests, but podzolisation is very variable in its effects and only occurs as a zonal process in the better drained areas of the forests and on high mountains; on coarse siliceous materials it can form intrazonal podzolised soils in any humid climate from the tundra to the equator; in some areas it has been introduced by man as a result of planting mor-forming trees.

Podzolisation

Podzolisation can commence only when leaching has removed the bases, particularly calcium carbonate, because in basic or neutral reaction conditions the soil colloids form a stable gel and only become mobile in an acid environment. This decalcification can occur both in very wet and in acid environments. The acid solutions derived from the vegetation and the humus can then mobilise the clay, humus and sesquioxides which are washed out of the A horizon and precipitated in the B. Much of the chemistry of this process is still obscure and it is still not known what agent mobilises the sesquioxides and what causes their precipitation. It has been commonly assumed that finely dispersed acid humus particles can mobilise the hydrated ferric oxide formed during the weathering process. It has also been suggested that the oxalic, citric and tartaric acids formed as metabolic by-products of the fungi in the mor layer are likewise capable of bringing sesquioxides into solution. In a series of laboratory experiments Bloomfield has shown that podzolisation can be produced by polyphenols extracted from leaves still on the tree and newly fallen. Bloomfield suggests, therefore, that the mor humus is not essential to podzolisation, and that it is merely produced simultaneously with podzolisation. Many Australian podzols have no mor humus and it may be that Bloomfield's laboratory experiment has demonstrated the true nature of podzolisation, but this has still to be tested extensively. A mor humus probably is significant in keeping the soil moist and in producing the reducing conditions in which ferrous iron is mobilised and converted to the ferric iron which is deposited in the B horizon.

The second problem concerns the cause of the deposition of the iron and aluminium in the B horizons. It is possible that the higher pH in the B causes the colloidal iron and aluminium hydroxides to be precipitated. It is also possible that silicic and humic acids could be deposited on the mineral particles of the B horizon during summer droughts, which would give the particles a negatively charged surface. During the wet season the positively charged iron and aluminium would be precipitated on to the particles to neutralise them. A third hypothesis suggests that during a summer drought, percolating solutions would become concentrated and the sesquioxides would crystallise out from them. It is clear that much has to be learnt of the processes operating to form podzolised soils.

Figure 33.i. The rooting habits of a tree are influenced by the soil. The deep A_2 of the podzol (P) is deficient in nutrients and the tree's feeding roots are confined to the A_1 and B horizons. In the brown soil (B) the nutrients are more evenly distributed through the profile.

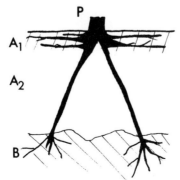

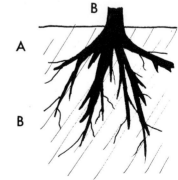

The Podzol Profile

There is a great range in the features of a podzol profile from the minimal development of a grey-brown podzolic to the maximum of a deep iron-humus podzol. A profile of the latter will indicate the main features:

10–20 cm	O	Mor humus, brown to black, few organisms.
2–5 cm	A_1 or Ah	Mineral horizon black and rich in humus but a single-grain rather than aggregated structure.
Variable thickness depending upon degree of development	A_2 or Ae	Bleached horizon. Its ashy colour and cinder-like structure led Russian peasants to assume that it is the result of forest fires. It is, of course, an entirely pedological feature and the Russian name *podzol* given to this layer is now applied to the whole soil. The sesquioxides and humus have all gone from this layer. The slight structures are the result of colourless organic colloids causing adhesion of the quartz grains. Irregular bottom tonguing into the B.
1–20 cm	B_1 or Bh	Black accumulation of colloidal humus with a fine pellicular or laminated structure.

Figure 33.ii. Profiles of podzolised soils. Brown forest and leached brown soils or grey-brown podzolic soils beneath deciduous forest. Brown podzolic soils beneath mixed forest. Iron podzols and iron-humus podzols beneath needle-leaf evergreen trees.

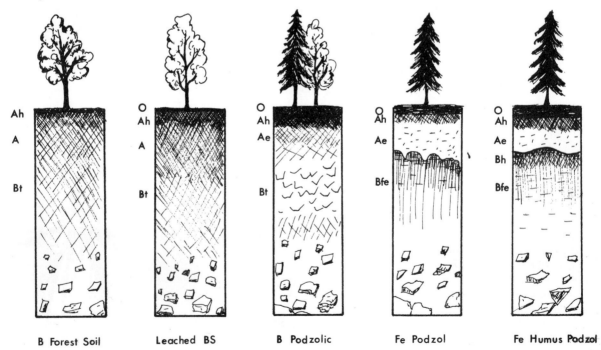

B Forest Soil   Leached BS   B Podzolic   Fe Podzol   Fe Humus Podzol
                  G-B Podzolic

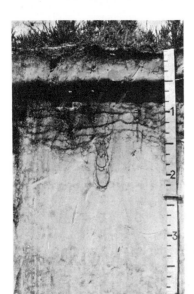

44. Profile of a humus podzol with a thick Bh horizon at a depth of 6–10 inches on the scale (15·2–25·4 cm).
Photo by Roy W. Simonson, US Soil Conservation Service

Figure 33.iii. Profiles of podzols developed under humid (oceanic) climates and under drier, colder continental climates.

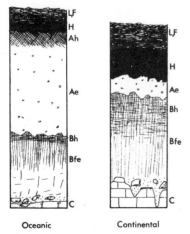

Oceanic   Continental

5–15 cm B_2 or Bfe   A reddish-brown accumulation of iron oxides. It may be hard, when it is called ortstein, or if deep and soft it is called orterde. Ortstein develops on sandy soils, and blocky structures form on clays.

The pH of the soil is lowest in the O horizon, where it is between 3·5 and 4·5, and increases down the profile. The A_2 horizon is very low in organic matter and very low in nutrients, but the Bh horizon may have 5–12% organic matter and the B horizon generally has more nutrients. Trees on these soils develop root systems with feeding zones which indicate the position of the nutrients (Figure 33.i). Converting podzols into agricultural land is very difficult. They have to be deep ploughed to mix the horizons and often to improve the drainage, for the Bfe horizon may form an impervious ironpan and the laminated structure of all parts of the B horizon can impede drainage. The soils have to be heavily limed to reduce acidity and encourage organisms, and most plant nutrients have to be added. They are generally best left in forest.

Types of Podzol

Figure 33.ii shows the range in the degree of podzolisation. The total depth of the A and B horizons varies more with parent material and site than with climate and vegetation. The O and A_1 are thin on well drained sites and the A_1 may not exist. The O and A_1 are thickest on ill drained sites. Ortstein is thin or absent on dry sites and in very coarse material. The whole profile is thickest on freely drained sandy sites. Age is also a factor in profile development. Tamm has suggested that 1000 to 1500 years are needed for a mature profile to develop on a well drained sand, and a 1 cm thick A_2 can be formed in 100 years.

There is usually a marked contrast between podzol profiles developed in oceanic and continental climates. In the oceanic climate, as in western Europe, a podzol is seldom the zonal soil and is the result of acid vegetation or parent materials. As a result podzolisation follows extreme *lessivage* and a very deep A_2 is formed. In the extreme continental climates of Siberia and central Canada the short summers, low rainfall and frozen soil of winter prevent deep leaching and profiles are therefore shallower (Figure 33.iii).

Where grass forms a conspicuous ground layer beneath a needle-leaf or broadleaf forest, as occurs when land is cleared of trees by man, the soil has a more granular and better structured A_1 horizon and microorganisms are more active than under forest alone. The grass is more demanding of nutrients than the trees and produces a richer and more stable humus, so that the A_2 may be scarcely discernible. Such soils are called sod-podzolic or turf-podzolic soils.

In general, podzolisation becomes more pronounced with increases in temperature and precipitation provided that the vegetation is mor forming. Usually, however, higher temperatures favour vegetation with base-rich litter. In the humid tropics podzols do form but only on siliceous parent materials such as quartz sands from which the vegetation cannot derive bases.

When Dokuchaev distinguished the five natural soil zones—tundra, podzol, chernozem, desert and laterite—he implied that the zonal soil is formed under conditions of adequate drainage. In the case of the tundra, however, drainage conditions are usually poor and tundra soils should not be treated at the zonal level of classification but regarded as intrazonal hydromorphic soils. Nor should it be assumed that a distinct soil forming process is operative in the tundra, but rather a podzolising process which weakens polewards from the zone of maximum podzolisation.

The cause of the wetness of tundra soils is the permafrost which extends downwards to depths of over 300 metres in many parts of the Arctic. Only the surface 0·5–1·5 metres thaw out in the 3 month long summer, and soil processes are therefore confined to this active layer and only operate at slow rates for this period. The permanently frozen ground is impermeable so that, although precipitation is seldom more than 2·5 cm a month for the summer period and there is a theoretical drought, the water released from melting snow and from the ice of the

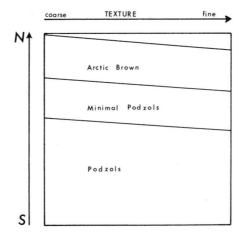

Figure 33.iv. Relationships between latitude, and hence climate, soil texture and soil type.
After Tedrow, Drew, Hill and Douglas

deepening active layer is enough to keep the soil moist even on well drained sites. On gentle slopes or in hollows, gleying and organic matter accumulation are therefore the usual conditions over very large areas.

A catena, as in Figure 33.v, shows the major soil types of the tundra. On rocky, steep areas skeletal soils or lithosols are formed; on well drained sites, where there is a thick active layer and winter conditions with little ice in the frozen soil profile (dry-frost), Arctic brown soils form. These may be regarded as the zonal soils of the tundra even though they occupy only a small area, for these alone form on the well drained

Figure 33.v. An Arctic catena.

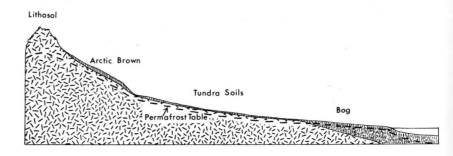

sites characteristic of zonal soils. On imperfectly drained sites with a shallow active layer, gleyed tundra soils form and these grade into bog soils as the accumulation of organic matter increases.

Arctic Brown Soil

An outstanding feature of an Arctic brown soil is the colour of the solum; it ranges from dark brown to dark reddish-brown in the upper horizon to more yellow hues lower in the profile. The colour changes in the profile are gradual. There is no translocation of clay in the profile, although a rather high clay content occurs near the surface. The surface horizon has a crumb structure which grades into a single-grain structure at depth. A small increase of iron and aluminium coatings on the clay surfaces of the B horizon indicates slight podzolisation. The low temperature, short summers and youth of many of these soils have prevented more advanced podzolisation. All Arctic brown soils have a very acid surface horizon of about pH 3·5 but lower in the profile pH is

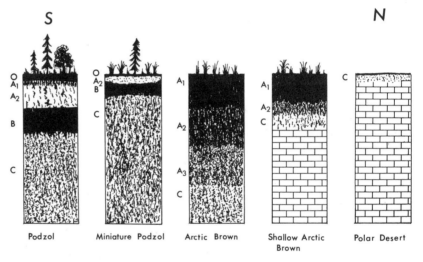

Figure 33.vi. Soil profile sequences from the zone of maximum podzolisation northwards.
After Tedrow, Drew, Hill and Douglas

usually controlled by the nature of the parent material. It is probable that low pH's are the result of leaching by meltwater, for precipitation is too low to cause it. In some soils calcium carbonate may even be precipitated at the bottom of the solum. In Figure 33.vi schematic profiles indicate the progressive shallowing of the zonal soil, and the weakening of podzolisation northwards.

Tundra and Bog Soils

Tundra soils are extremely variable because of the complex pattern of ground ice and the amount of organic matter on the surface. The profile is acid in reaction, yellow to brown in colour, silty in texture, mottled as a result of gleying and usually saturated with water. The active layer is seldom more than 60 cm deep. The low annual increment in organic matter is offset by the very slow humification resulting from an anaerobic environment and a lack of organisms, as well as predominant reduction processes which give the grey and blue colours of the profile. A thin O horizon forms on a very thin Ah horizon over a gley horizon.

Bog soils are very common on all gentle slopes where the waterlogged condition of the soil prevents organic matter decomposition and peat tends to accumulate. The peat is seldom more than 1·2–1·8 metres deep and is usually fibrous. (See colour plate 21.)

Antarctic Soils

About 3% of the Antarctic continent is free of ice, and in parts of that area soil forming processes occur. There are few plants and those are only mosses and lichens so the biological factor of soil formation is of only minor significance, although bacteria, rotifers and nematodes have been found in Antarctic soils. The absence of plant debris, however, makes it justifiable to regard them as devoid of humus (ahumic) but having life (biotic). Small quantities of organic matter are locally supplied by birds such as penguins around their nesting places. The main processes of soil formation are chemical and physical weathering which have progressed far enough to produce clay minerals vermiculite and montmorillonite. Soils do not form on extremely coarse materials and recognisable soil structure is developed only on parent materials which are fine-textured, such as clays and silts.

Because of the low precipitation and lack of organic activity, profile differentiation is limited. There is little change of texture or colour, and the latter is controlled by the colour of the parent materials, but in some soils there is an accumulation of calcium carbonate or of gypsum as either a horizon or a surface crust, or throughout the profile. Typical profile features are shown in Figure 33.vii. The surface is protected by a lag gravel presumably left by deflation of the fines; beneath that during the summer is the dry soil, then the firm accumulation zone with moist soil beneath it. The whole soil is frozen during the winter but the

45. Nesting area of Antarctic penguins showing the accumulation of organic matter around the nests. *Royal New Zealand Navy*

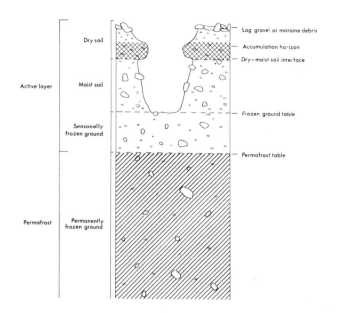

Figure 33.vii. Features of a soil profile from Taylor Dry Valley, Antarctica. *After McCraw*

frozen ground table sinks during the warmer season until at a depth often around 30 cm the permafrost table occurs. Water for chemical processes is largely derived from melting of ground ice and from snow melt. It is thought that there is little leaching, except perhaps on slopes, and that the salt accumulations result from upward movements of soil water in the summer.

The zonal soils of the cold deserts of the northern and southern hemisphere seem to have an accumulation of salts and lack of organic matter much as do the soils of the hot deserts. The soils of the penguin rookeries and other local conditions should be regarded as intrazonal. (See colour plate 22.)

34 Azonal and Intrazonal Soils

Concentration of discussion on zonal soils is apt to underestimate the importance of those soils which develop under the special conditions which limit profile development—azonal soils, and of those which have a distinctive profile developed under peculiar local conditions—intrazonal soils. In the new US classification some of the soils which owe their characteristics to drainage conditions are considered as aquic suborders of each order. This is a useful reminder that many intrazonal soils which have the same genesis are really local modifications of the zonal soil. Because, however, many of the modifications are repeated around the world—gleyed and organic soils for example—it is useful to consider them as a group, for their genesis is the same whether they occur within the zones of podzols or of chernozems. In some areas the non-zonal soils may occupy larger areas than the zonal soils: in the tundra, for example, the gleyed and organic soils occupy most of the area and the zonal Arctic brown soils only the few well drained sites; in northern New Zealand, parts of Indonesia and Japan, recent volcanic activity has laid down such fresh ashes that the zonal soils have not yet had time to form on the ash; in most steep mountains, slope conditions hinder the formation of mature soils. (See colour plates 16, 17, 23.)

Azonal Soils

All azonal soils have shallow A horizons, coloured by organic matter, and C horizons but not B horizons. Because soil forming processes have hardly begun to act in them, they are usually classified according to the nature of their parent materials. There are three main groups of azonal soils: lithosols are thin, stony soils formed on hard, resistant parent materials; regosols are formed on unconsolidated parent materials; alluvial soils are formed on active floodplains which receive additions of material during high floods.

Lithosols

Lithosols usually occur on slopes where there is rapid runoff and such active erosion that fine material is rapidly washed away. The sites are therefore dry, and although a thin A horizon forms beneath vegetation, rapid lateral leaching in the soil and the washing away of decaying litter

Figure 34.i. Soils on a high mountain.

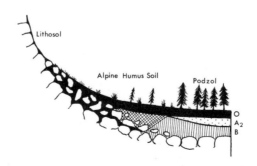

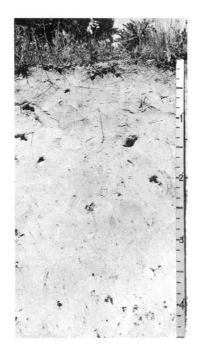

46. Profile of a regosol, showing little horizon differentiation. The scale is in feet.
Photo by Roy W. Simonson, US Soil Conservation Service

47. A regosolic yellow-brown pumice soil formed on pumiceous volcanic ash which was erupted in 130 AD. The dark A horizon is caused by bracken melanisation. The scale is in feet.
NZ Soil Bureau

prevent it increasing in thickness. Because of the thin solum, the development of a B horizon is inhibited. As the slope lessens, so these inhibiting factors diminish and a thicker zonal or intrazonal soil develops. In Figure 34.i a lithosol on a high mountain grades downslope into an alpine humus soil developed under alpine grasses. The alpine humus soil has a thick A horizon rich in organic matter, some of which is fibrous; the B horizon is thin and contains mainly gravel; the total depth of the soil is seldom more than 50 cm. If such soil deepens and the grass gives way to a needle-leaf forest, a true podzol may form.

Regosols

Regosols form on deep incoherent or soft materials such as dune sands, loess and glacial till. Because of their depth and permeable nature, and also because they frequently contain partly weathered minerals, the parent materials of regosols are more rapidly converted to zonal soils than those of lithosols. By definition, regosols have only A and C horizons but, except under conditions of extreme leaching and soil dryness found in some sand deposits, B horizons form readily in humid climates. Thus in Indonesia soils on volcanic ash change to brown then to red-brown and finally red tropical soils; in northern Japan they change to brown forest soils, then to podzols. In central Europe many loesses have grey-brown podzolic soils, brunizems or chernozems. Glacial tills vary greatly in texture, depth and topography, for those laid down by advancing ice are usually compacted, well mixed and fine-grained, but those deposited by dead ice are poorly sorted, hummocky and less compact. The various origins of tills can also affect the soil development, for some are lime-rich, others acid, depending upon the nature of the minerals in them. The soils on tills vary from regosols to chernozems and podzols, depending upon age and parent material.

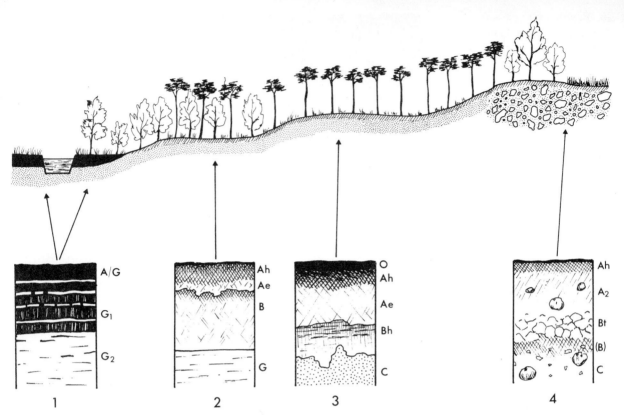

Figure 34.ii. Soils on a flight of terraces in north Germany: (1) gleyed alluvial soil of the present floodplain; (2) podzolised soil on sands and gravels of the lowest terrace; (3) a podzol on the older middle terrace; (4) on the highest terrace a leached soil (*sol lessivé* or grey-brown podzolic) has formed on glacial moraine.
After Ganssen and Hädrich

Intrazonal Soils

Alluvial Soils

Alluvial soils are extremely varied in colour and parent materials. Those formed on coarse debris of the type deposited by many mountain streams or of fluvio-glacial origin are pale-coloured and have textures varying from boulders to silt; those of high order streams may have fine, silty materials rich in organic matter. These soils have many thin, discontinuous beds laid down by floods and many buried topsoils. Alluvial soils frequently have only A and G horizons. The gleying occurs because of the high water tables near the stream. If stream discharge falls permanently or the stream becomes incised, the old floodplain is converted into a terrace and the soil is gradually transformed towards a zonal soil. Figure 34.ii shows this happening.

There are three types of intrazonal soils—calcimorphic, hydromorphic and halomorphic, or very saline. Of these the halomorphic soils were discussed in Chapter 31.

Calcimorphic Soils

Soils formed on highly calcareous rocks usually have an AC profile because the humus and sesquioxides remain stable in the presence of the calcium carbonate and are not dispersed and translocated to form a B horizon. With time, the lime is leached from the surface soil and the thin initial A horizon thickens; in very humid areas, as on some of the Carboniferous limestone hills of western Britain, the A horizon can become acid. The leached lime is either removed in the drainage water or, in dry climates, redeposited to form a lime pan or caliche as a Cca horizon.

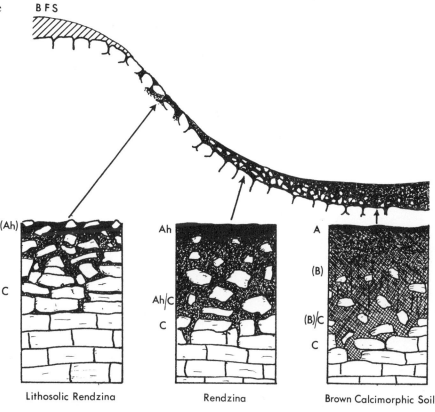

Figure 34.iii. Soils on a slope formed on hard limestone. The brown forest soil on the plateau gives way to rendzinas and at the foot of the slope to a brown calcimorphic soil which will eventually develop into a brown forest soil.

B F S

(Ah)	Ah	A
C	Ah/C	(B)
	C	(B)/C
		C

Lithosolic Rendzina   Rendzina   Brown Calcimorphic Soil

The parent material for calcimorphic soils is highly variable and ranges from limestone and chalk to calcareous sandstones. The parent material greatly affects the texture and colour of the profile. In temperate and cold climates, rendzinas with dark grey or black A horizons form. Initially they are lithosolic and a very thin A horizon occurs above an AC with fragments of rock and humic material mixed. As the rendzina develops, the A horizon thickens and a C horizon of rock fragments becomes more prominent. A rendzina has an alkaline reaction through the profile and a crumb structure formed by high earthworm activity. Once it becomes very thick, the A horizon may become leached and humus may then be mobilised and moved lower into the profile. If clay is available from the parent material and if this degradation of the rendzina continues for long enough, a brown calcimorphic soil may form and this could eventually be podzolised in a humid climate. The soil at the foot of the slope is thickened by colluvium. In many of the chalk areas of southeastern England sequences of soil are repeated, with slightly podzolised brown earths being formed on interfluves on the clay-with-flints derived from capping Eocene rocks and from residues from the chalk, and on the colluvium in the valleys. Rendzinas form the shallow soils on the upper convexities, and thicker brown calcimorphic soils occur on the longer mid-slopes.

Figure 34.iv. Topography and genetic soil types. Brown earths (BE) on flat interfluves and valley floors, brown calcimorphic soils (BC) on slopes and rendzinas (R) on steep eroded slopes. In areas of chalk lithology in southern England.
After Avery

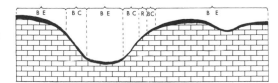

Hydromorphic Soils

The two main kinds of hydromorphic soils are gleys and organic soils. Within these groups there is an extensive range of properties.

Gley soils are formed in conditions of temporary or permanent waterlogging of the soil. The exclusion of oxygen from the soil produces an anaerobic environment and prevents oxidation, particularly of iron; reduction becomes the dominant process, reducing the red ferric iron to blue ferrous iron. The resulting gleyed horizon, G, is blue, grey or olive; sticky, structureless and compact when wet, it may dry out to blocky structures and show rusty mottles when dried. Within deep G horizons all of the ferric iron is reduced, sulphates form sulphides and nitrates form ammonia, so that a gleyed horizon can be smelled as well as seen.

Gleying can occur at any depth within a profile. If it is caused by a permanently high water table, a typical profile shows a thin, oxidised A horizon, maybe a mottled Bg if the water table level varies, and a G horizon of permanent waterlogging. Such a soil is called a ground water gley.

Surface water gleys develop in impermeable soils, such as many clay soils. Rain and drainage water cannot pass through the surface horizons readily, and they suffer alternate oxidation and reduction as the supply of oxygen is available or cut off by the water. The topsoils usually have a moder humus form and mottled Ag and Bg horizons, but C horizons normal for the zonal soil. Such surface water gleys are called stagnogley soils if the wetting is prolonged.

Those soils which have permeable surface horizons and well drained C horizons but impermeable B horizons, like some podzols, have a temporary perched water table above the compact horizon and consequent gleying—such a podzol would have an A_2g horizon. Soils of this type are often called pseudogleys. They frequently develop a mor humus which may be replaced by a blanket of moss. In upland Britain, the formation of iron podzols with an ironpan frequently causes pseudogleying and the development of such a thick blanket of moss that a blanket peat bog is formed (see Chapter 48 for further information on peat).

Organic soils are developed on the partially decayed organic remains which are usually called peat. Traditional usage would confine the term organic soil to the mineralised surface of peat deposits, and peat would be regarded as inert organic debris. Such a point of view is hardly justifiable, for peat has plants growing on its surface and chemical or organic activity in its profile. Peat may therefore be regarded as a soil with greater than 50% organic matter.

The accumulation at the soil surface of semi-decomposed plant residues occurs because of the slow rate of humification and mineralisation of organic substances in the presence of a high water table, which produces an anaerobic environment in which there are few micro-organisms. In thick moss bogs the surface horizons are nearly sterile, so that when the peat is drained for farming it may be necessary to inoculate it with micro-organisms.

Most of the world's peat occurs in the boreal zone of the USSR (60%) and North America (*c.* 20%). Its development is particularly favoured by the extensive flat shield areas of Canada and USSR with

48. The profile of a shallow rendzina.

49. Profile of a thin peat developed over impermeable gleyed silt.

their discontinuous cover of glacial tills. The distribution of peat is very uneven (Figure 34.v); the largest single area is in western Siberia, but all of north European Russia and Scandinavia have extensive deposits, and in Finland about 70% of the land surface is covered by peat. The flora of a peat bog depends upon the origin and status of the nutrients. In an ombrotrophic bog the nutrients are all derived from the atmosphere and the bog is therefore very acid and deficient in nutrients; the vegetation is then usually dominated by the undemanding peat-mosses (*Sphagnum*). In minerotrophic bogs where the minerals are derived from ground waters, the nutrient supply is greater and the flora richer so that cotton-grasses (*Eriophorum*), sedges (*Larex*) and some willows (*Salix*) occur. The pH of an ombrotrophic bog may be as low as 3·0, but minerotrophic bogs have a range upwards to the neutral bog (pH 7–8), developed in the presence of base-rich water, known as a fen.

There are many classifications of peat bogs but the division into fen, low moor, high moor and blanket peats is probably the simplest and most useful. In a depression where a shallow lake can form muds, reeds and other deposits will gradually fill the area until low shrubs and trees can colonise their surface. At this low moor stage the surface of the bog is almost level, although the margin may be raised and carrs or mounds with forest trees exist. Further development of the bog usually involves

Figure 34.v. Major areas of peat in the boreal zone. *After Sjörs*

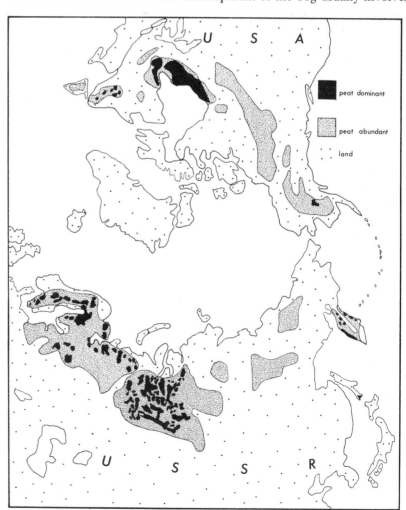

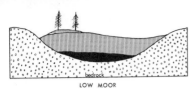

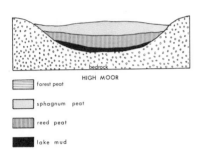

LOW MOOR

HIGH MOOR

forest peat

sphagnum peat

reed peat

lake mud

Figure 34.vi. Basin peat development showing low and high moor stages.

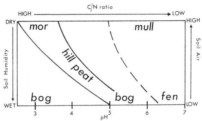

Figure 34.vii. Upland soil organic matter relationships. *After Pearsall*

50. A peat in which the top layers have become pale in colour, dry and fluffy. The still moist layers are on a level with the blade of the spade.

the extensive growth of sphagnum mosses, which can obtain sufficient nutrients from the atmosphere and rain and hence form self-perpetuating domes near the centre of the bog, leaving the margin of the bog with its higher nutrient status as a depressed margin (Figure 34.vi). Peat grows very slowly and although the rate is very varied, it is of the order of 30 cm in 500 years.

The texture of peat depends upon the nature of the original plants and the extent to which they have decomposed. Peat decomposes in two main ways—by humification and by anaerobic decay. Humification is the oxidation of the plant residues, in the presence of oxygen either in water or in the air, to leave behind mineral matter and loose carbon. Anaerobic decay takes place in the absence of oxygen in submerged peat. Each layer of peat is subjected to aerobic decay as it forms the surface and later to anaerobic decay as it is buried beneath the next layer of fallen vegetation. Humification is usually greatest in warm climates and least in cold ones. The amount of decomposition is gauged in the field by squeezing the wet peat in the hand. Von Post proposed a ten point scale in which $D1$ refers to undecomposed peat, which yields only clear, colourless water when squeezed and in which the plant structure is unaltered, and $D10$ is completely decomposed so that plant structure is unrecognisable and the peat is a jelly-like material which all escapes between the fingers.

A typical peat profile shows a surface horizon of active humification, a transition horizon of alternate humification and anaerobic decay, and a deeper horizon of anaerobic decay. There is sometimes a difficulty in deciding whether a forest-floor deposit is a peat or a thick mor humus. In boreal zones and areas of blanket peat, where peat may be regarded as a zonal form, the distinction is difficult, but elsewhere it can be assumed that peat will show maximum humification at the surface and mor will show it at the base.

Fen peats form very rich soils when drained, but ombrotrophic peats are acid and deficient in nitrogen and other nutrients. They do however form excellent farmland when top-dressed with fertilisers. Care must be taken not to overdrain peats, for they may shrink excessively and the light topsoil will then be easily blown away or have a resistance to re-wetting which will cause water deficiency in the root zone of crops.

Peat and humus forms are part of a continuum of organic deposits and this is illustrated by Figure 34.vii. It should be emphasised that both gley and peat soils can be zonal if the climate is sufficiently wet.

35 Soils and Man

Man has so modified the soil that over large areas of the earth it is not possible to find soil in its virgin state. The modifications are too many to discuss in one short chapter but Table 35.1 attempts to show some of the effects of man on Dokuchaev's five classic factors of soil formation. Man's impact on the soil began with the earliest societies: the use of fire to clear vegetation or herd animals; the domestication of plants and animals; the invention of irrigation, terracing and swamp draining; the breaking in of grasslands for agriculture after the introduction of iron ploughshares; the spread of cash cropping and European colonisation of the New World; the mechanisation of agriculture; the spread of towns, roads and factories over farmland; and more recently the growth of industrialised 'factory' farming. Each has had an increasing impact on soils.

Irrigation and Terracing

Between about 3000 and 1000 BC ancient civilisations flourished in the riverine and delta regions of the Nile, Tigris, Euphrates, Indus and Hwang Ho. In these areas flooding produced an increment of fertile alluvium which allowed intense agriculture to be continued without exhausting the soil for hundreds of years.

Furthermore the delta areas, particularly of the Nile, were so flat and low-lying that control of water during periods of low river level was relatively easy and irrigation could be practised to give as many as four or five crops a year. These early civilisations not only used natural soil conditions but, by controlling the soil water, greatly increased soil productivity to provide the surplus of food which is the basis of a complicated and advanced society. The control of soil conditions was

51. Open paddy fields growing rice in western Johore.
Photo by D. W. McKenzie

Table 35.1. Suggested Effects of the Influence of Man on Five Classic Factors of Soil Formation (After Bidwell and Hole)

BENEFICIAL EFFECTS*	DETRIMENTAL EFFECTS*
(1) Parent material	
(a) Adding mineral fertilisers (b) Accumulating shells and bones (c) Accumulating ash locally (d) Removing excessive amounts of substances such as salts	(a) Removing through harvest more plant and animal nutrients than are replaced (b) Adding materials in amounts toxic to plants or animals (c) Altering soil constituents in a way to depress plant growth
(2) Topography	
(a) Checking erosion through surface roughening, land forming, and structure building (b) Raising land level by accumulation of material (c) Land levelling	(a) Causing subsidence by drainage of wetlands and by mining (b) Accelerating erosion (c) Excavating
(3) Climate	
(a) Adding water by irrigation (b) Rain-making by 'seeding' clouds (c) Release of CO_2 to atmosphere by industrial man, with possible warming trend in climate (d) Heating air near the ground (e) Subsurface warming of soil, electrically, or by piped heat (f) Changing colour of surface of soil to change albedo (g) Removing water by drainage (h) Diverting winds	(a) Subjecting soil to excessive insolation, to extended frost action, to exposure to wind, to compaction (b) Altering aspect by land forming (c) Creating smog (d) Clearing and burning off organic cover
(4) Organisms	
(a) Introducing and controlling populations of plants and animals (b) Adding organic matter (including 'nightsoil') to soil directly or indirectly through organisms (c) Loosening soil by ploughing to admit more oxygen (d) Fallowing (e) Removing pathogenic organisms, as by controlled burning	(a) Removing plants and animals (b) Reducing organic matter content of soil through burning, ploughing, overgrazing, harvesting, accelerating oxidation, leaching (c) Adding or fostering pathogenic organisms (d) Adding radioactive substances
(5) Time	
(a) Rejuvenating the soil through additions of fresh parent material or through exposure of local parent material by soil erosion (b) Reclaiming land from under water	(a) Degrading the soil by accelerated removal of nutrients from soil and vegetative cover (b) Burying soil under solid fill, water, asphalt or buildings

* The terms 'beneficial' and 'detrimental' imply a value judgement, and the table is admittedly over-simplified and patently biased

also a characteristic of many societies in southeast Asia, and also in Peru, where great stone terraces kept erodible soil in place to make agriculture possible on steep slopes, and where irrigation canals carried water long distances. In China nightsoil, and in Peru guano, were used to maintain fertility.

Attempts to increase production by the use of irrigation have not always been successful. Recent irrigation schemes in Turkey and Pakistan have been just as unsuccessful as much older schemes because of failure to prevent accumulation of salt in the soil. In arid and semiarid climates, salts accumulate in the soil instead of being leached into the drainage water, so that when the water table is raised by irrigation they are carried into the root zone and even left as a crust on the soil surface when the water evaporates. The salt can be controlled by adequate drainage and by frequently flushing the soils with clean water, but not all irrigation schemes have had the necessary drainage and adequate water for flushing.

Clearance of the Temperate Forests

About 2000 BC when Neolithic culture was becoming widespread in western Europe, the area was largely covered with deciduous forest in the southern areas and coniferous forests to the north. The deciduous broadleaf forests protected the soil from direct insolation and rainbeat, and also reduced evaporation from the soil surface. The trees transpired water which was derived from the ground-water of the root zone. The fires and clearances of the early settlers caused direct insolation to increase and allowed rainbeat and washing to remove fines from the soil surface. The surface soil became more liable to drying out and in many areas there was a rise in ground-water levels. The increased

52. Clearance of native forest by burning. East coast area of the North Island of New Zealand.

exposure and drying allowed a more rapid oxidation of organic matter, so that the water-holding capacity of the soil was reduced. The loss of organic matter and the absence of the annual leaf-fall with its supply of bases caused a decline in soil fertility. The lowered organic matter content of the topsoil also reduced the earthworm activity so that organic matter was no longer rapidly incorporated into the soil and a mor humus tended to form. The result of all these changes was a tendency for greater podzolisation on sandy soils, especially when heath vegetation became established, but on clay soils the turning over of the topsoil by cultivators brought the bases back to the surface. In addition, increased evaporation tended to bring salts to the surface, and man and animals added organic matter to the soil. The final result of clearing broadleaf deciduous forests was usually to turn brown forest soils into a kind of prairie soil.

The induced prairie soils frequently suffered a decline in fertility, once the nutrients in the wood ash from the first fire had been exhausted, so that shifting cultivation became necessary. In an area of slow-growing trees, forest destruction became rapid and widespread. Mixed farming, with animals being folded on the stubble of the crops, and liming and manuring being practised allowed stable farming to become the pattern in medieval Europe, but yields were low and fallowing was necessary until the agricultural revolution, when crop rotation, better manuring, and the introduction of new crops allowed fertility to be raised.

During the clearances of the European forests the soil had been protected from erosion by a nearly continuous vegetation cover, and by the characteristically gentle rainfall. Wind-blowing and some rilling with downstream flooding had occurred but these effects were neither widespread nor disastrous. The introduction of European technology to North America was not so successful. Tobacco planting had exhausted many soils in Virginia by the middle of the nineteenth century and the upland cotton lands of Georgia were dissected with gullies. The spread of a forest-based agricultural technology to the grasslands of the west was even more disastrous. Once the sod of the chernozems and other soils of the Plains was broken, wind could carry away the dried humus

53. Scene on the High Plains of Texas. Wind-blown dust accumulates round an abandoned farm.
US Dept of Agriculture

54. Gully erosion in the pumice-lands of the central North Island of New Zealand.

and silt; the intense summer rainstorms could wash the fine soils away, leaving behind exposed B horizons of low fertility, and cause downstream flooding by silt-laden waters. For many years little notice was taken of this 'rape of the earth' for there was always more cheap land further west. The occupation of the High Plains took place in wet periods, but characteristically these are followed by dry periods and it was the drought of the 1930's which caused North Americans to realise that the irreplaceable fertility of their land was being washed away. The fine soil from the Dust Bowl of the west which settled on the Capitol at Washington caused the US Soil Conservation Service to be established and ended an era of soil exploitation.

Overgrazing

Overgrazing and over-cultivation lead to patches of bare soil which are susceptible to erosion. Although cattle do not graze the vegetation as closely as sheep or goats, they tread heavily and compact the soil so that it becomes impermeable, or break it up so that the grass dies out and wind and water can erode it. These processes go on wherever there are pressures on the land and they are particularly common in semiarid regions where animals concentrate round water holes. Many savanna watering places have now become dry and severely eroded. The goat has done particular harm wherever it has been free to browse in forests. In the Mediterranean lands goats and the constant demand for timber have denuded many areas so that infiltration rates have declined, runoff has increased and erosion has been accelerated. These effects were widespread by late classical times, and many harbours became silted up, valley floors became marshland and slopes were bared of soil.

Shifting Cultivation

Shifting cultivation has ceased to be a feature of advanced agricultural technologies or of any large area outside the humid tropics, but within the tropics over 200 million people, thinly scattered over 36 million square kilometres, obtain the bulk of their food supply by this system.

They form nearly 10% of the world's population and are spread over 30% of its cultivable soils.

In regions as widely separated as the tropics of Africa, America, Oceania and southeast Asia, areas are cut and burned out of the forest, scrub or savanna, cultivated for a few years and then abandoned for another area as yields fall. This process has been going on for hundreds of years and where there is no pressure of population, soil fertility and stability and crop yields have been maintained. When a fresh clearing is made, its boundaries will not necessarily coincide with those of an older clearing so that the fallow periods of one patch may be 5, 10 and 20 years. In many areas 3 years of cropping followed by 8 years of fallow appear to be sufficient to maintain fertility, but where there is pressure of population the fallow periods have become too short, and soil deterioration followed by a fall in crop yields and erosion has become common.

At the start of a fallow period under forest, the bush grows rapidly from new seedlings and old roots still in the soil. The nutrients taken up from the root zone will be partly stored in the vegetation and partly returned to the surface soil as litter, or in rainwash from the leaves. Humification is so rapid that the litter layer will never be greater than about 13 mm thick. As the vegetation re-establishes itself, the soil is once more protected and after a few years the system is again in equilibrium, with weathering and losses by erosion being balanced. At the same time roots, earthworms and other organisms will make the soil friable, permeable and aggregated.

When the forest is again burned, nitrogen and sulphur are lost as gases but the other nutrients are deposited in the ash on the soil surface. Some ash is washed or blown away and the loss of humus causes a deterioration in soil structure. Nutrients are also lost by increased leaching. It is usually only a few weeks before the soil is again covered by vegetation, but removal of a crop and the reduction of humus rapidly leads to a decline in productivity, so that it is seldom worthwhile cropping the same land for more than 3 years. The patch is then abandoned to the forest.

55. Sheep grazing on a pasture which has been treated with bipyridylium herbicides and then drilled with new grasses. The dead stubble can be seen between the rows of new grasses.
Plant Protection (ICI) Ltd

In savanna areas, when the land is abandoned after burning, grasses do not establish themselves quickly so the soil is left bare and liable to erosion. Unlike the forest fallow, the grasses do not increase their store of nutrients year by year and the annual burning reduces the humus buildup and the amount of litter available to soil organisms. The thorough cultivation of the soil necessary to remove the tussock grasses also leads to rapid oxidation of humus. The savanna recovery is therefore slow, and the rotation period should be longer than that of the forests, but as the savannas are constantly encroaching on the forests the pressure for short rotations is increasing.

It is clear that in forest areas shifting agriculture is an effective response to the problems of soil use and fertility in the humid tropics. Even in the forests of the eastern USA, the European settlers of the seventeenth century adopted the shifting cultivation methods of the Indians and continued to use them for two and a half centuries. Their replacement by rooting and stumping, with ploughing and row cropping, was rapidly followed by gulleying and topsoil destruction. Harm is only done when the tropical forests are needed for timber production. In the savanna, however, annual burning, which prevents the growth of tree and shrubs and encourages erosion, causes a permanent loss of nitrogen and so prevents the rapid regrowth of the grasses. Nothing but soil deterioration and loss is the result.

There is, as yet, no sign of a type of farming which will replace shifting cultivation in the forest lands, but some form of replacement, possibly using mechanisation, artificial fertilisers, mixed animal-crop methods and planted fallows, will have to be devised. The animal must replace the natural fallow as the source of organic matter if fertility is not to decline.

Modern Developments

So much research has been done in the last 20 years that it is not possible to mention more than a few examples of how modern technology can offset many of the traditional hindrances to higher production.

Soil erosion usually occurs because soil structure is lost during cultivation or because the soil is left bare during certain times of the year. A recently developed herbicide called bipyridylium, developed by a British company, appears to offer an answer to these problems. Bipyridylium kills green vegetation upon contact and then passes into the soil, where it is adsorbed on to the clay micelle and inactivated. By spraying stubble, weed beds or old pasture with bipyridylium and then resowing with a machine which merely rips a small channel and plants the new seed in it, it is possible to avoid breaking the soil structure or exposing the soil. The dead vegetation acts as a mulch and prevents weed regrowth while the planted seeds can establish themselves rapidly because there is no competition from weeds. The roots of the old crop are left in the soil to supply organic matter and there is no increase in the rate of humification such as occurs after ploughing. It seems probable that the use of compounds like bipyridylium will completely revolutionise farming and prevent soil losses.

56. Attempts to stabilise sand dunes in Libya with grass.
Shell Oil Co. Ltd

A second example is of soil creation. In parts of Israel shifting sand dunes have been transformed into fertile fields by irrigating them with sewage. As an experiment 60 square kilometres of shifting dunes were sprayed with the sewage from one town. Not only did the dunes produce good crops of alfalfa and grass but the high nitrogen and organic matter content of the sewage encouraged a great increase in the micro-organism population of the dunes and an increase in the clay fraction. In addition, an effective and relatively cheap way was found to dispose of and purify the sewage.

Man has modified and often destroyed soils when he has attempted to increase their yield or to produce crops from them. He has the capacity to improve rather than destroy and as world population continues to grow modern technology will have to reverse the historic process of soil degradation.

PART III   BIOGEOGRAPHY

36 Introduction to Biogeography

The study of the origin, distribution, adaptation and association of plants and animals which make up the complex components of the biosphere is called biogeography. Traditionally the study of plant distributions—phytogeography—has been distinct from animal geography—zoogeography. The distinction is artificial for the two are not separate in nature. In the following chapters, therefore, separation is made only for the sake of clarity.

The origin of organisms must be studied in the light of the climatic and geological events which have influenced the evolution and distribution of floras and faunas. Adaptation includes a study of the relationship of the flora and fauna with the environment and association includes the relationships found within the communities of plants and animals.

There are three levels at which these relationships can be studied: (1) at the individual level (autecology); (2) at the population level (synecology); (3) at the ecosystem level. An individual plant or animal is a unique entity, although in the case of plants such as strawberries which multiply by sending out runners on which new plants grow, and which are genetically identical to the parent, the distinction can be difficult. The population is an interbreeding group of individuals which has become adapted to its local environment and may be differentiated from other populations of the same species by small genetic differences. The individuals making up the population may be genetically slightly different from each other because each is adapted to its own immediate environment. Such differences give the population a greater chance of survival because if there is a change in environmental conditions—such as an exceptionally cold winter—the chances of a few individuals surviving it is greatest in a heterogeneous population. Diversity, then, is a kind of genetic insurance which enables the population to become adapted, by natural selection or survival of individuals with certain genes, to changing conditions. Individuals and populations do not live alone in nature but are grouped or aggregated into communities of plants and animals called ecosystems. An ecosystem can be of any size, from the Amazon rain forest to a tank in an aquarium. Whatever its size, the individuals and populations which make up the ecosystem are so enmeshed in their functions that it is difficult to separate them. The ecosystem properly includes the physical environment and the biological community occupying it.

The zone of the earth's surface within which organisms exist is called the biosphere. The biosphere is a thin veneer on the surface of the earth extending little higher than the height to which birds can fly. Even in the sea most life is to be found within 150 metres of the surface. Within the biosphere there is a great complexity of living things, with upwards of 1,300,000 kinds of plants and animals. Most of these organisms are limited in their distribution for only man has spread widely over the earth, and even his settlements are largely confined between latitudes 60°N and about 40°S. With man have spread his congeners—the housefly, cockroach, house mouse and a few other species. Most animals have not even spread into areas which are suitable for them. Thus Australia has no monkeys, Africa has only monkeys with weak tails, and only South America has monkeys with tails sufficiently strong for them to be used to cling to branches.

For the sake of precision the biosphere is divided up into areas of various magnitude by ecologists and biogeographers: the terms used here are those of Dansereau. Some of them are used in different senses by other workers, so the reader may often have to define his terms when using them.

The biosphere may be divided into three biocycles—the salt water biocycle, the land biocycle, and the fresh water biocycle (Figure 36.i). The land biocycle can be divided into biochores, or areas within which the climate is such that a major type of vegetation—desert, savanna, grassland or forest—occurs. Each biochore contains many different formations. Thus the forest biochore has several types of forest within it, e.g. tropical rain forest, temperate rain forest, temperate deciduous forest, coniferous evergreen forests. Within each formation of mature trees there will be areas dominated by one or more species of tree, e.g. oak forest or maple forest, and the area occupied by this species may be called the climax area. Within each climax area there will be areas of uniform topography and each such area will be the habitat for a particular community of plants. Within each habitat will be a number of small units of space such as those occupied by a forest pool or a clump

Figure 36.i. Subdivisions of the biosphere.
After Dansereau

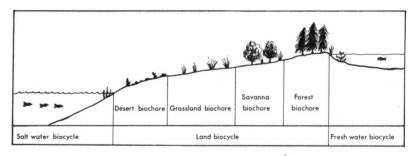

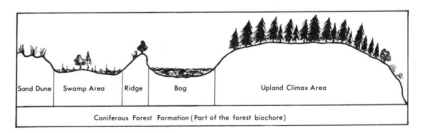

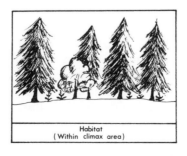

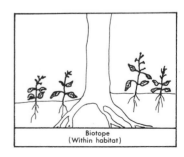

of grass. This smallest unit is the biotope. The reader is recommended to study Figure 36.i to see these relationships, which may be summarised by a list in descending order of magnitude:

Biosphere > biocycles > biochores > formations >
climax areas > habitats > biotopes

Within any part of the biosphere, each organism has a particular function which is called its niche. The habitat of the organism is its place of work, its niche is its professional occupation. In grassland areas kangaroos in Australia, bison in North America, zebra in Kenya and beef cows in Argentina all occupy the same ecological niche of herbivores and are thus ecological equivalents. The same species may, however, function differently in different habitats or geographical environments; thus man himself may be a herbivore, a carnivore or an omnivore.

Most natural ecosystems contain a variety of species, some specialised and some generalised, each of which occupies its own niche, but in different ecosystems the same type of niche may be occupied by different species—thus the owl, the hawk and the eagle may occupy similar niches but in different ecosystems.

Vegetation may also be classified into formation-types, formations and associations. The formation-type is a world vegetation type dominated throughout by plants with a similar morphology or life-form. Thus the tropical rain forest or the tundra is a formation-type, but within these types there are geographically distinct units such as the Amazonian, African and Indo-Malaysian rain forests, or the Eurasian and North American tundras. A distinct plant community dominated by two or more distinct species is called an association, or if the community has one species only as a dominant it is called a consociation.

37 Plants and their Relations with the Environment

The place where an organism or community lives is its habitat. In such phrases as dune habitat or swamp habitat is implied a set of environmental conditions which are sometimes called factors. The factors listed below in Table 37.1 exist within the dimensions of space and time, so that space and time should not be considered as factors.

Table 37.1. Environmental Factors

Physical factors		Biologic factors	
Edaphic:	Soil	*Biotic:*	Plants
Geologic:	⎰ Topography ⎱ Rocks		Animals
			Man
Climatic:	⎧ Water ⎪ Temperature ⎪ Light ⎨ Atmosphere ⎪ Wind ⎩ Fire*		Fire*

*Fire may be natural or induced by man.

A simple list of factors gives a false impression because the factors are intermeshed and are constantly changing, both in the short term and in the long term. The separation of the factors is part of the analytical procedure of the observer and not a natural characteristic.

Adaptations

Plants acquire physiological and morphological features by selective evolution which enables them to withstand adverse factors in the environment. An organism has been described as 'a bundle of adaptations'. Adaptation confers benefits on a plant, but by being specialised so that it can withstand the factors in a particular environment only, a plant loses its adaptability to changing factors. Over-specialisation is the first step to extinction.

In the early literature it was customary to use anthropomorphic terms ascribing human responses and attributes to plants. This should be avoided for it gives a false impression of the actual processes of evolution and adaptation. It is more accurate to regard the environment as a selecting agency by which plants with advantageous mutations survive, and those without necessary adaptations fail. Such a condition also implies that a plant may have morphological and physiological features which it does not require for success in its natural habitat. Many tropical plants, for example, are known to possess adaptations which allow them to withstand cold and drought but, unless the plant is moved by man, such adaptations are not used. Plants, because they are essentially immobile, have to be adapted to the environment to survive. Animals, by contrast, can move and so have less need of such adaptations, although many animals do have them.

Geological Factors

The rocks of the earth's crust have to be weathered and a soil formed before the higher plants can grow. The immediate environmental factor affecting plants is therefore the soil, but by influencing the characteristics of soils, especially the azonal soils, rock structure and composition influence plants. Rocks such as limestone, dolomite, serpentine, alluvium and dune sands offer environmental conditions favourable to some plants and unfavourable to others. In some areas of Australia and Canada, prospectors for minerals have used plants as indicators of ores beneath the soil. The effect of the substratum on plants is most noticeable in dry and in cold climates, for here soil development is so slow that the mineral composition of the bedrock strongly controls the soil minerals. Under climates with slow weathering and little clay formation, the availability of water is reduced by the occurrence of sandy soils. Such influences have marked control over the distribution of plants. Even animals may be restricted in their distribution if bedrock is near the soil surface and prevents burrowing, or the growth of suitable plants for food. In such cases, plants and animals have distributions controlled by geologic factors.

Edaphic Factors

These are numerous and so interrelated that it is often difficult to separate them. The soil itself is a physico-chemical-biological complex which provides the higher plants with an anchorage, with nutrients and with water. The plant in its turn supplies the soil with humus and assistance in forming soil structure, so plants and soils exist in an intimate relationship.

The two extremes of soil texture—coarse and fine—both have advantages for plants, so that when man is attempting to form an environment most favourable for crops he attempts to produce a soil with an intermediate texture which he calls a loam. A good soil structure makes the soil permeable to water, air and rootlets. At the same time the aggregates have high water and nutrient-holding capacities because they are bound together by colloids. A well aggregated, fine-textured soil may possess many of the best properties of poorly structured sandy or clay soils.

In both coarse-textured and well structured soils, soil aeration is provided by the interstitial spaces required for allowing toxic carbon dioxide to be replaced by essential oxygen. The conditions under which good aeration exists do, however, have the disadvantage that they also allow the rapid oxidation of humus. Poor aeration is, of course, often associated with poor drainage, so these factors operate together in being limiting factors in plant growth. Because of their freer gas exchange and lower moisture-holding capacity, coarse-textured soils usually warm up more quickly in the spring than fine-textured soils. If the coarse soils are also pale-coloured, the high albedo of the soil may reduce the rate of warming up. This effect can inhibit plant growth on sandy soils in the winter and early spring.

Organic matter and biotic activity improve both the soil fertility and its structure. Many of the plant nutrients taken from the soil are returned to it in the form of decaying litter, and by removing nutrients from depth and returning them to the surface soil, plants can actually increase the fertility of the surface soil. The nitrogen-fixing bacteria (such as *Azotobacter* in aerated soils, *Clostridium* in unaerated soils,

Rhizobium in nodules on legumes) also help to improve fertility by fixing nitrogen, which becomes generally available in the soil after the plant decays. The most important function of the soil flora and fauna is their improvement of soil structure. Roots grow and animals burrow into the soil, allowing water and air to permeate. Humus itself improves structure by promoting the aggregation of particles. Further, it has a high water-retaining capacity and so makes a better environment for plants. The fertility of a soil will clearly affect plants, for any nutrient deficiency can limit plant development. Sulphur and magnesium are frequently limiting factors in natural conditions, although under cropping systems nitrogen, phosphorus and potassium are the nutrients most readily depleted. Fertility is also affected by pH—a point discussed at greater length in Chapter 24.

Under conditions of high or moderate rainfall for at least part of the year, material eroded from the land is carried away by streams, in solution or as solid particles. In arid climates drainage is frequently endoreic (that is, limited to internal drainage basins), so the soluble salts are precipitated in ponds and lakes to form salt flats or playas when all the water has evaporated. To live in such conditions plants have to be tolerant of salt, i.e. halophytic. Halophytes occur around most salt flats and on many coasts, with the most tolerant species occurring in the saltiest soil.

In brief, by providing an anchorage for the whole plant and the environment for its roots, the soil must have a marked influence on all plants.

The Water Factor

Water is of the utmost importance to plants. As an almost universal solvent, it carries the nutrients a plant removes from the soil; it is essential for chemical reactions within the plant; it is a raw material in photosynthesis; it is imperative for the very existence of protoplasm and for the maintenance of turgidity; and finally it helps to maintain an equable soil climate because it can absorb much heat with relatively little temperature change.

Not all of the water which falls on an area as precipitation becomes available to plants. Depending upon the local conditions of the precipitation and the distribution of plants, some of the rainfall will be intercepted by leaves and branches and evaporated, some will drip to the ground, some will run down trunks as stem flow, and some will reach the ground directly as throughfall. Light showers may be completely intercepted and never reach the ground.

Most plants depend upon capillary water for their needs, but a few, such as willows, depend directly on the ground-water. Such plants are called phreatophytes. In arid areas, phreatophytes may still be thriving when all other plants are dormant. A few plants can absorb their moisture directly from the air. Some orchids, for example, which grow on tree branches in tropical forests, take up water directly from the air when the relative humidity rises above 85%. Some desert plants like the Sahara caper plant (*Capparis spinosa*) have a similar capacity, but even more notable are the forest plants of the rainless coast of Peru, which get all of their moisture requirements from cloud and fog. Some of the fog is utilised because water condenses on leaves and then falls to the

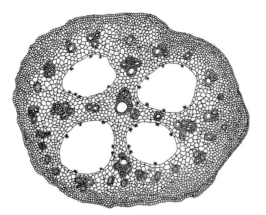

Figure 37.i. Cross-section of the leaf-stalk of a water-lily showing the large air passages which give it buoyancy in water.
After Polunin

ground, but most of these plants obtain their moisture directly from the air.

Primitive plants are thought to have originated in the seas, where transpiration, wilting and drought are absent. When they emerged on to land, therefore, the successful plants were those which evolved organs capable of maintaining the water content of the protoplasm at the required level. Landward migration was controlled by the rate at which adaptations to water loss were evolved. Two main types of adaptation developed: one was the growth of cutin and suberin (skin and cork) tissues which reduced transpiration, and the second was the formation of roots and rhizoids which could extract and retain soil moisture.

According to their responses to their water requirements, plants can be classified into three main groups:

Hydrophytes: Plants which cannot withstand drought. Many of them, like the water lilies and mangroves, have their roots permanently in water. Others, like the ferns of rain forests, live in a permanently humid atmosphere.

Mesophytes: Those species which cannot inhabit water or saturated soil, yet cannot survive prolonged deficiency of water. Most temperate region plants come into this category.

Xerophytes: Plants which can withstand lack of ground-water for long periods. All desert plants come into this category, and in higher rainfall areas shallow-rooted plants on sand dunes, or many mosses and lichens, are good examples. The xerophytes are of four main kinds:

(1) Annuals are actually drought evaders because they get through a drought in seed form.
(2) Phreatophytes tap ground-water.
(3) Succulents store water in their stems or leaves. Cacti are perhaps the best-known succulents, but because they are susceptible to frost, they are confined to warm climates.
(4) The major groups of true xerophytes withstand drought by a great variety of structural adaptations. Some have thick cuticles on their leaves, or waxy coats on them, others have very small leaves, and ones which curl up in dry periods, or even drop off while photosynthesis is performed

57. The trunks of these Australian trees are thick and capable of storing water.
Photo by John Crowther, Australian News and Information Bureau

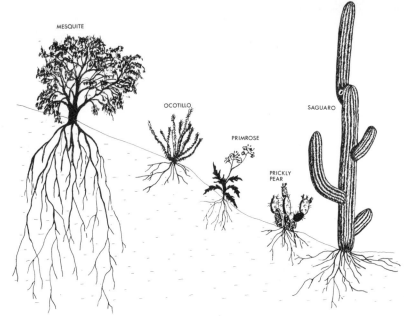

Figure 37.ii. Plants adapted to an arid environment: the mesquite is deep-rooted; the ocotillo sheds its small leaves during dry periods and flowers when it rains; the primrose is an annual; prickly pear and saguaro are succulents.

by the green stems. Still other plants store moisture and nutrients in bulbs and tubers. Some examples of adaptations are shown in Figure 37.ii. The shallow root systems of many plants allow them to make maximum use of light precipitation; the storage of moisture by succulents and tubers is shown; the mesquite of the USA has roots which can reach 30 metres below the ground surface.

In summary, their adaptations allow plants to overcome shortages of water, to reduce losses caused by transpiration, to avoid some of the heating effects of sunlight, to increase water storage, and to suspend life during unfavourable climatic conditions.

The Temperature Factor

The effect of climatic factors on plants has to be considered at the continental and the microclimatic scale. On the continental scale there is clearly a marked difference between the plants of the tundra and those

58. The leaves of the book-leafed mallee (*Eucalyptus kruseana*) are adapted to drought.
Photo by R. Woldendorp, Australian News and Information Bureau

of the tropical forests. These differences are mainly the result of temperature.

Biological activity is largely confined to the temperature range of 0–50°C, but each plant has its own optimum growth confined to a particular temperature range. Some temperatures at which growth begins may be listed:

Evergreen conifers of temperate areas	-3°C
Many Arctic plants	0°C
Peas, rye, wheat	-2° to $+5$°C
Date palms	15° to 18°C
Hot spring algae	77°C

In addition to the lowest temperatures at which growth can occur, there is an optimum range which is most conducive to particular functions. Some plants require rhythmic daily changes of temperature with cool nights and warm days—thermoperiodism. Tomatoes, for example, fruit best when the average daily temperature is 26·5°C and the night temperature is 17–19°C. Because of such requirements, figures of annual average temperature seldom have any ecological significance. An extreme continental climate may, for example, produce the same average annual temperature as a coastal climate but the plants will be totally different in these two areas.

Many fruit trees such as peaches, pears and apricots require a cold season in which they are dormant. Without this the fruit crops may be small or the tree can die. A similar response to cold is shown by many spring flowering bulbs which shoot only after a cold period.

Injury to plants by extreme temperatures includes freezing and death of tissues, or exposure of roots by frost heaving. Heat can cause sunscald or wilting. Plant temperatures are very close to those of the immediate environment so shade, soil moisture, slope and other factors are very important. Soil temperature is partly controlled by the albedo of the surface—thus dark-coloured soils warm up most rapidly—and also by moisture content. The wetter the soil, the slower its temperature changes, because about 5 times as much heat is required to raise the temperature of water in a pore space as would be required to raise the temperature of an equivalent volume of soil particles.

Insolation received on a slope is significant because a poleward slope of 5° can have the same effect on soil temperatures as a latitudinal poleward displacement of 480 km. The aspect, drainage, shade and other local factors can thus favour quite different plant communities on slopes, ridges and valley floors.

The Light Factor

Light makes up about half of the solar radiation reaching the earth's surface. It is of some importance in heating the environment but its chief significance is as the effective radiation in photosynthesis. Atmospheric gases, chiefly oxygen and nitrogen, absorb small quantities of light, so that areas at high altitudes receive more radiation than those at low altitudes, but of more ecological significance is the screening effect of moisture in the air. Light intensity is at a maximum in the dry tropics, and is severely reduced in all cloudy or foggy climates. On a cloudy day it may be reduced to 4% of the normal intensity. On clear days about 10% to 15% of total light may be diffused or sky light

Figure 37.iii. The ivy, I (*Hedera*), climbs using special roots; the clematis, C (*Clematis spp.*), has twining leaf tendrils.
After Polunin

(light scattered by gases and water droplets), but on overcast days it may be 100% of total light and thus no direct light would be received.

Latitude and aspect also affect light received. At low latitudes the light passes through only a thin layer of atmospheric gases, but at high latitudes the light rays cut obliquely through the atmosphere and are spread over a large area. Slope direction is probably not as important in affecting light as temperature, but it is none the less significant. The shady sides of many mountain valleys are avoided by cultivators and some poleward-facing slopes may never receive the noon sunlight, so that plants have to rely entirely on sky light.

At the equator daylight prevails for about 12 hours out of every 24 at all seasons, but towards the poles the length of day (photoperiod) becomes increasingly longer in the summer and shorter in the winter, until poleward of 66° effective midsummer daylight is 24 hours, and effective midwinter daylight is nil. It is because most plants flower or drop leaves in response to light, that tropical plants flower and fruit throughout the year, but Arctic plants flower only in the long days of Arctic summer.

Local conditions can often affect the amount of light received. In a clear stream as much as half the light may penetrate to the bed but in a muddy stream only 5% may penetrate. Under snow, light penetration may be sufficient for growth to begin before the spring melt. In industrial areas as much as 90% of the light may be cut off by pollutants and for many plants the remaining light may be insufficient for photosynthesis. In an evergreen rain forest with a dense canopy, only the

Figure 37.iv. A prickly lettuce as viewed from the east or west (a) and as viewed from north or south (b).

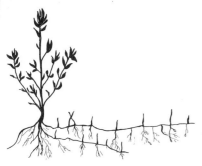

Figure 37.v. A black mangrove showing the pneumatophores on the roots.

tallest trees will receive full insolation, and the understories may never receive direct light and possibly as little as 1% of full sunlight. Few plants can survive under these conditions, and the ground remains bare of vegetation until a break in the canopy improves lighting conditions. Some plants such as cave-dwelling mosses and algae can effect photosynthesis under light intensities no greater than that of moonlight. Such plants are called sciophytes. Plants which grow best under intense light are called heliophytes. In moist climates the superimposed layers of the forest vegetation favour sciophytes except for the canopy species. In dry climates heliophytes abound. Because the light requirements of trees are most critical at the seedling stage, and those requirements vary from one type of tree to another, some plants succeed where others fail. On ground newly cleared by fire or on an abandoned field heliophytes colonise first and in their shade sciophytes develop. After a time quicker-growing heliophytes may shade out the first heliophytes and so dominate the community. This type of succession is described in more detail in Chapter 38.

Some plants have developed notable adaptations which enable them to seek light. The climbing plants are the best known. The ability of peas and runner beans to climb by turning themselves is well known; other climbers like the ivy have suckers or climbing roots which enable them to cling to smooth surfaces (Figure 37.iii). An unusual plant is the prickly lettuce. This weed grows so that its greatest spread of leaves is aligned north to south; so that viewed from north or south it is flat and viewed from east or west it is broad. This allows the plant to benefit from almost maximum sunlight for most of the day (Figure 37.iv).

The Atmospheric Factor

59. The buttressed trunk and knees characteristic of the bald cypress. (*Taxodium distichum*). *US Dept of Agriculture*

The earth's atmosphere envelops all plants and their root systems, providing them with carbon dioxide and water vapour for photosynthesis. The gaseous envelope of the earth decreases in density with altitude, and although this is perhaps of more significance to oxygen-breathing organisms than to plants, the reduced availability of carbon dioxide may be an important factor in the environment of alpine plants. Green plants release oxygen into the atmosphere and absorb carbon dioxide. Animals do the reverse, thus helping to maintain an atmospheric balance between these gases. Small soil animals release large volumes of carbon dioxide, and in some forests this may raise the carbon dioxide concentration near the ground to 6 times the average concentration, thus largely compensating for the low light intensity.

The release of carbon dioxide in the soil can reach toxic proportions unless a soil is well aerated, so this factor can be critical for plant growth. Other toxic substances in the air are derived from pollutants from industrial processes and domestic coal fires or motor vehicles. In some industrial areas, plants have been completely destroyed and a man-made desert formed (see Chapter 10). Some plants which grow in poorly aerated environments have special morphologic adaptations. Shallow root systems are quite common, but more noticeable are the 'breathing tubes', or pneumatophores, of the black mangrove, which stick above the water at high tide level, and some trees such as the bald cypress of Florida have knees which project above the level of high water (Figure 37.v and Plate 59).

60. Increasing exposure with altitude reduces the height of the vegetation.

The Wind Factor

Wind is most significant as an ecological factor on flat plains, along coasts, and at high altitudes. It affects plants by increasing transpiration, by mechanical damage and by scattering pollen and seeds.

Wind damage is most obvious on trees which are exposed to strong winds from a dominant direction. They may have a flag shape or be permanently bent away from the wind. Flag shapes are usually caused by wind-carried sand, ice or salt killing off the buds on the exposed side of the tree and leaving those on the protected side. In areas with winter snow the base of the trees may be protected in winter from the driven ice so that a basal 'krummholz' of bushy twisted branches is formed (Figure 37.vi). Near the tree-line, stunted and flattened trees are common. Wind may actually be a more significant control of the tree-line than any other factor in some mountains. Only shrubs and pliant grasses can grow in very exposed places. Desiccation of plants by the increased evaporation in moving air may be even more important in controlling the form of plants than actual damage. Cushion plants and dwarfed trees have a shape which is particularly suited to reduce evaporation and the specimens forming the outer individuals of many forests have this stunted form. Föhn winds may not be tolerated by trees, but grass will survive—a factor of significance in many prairies. A cover of trees and shrubs greatly reduces wind velocity near the ground. Shelter belts or wind breaks of trees are therefore of considerable value for protecting crops (see Chapter 10).

Most cool and cold climate trees and shrubs are pollinated by wind-carried pollen. This is a very inefficient process which requires the production of large quantities of pollen, but it is clearly successful. In areas with a constant wind direction, pollination is only readily possible downwind so colonisation can only take place in that direction. The shrubs of a forest understory living in a largely windless environment are seldom wind-pollinated types.

The Fire Factor

Fire is not an environmental factor of continuous importance, either in time or space. Natural fires started by lightning, or more rarely by volcanism, have always been environmental factors except in the very

Figure 37.vi. The flag-form and basal krummholz of wind-trimmed spruce trees near the tree-line.

cold and very wet regions. In the forests of the western USA, lightning causes hundreds of fires every year. Elsewhere, man has used fire as a most effective agent for controlling and altering the vegetation. Fires may be confined to accumulations of litter or peat—ground fires; they may sweep over the ground surface and through herbs and shrubs—surface fires; or they may travel through the tree canopy—crown fires. Ground and crown fires are usually disastrous, but some plants survive surface fires.

In environments with a marked dry season, some plants have become adapted to the fire factor. Some shrubs have seeds which remain dormant until their hard pods are cracked open by fire and seedlings can emerge. Some woody plants pass through the life cycle rapidly so that they produce cones or seed before sufficient ground litter has accumulated for another fire to start. Cork oaks and some pines have such thick barks that they are protected from all but the hottest fires. Some trees produce buds and woody tubers after fire destroys the old ones. Some pines like *Pinus contorta* have cones which are opened after fire and the seeds allowed to scatter.

Fires kill off many species but favour others. Thus fire-resistant species increase in abundance after a fire because of the reduced competition. Fires also open up the canopy, so that heliophytes colonise the ground after a fire and only later can sciophytes establish themselves. Many plant nutrients become locked up in litter which is too dry for bacteria and fungi to attack. Fire releases these minerals and makes them available to plants; hence occasional burning can raise fertility. Fires also burn off the dead litter on grazing lands and allow animals access to new green shoots. The destruction of litter, however, can result in increased runoff and erosion.

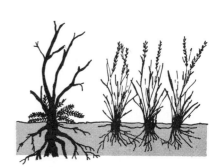

Figure 37.vii. (Left) In the absence of fire, mesquite chokes out the grass. (Right) Fire favours grass over mesquite.
After Odum

Carefully used, fire can be a most valuable tool for man. Many grasslands owe their continued freedom from forest to burning, and the grazing areas of many savannas have to be constantly maintained by fires. Many shrubs, like the sage-brush and mesquite of the semideserts of western North America, are only kept out of grazing land by burning. The shrubs are thus destroyed but the grasses survive, so that as long as grazing is not permitted until the grass has formed a complete cover, permanent pastures result.

The Biotic Factor

No plant is independent of the activity of other organisms. All depend upon fungi and bacteria to break down humus and release nutrients, and on bacteria or legumes to fix nitrogen. Many plants depend upon

249

insects for pollination, and upon birds and mammals to disperse seed. Even much of the carbon dioxide used in photosynthesis has been released in respiration by animals. One plant shaded by another, or any two plants competing for nutrients, water or space, are being affected by the biotic factor. There is, then, no such thing as an independent plant.

Herbivorous animals which graze on herbs or browse on shrubs and trees have a marked influence on vegetation. By eating those plants which they find most palatable, by preventing plants from reaching the seeding stage, by eating the photosynthetic organs, by trampling out some plants, and by depleting vegetation so that erosion can occur, animals can modify or destroy plant communities over vast areas. In areas where there is a reasonable balance between food supply and animal population, modification of plant communities depends only on the eating habits of the herbivores. Thus sheep prefer forbs, horses and cattle grasses, goats and deer broadleaf woody shrubs, so that flocks and herds usually leave an area depleted in the plants they prefer.

Constant defoliation by grazing animals causes the carbohydrate reserves of grasses to be depleted, so that there is a reduction in leaf and root size. This in turn reduces the intake of water and nutrients, making the grass more susceptible to drought, heat and frost. The tall grasses are the most seriously affected, so under heavy grazing there is always a tendency for flat-growing grasses and forbs to assume dominance.

Overgrazing, caused by man's carelessness or by his introduction of such animals as rabbits into areas where they have no natural predators, turned large areas of Australia and the South Island of New Zealand into deserts until poisoning and the disease of myxomatosis brought them under control. Rabbits eat herbs so close to the ground that the plants are completely destroyed. Regrowth in many areas has only been made possible by top-dressing large areas with seed and fertilisers in an effort to reduce erosion and once more make the land productive.

Cross-pollination of most colourful and odorous flowers is effected by flying insects, chiefly bees, moths, butterflies and flies. Some birds and bats are also effective pollen carriers. Most fruit trees are so dependent upon insects that orchardists keep hives of bees for the purpose. Red clover, which could not be successfully established in Australia until bumble-bees were also introduced, is but one example of the almost complete dependence of some plants upon pollen-carrying insects.

Animals are also very effective in dissemination. By eating fruits and excreting the seed, animals can spread some plants widely. Anyone who has bought old sewage waste for his garden and noticed the large number of tomato plants which spring up in it can testify to the effectiveness of this process. Other animals and birds carry sticky or hairy seeds which cling to them, and many spores are carried in like manner. Rodents and birds which store seeds are probably responsible for spreading many plants such as oak trees which could not migrate upslope unless the large and heavy acorns were carried there. Destruction of some predatory animals, such as foxes, can allow such an increase in rabbits and rodents that the increased grazing may destroy an ecological balance and prevent forest regeneration.

Some plants are entirely dependent upon others for providing support. The best known are the vines or lianas and epiphytes. Lianas are rooted in the ground and maintain a more or less erect position by using other plants for support. Although true lianas do not have a nutritional relationship with their supports, some do have a very close relationship. Those that twine themselves round the supporting tree can so restrict the growth of its trunk that the support may die. Other lianas, like the strangling fig, have such a luxuriant vegetation at the crown that they shade out and kill the support. Epiphytes are plants which grow entirely on other plants and have no contact with the soil. They are not parasitic upon their support but obtain their moisture from precipitation, and nutrients from dust, rainwater and decaying bark. Being susceptible to drought, they are most common in rain forests. In cold or dry climates they consist mostly of mosses and lichens, but in the tropics epiphytes like orchids can develop luxuriant foliage. Some epiphytes live on rocks and telegraph wires, and there are numerous other special adaptations found amongst them.

Parasitic plants, unlike the epiphytes, have structures, called haustoria, which penetrate the bark of the host and tap its supply of nutrients and water. A well known example is the mistletoe (*Viscum album*). The adaptations between a host and a parasite tend to become balanced, so that the parasite derives maximum benefit from the host without interfering with its life cycle.

Other well known relationships are the symbiotic relationship of nitrogen-fixing bacteria with the roots of legumes and the mycotrophic relationship of fungal mycelia with plant roots to form the compound structure called a mycorrhiza (see Chapter 22). The significance of man as a biological factor will not be discussed here, as Chapter 50 has been devoted to this subject.

Figure 37.viii. Haustoria of a mistletoe penetrating a branch of its host.

Limiting Factors

In the earliest studies of ecology many generalisations were made, based upon superficial observations and crude assumptions of cause and effect relationships. Modern ecology is increasingly based upon quantitative observations, and the complexity of plant and environment interrelationships is acknowledged. It is now realised that there is no fixed optimum intensity of any one factor but that for each change in one factor there is a change in the effectiveness of all other factors. Thus in a dry climate the lack of bumble-bees to pollinate red clover may be compensated for by smaller flowers which can be pollinated by smaller bees.

As one plant extends its range so that one factor of the environment controls its existence, that factor is said to be the limiting factor. The principle of limiting factors was enunciated by the nineteenth-century agricultural chemist Justus von Liebig, who saw that if a deficiency in any one factor limited crops, then that one factor limited all growth. Additions of nitrogen, for example, could eliminate nitrogen as a limiting factor, until further growth was limited by another factor—perhaps water. Because a surfeit of any one material could destroy the plant, Liebig spoke of the 'Law of the Minimum' and the 'Law of the Maximum'. Today we speak of limiting factors because at any stage there is an optimum balance of factors.

38 Plant Communities

At the autecological level individual plants are studied, but plants do not usually exist in nature on their own but as part of a community. The study of communities is often complex because the individuals within it act upon each other. The techniques of detailed study of communities are only just being worked out and, as is shown in the section on ecosystems (Chapter 39), ecologists have so far concentrated either on very simple communities or upon very generalised study of large communities.

A community is the basic vegetation unit of the ecologist. It may range in size from a mat of lichens on a boulder to a forest stand of 400 hectares or more, with the outstanding feature of a clearly defined uniformity of structure and floristic composition. It is rare for a community to have a definite boundary, and a transition, or ecotone, commonly occurs between one community and the next.

Mutual Relationships Among Organisms

Within a community there is a full range of mutual relationships from competition to parasitism. Some of these situations were discussed under the heading of *the biotic factor* in the previous chapter.

Competition occurs whenever two organisms require the same things in the same environment, and the intensity of the competition is determined by the amount by which the demand exceeds the supply. Competition is often most noticeable in stands of the same species because, being alike, the individual plants have identical requirements. In young, crowded forest stands this leads to the elimination of smaller or weaker saplings and so ensures a larger supply of required nutrients and light for the stronger survivors. For this reason, thinning of forest trees or weeding amongst crops usually pays with a higher yield. In entirely natural communities the number of individuals and species becomes adjusted through competitive demands to use the resources of the environment to near-capacity. This results in stable communities which are usually immune from invading species unless a disturbance breaks the stability. This stability provides a kind of mutual support for members of the community. A few plants such as the black walnut, which secretes poison from its roots and keeps other plants away, are well equipped for competition but this is not usual.

Dependence is a feature of most communities. Even in very simple or small stands there are usually dependent plants such as mosses and fungi, which rely upon the shade, moisture or other environmental conditions produced by the rest of the community. Epiphytes growing on the branches, trunks or foliage of larger plants are obvious examples. Less noticeable but equally important are the symbionts, parasites and saprophytes. Many bacteria and fungi live in a close symbiotic relationship with the roots of trees, so that both the host and the symbiont benefit. Parasites, however, reduce the vitality of the host, and many fungi cause diseases which can destroy the community, e.g. *Phytophthora infestans*, which caused the Irish potato blight in the late 1840's.

Saprophytic fungi living in the soil and litter are also dependent upon the community, although the rest of the community is dependent upon them for recycling the nutrients in the litter.

Structural Features of Communities

To study and record the features of a community completely, it would be necessary to list and map the position of every plant. For any community of size this is clearly impossible and mapping and sampling techniques are used which seek to record characteristic features and examples from the community (see Chapter 45). A valuable and useful record for comparative purposes can also be made by studying the vertical structure, or stratification, of a stand and the life-forms of the plants composing it.

Stratification in vegetation causes distinct layers each with its own assemblage of plants. It is useful then to categorise and speak of one of the following classes:

Dominant or overstory tree
Secondary or understory tree
Seedling tree or sapling
Tall shrub
Low shrub
Perennial herb
Annual herb
Moss or lichen

All types of communities, including those of deserts and grasslands, are stratified and on a miniature scale might be very complex. A tropical rain forest, with its epiphytes, lianas and multiple understories, is among the most complex types but even temperate forests or desert communities often have four or five layers. Each layer has its own features, and to many ecologists each is known as a synusia. A typical profile is shown in Figure 38.i.

Figure 38.i. Stratification of a woodland.

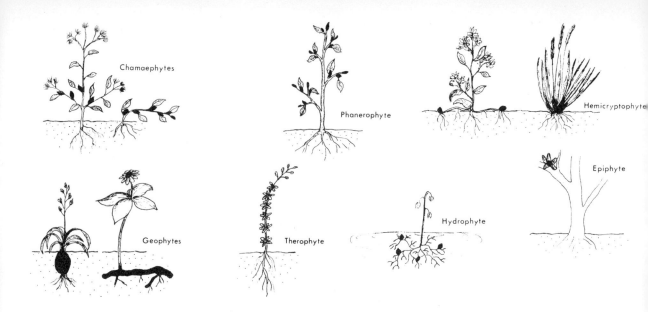

Chamaephytes

Phanerophyte

Hemicryptophyte

Geophytes

Therophyte

Hydrophyte

Epiphyte

Figure 38.ii. Characteristic life-forms according to Raunkiaer's classification. The perennating bud is shown in solid black.

Life-forms described along the lines of the profile in Figure 38.i may contain several forms in each category. The life-form of the dominant plants thus gives the essential character to a plant formation. One of the most obvious characteristics is the nature of the foliage, which may give strong indications of the type of environment to which each plant is adapted. Another feature is the number and nature of the shoots which the plant produces. Thus the forest tree has a woody stem with branches high above the ground. Shrubs are also woody but less lofty and have many stems. Herbs are soft-bodied, small and have a great variety of forms. Herbs may have perennating organs which spread through the soil, or surface runners. Trees and shrubs have their buds above the ground.

The position of the buds, which give rise to new shoots and thus carry on life from year to year, in relation to the ground was used by

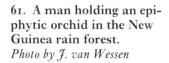

61. A man holding an epiphytic orchid in the New Guinea rain forest.
Photo by J. van Wessen

Professor Raunkiaer of Copenhagen as the basic principle for a classi-fication of plants by their life-form. The distinction is valuable because height of the perennating bud above the ground indicates the extent to which the bud is protected. In most climates there is a season of weather —either cold or dry—unfavourable to plant growth, in which the bud needs protection. Buds which are buried are protected from extremes of heat and drought, and those above ground are protected by bark or are exposed. Raunkiaer's classification has seven main categories (Figure 38.ii).

> *Phanerophytes (Ph)—exposed plants* have their buds high above the ground, where they are exposed to cold, drought and wind. This category includes virtually all trees and shrubs.
> *Chamaephytes (Ch)—ground plants* are herbaceous or low, woody plants with buds close to the soil, where they are protected by dead leaves or by snow. They are most common in dry and cold climates.
> *Hemicryptophytes (H)—half-hidden plants* undergo considerable growth during the favourable season, but at the end of it they die back to soil level, where the perennating bud is found. They are common in cold, moist climates. The grasses, herbs, rushes and sedges belong to this category.
> *Geophytes (G)—earth plants* die back below soil level because of cold or drought. The bud is then entirely protected. Bulbs, corms, tubers, rhizomes and rootstocks are characteristic of the plants in this category.
> *Therophytes (Th)—summer plants*, or annuals, produce seed and thus can remain inactive during unfavourable seasons. They occur in large numbers in deserts.
> *Epiphytes (E)* exist above ground level growing on the trunks or branches of woody plants.
> *Hydrophytes (HH)—water plants*, or aquatic plants, get their protection by immersing the bud in water.

> *Geophytes and hydrophytes may be grouped together under the term cryptophytes (i.e. hidden plants).

The relationships between abundance of life-form types and climatic regions is shown in Figure 38.iii. The very marked preponderance of hemicryptophytes in the tundra regions, of therophytes in deserts, and phanerophytes in the humid tropics is demonstrated. This type of analysis gives a biological spectrum, or life-form spectrum, for each region compared with that for the world as a whole.

Raunkiaer's scheme has a number of disadvantages for, although it is useful and accurate, it is highly selective of the features it uses. It does not, for example, take into account that in some parts of the world continuous growth of plants is possible, nor that the adverse season

Figure 38.iii. Life-form spectra diagrams for each of the major climatic zones.

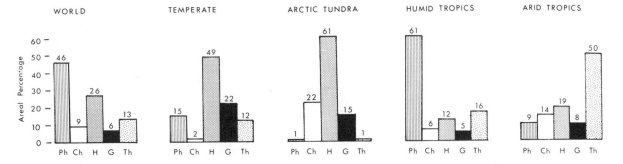

62. The camera rests on a patch of scab weed which is one of the early colonisers of eroded soil in central Otago, New Zealand.

may be due to either cold or drought or both. As a result, he placed in the same vegetative form the trees of the equable humid tropics and those of temperate areas with a marked cold season. Raunkiaer's scheme is based solely on the number of species—not the number of plants—so it takes no account of the rarity or abundance of particular species, nor of the dominant species of a community and the effect it may have as a competitor, preventing other plants from establishing themselves. In spite of these disadvantages the scheme is of considerable value to biogeographers and ecologists because it does help to establish relationships between life-forms and climatic zones.

Successions

No plant community is stable for ever because environmental conditions are not stable. Furthermore, plants spread on to newly available areas whenever they can, so the concept of a succession of plants and communities becomes an important one in community studies.

On bare rock newly exposed by processes such as mass movement or recent volcanism, the first colonisers are algae and lichens. These are followed by mosses, grasses, shrubs, and then small trees. This is a primary succession. It is a feature of such successions that the vegetation becomes more complex with time as lower life-forms create suitable conditions for more highly evolved forms, and open conditions are replaced by closed conditions. Such successions involve the creation of soil and general improvement of the environment for higher plants, and therefore require long periods of time for their development.

The whole group of plant communities which successively occupy the same site from the pioneer to the final stage is called a sere. The relatively transitory communities which occupy the site at a given time are seral stages, and the final plant community which then perpetuates itself indefinitely is the climax. The complete sere from bare ground to the climax community is called a primary sere or prisere.

By contrast, secondary successions are rapid because they occur in a soil which is already formed. After a fire, a killing drought or cultivation, plants recolonise bare soil rapidly and a succession of plants to the highest life-forms can occur. The time required for attaining a climax is extremely variable. In a humid temperate environment, at least

63. A primary sere: in the foreground a floodplain is being colonised by mosses, grasses and low shrubs; behind on the right, near the man, tall shrubs occupy the older first terrace and to the left of them mature trees occupy the higher terrace.

1000 years may be required for the development of a primary forest climax, although a secondary succession on cut-over or abandoned agricultural land may require only 200 years to reach the climax. In the plains of the central and western USA, the climax grass vegetation may re-establish itself in 20–40 years after land abandonment.

The principles of succession, and systematic development of seres were first studied in Denmark and Lake Michigan, USA during the first few years of the twentieth century. The study began a new era in ecology. Lake Michigan was a particularly suitable site because the lake was once much larger than now, and as it shrank it left chains of sand dunes with the oldest dunes furthest inland, and the youngest near the present lake shore (Figure 38.iv). In the climax forest there is a deep, humus-rich soil while towards the lake the soil is progressively less mature. Each dune chain also has an associated community of animals.

According to the environmental conditions, seres can be classified thus: hydroseres are initiated in fresh water; haloseres are initiated in saline water; dry places give rise to xeroseres, of which lithoseres are initiated on bare rock and psammoseres are initiated on dry sand.

Figure 38.iv. Ecological succession of plants on the Lake Michigan dunes. On the youngest dunes cotton-woods (C) and beach grass (BG) are established on dry sandy soils, and on the older dunes jack pine (JP), black oak (BO), oak and hickory (OH), until on the oldest dunes climax beech and maple (BM) forest occurs on a deep, humus-rich soil.
After Shelford

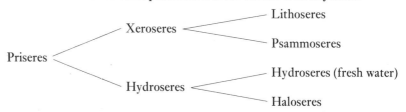

Examples of such seres are shown in Figure 38.vi.

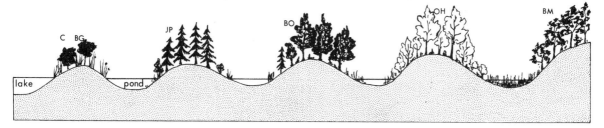

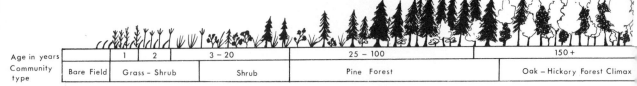

Age in years		1	2	3 – 20	25 – 100	150 +
Community type	Bare Field	Grass – Shrub		Shrub	Pine Forest	Oak – Hickory Forest Climax

Figure 38.v. Secondary succession on the piedmont region of the southeastern USA on an abandoned cornfield.
After Odum

A major feature of the successions so far described is competition between plants for nutrients and space. This is most noticeable in the very rapid successions which occur in the humid tropics, where an abandoned field may be occupied by a jungle within 4–5 years, but in the tundra and the deserts competition is not necessarily an obvious feature of existence. Went, in particular, claims that desert plants share the available water and nutrients, although he comments that he is mainly talking of annuals which have little time for competition. Some desert plants like the creosote bush are such effective competitors that they appear able to excrete a toxic substance which prevents other plants from growing within the range of their roots.

64. A field in New York State 14–15 years after being abandoned.
Photo by E. M. Stokes

Climaxes

Because the state of equilibrium implied in the concept of the climax can be disturbed, and a retrogression or complete change of vegetation occur, some authorities have denied the whole concept. In spite of this denial, it is clear that many forests, like the Amazon rain forest or the coniferous forests of northern unglaciated Russia, can retain their characteristic composition for long periods of time, and that within them there is a maximum utilisation of resources by plants and animals, which makes the climax idea useful. At its most extreme, the idea implies that any area of given uniform climate will have a uniform climax vegetation, after a period of time, which is called the climatic climax or, preferably, the monoclimax. More frequently the climax is modified by edaphic or topographic factors which affect the vegetation, so that a number of different stable associations exist in the same area; such a situation is called a polyclimax. A further distinction can be made to indicate the arrest of a natural seral development by such interventions as grazing by animals or mowing. In these cases the grazed and mown areas develop stable communities which may be called plagioclimaxes (by English ecologists) or disclimaxes (by American ecologists). The short sere which leads to a plagioclimax may be called a plagiosere.

Figure 38.vi. Priseres: (A) lithosere; (B) psammosere on sand dunes in New Zealand (soil begins to form in the marram grass area); (C) fresh water hydrosere.

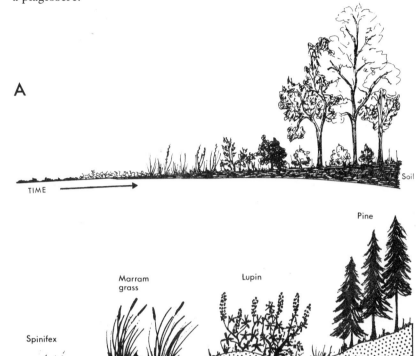

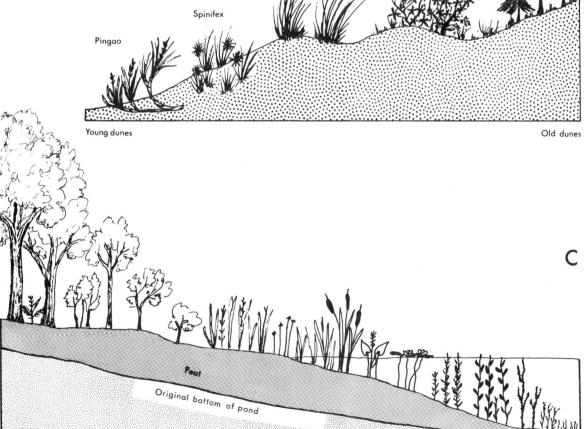

In brief, the chief characteristics of an ecological succession are: an orderly and continuous process of change in the plant and animal components of the succession until the climax is reached; a progressive modification of the physical environment by the plant communities—notably microclimate and soil; culmination in a steady state or stable ecosystem. In the early stages of a successional sequence, the energy fixed by the plant community exceeds the losses from the system and the energy accumulates in the plants and animals, but at the climax the energy within the system is almost constant (Figure 38.vii).

The final stability is extremely important to man because at that stage the community of plants has its maximum effect upon the physical environment, providing the greatest resistance to changes in micro-climate, or modifications by storms and floods, etc. The actual productivity of the community is greatest in the young stages of the succession, so to provide an environment which has the advantage of stability and productivity man should aim at forming a mixture of climax and young communities. In many parts of rural Europe a long-term stability has been achieved in the countryside, where cropping, grazing and old and young woodlots are all components of the landscape. Here natural calamities are buffered, and production is high. By contrast, large scale destruction of the original vegetation in the new areas of the Great Plains of the USA and the grasslands of southeast Russia has been followed by rapid soil erosion in periods of drought.

Figure 38.vii. Stored energy within an ecosystem is dependent upon the stage in a succession. It builds up to a steady state at the climax. *After Woodwell*

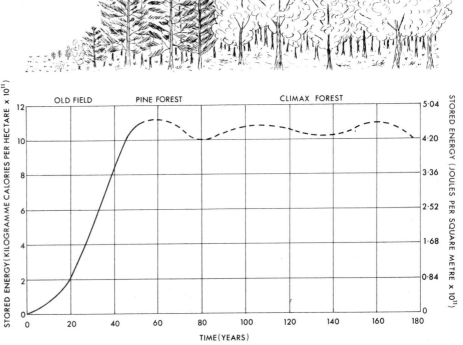

39 Ecosystems

The term *ecosystem* was proposed by Tansley in 1935 to describe the whole complex of plant and animal organisms naturally living together as a sociological unit together with their habitat. More recently Fosberg had developed the definition:

> An ecosystem is a functioning interacting system composed of one or more living organisms and their effective environment, both physical and biological. . . . The description of an ecosystem may include its spatial relations; inventories of its physical features, its habitats and ecological niches, its organisms, and its basic reserves of matter and energy; its patterns of circulation of matter and energy; the nature of its income (or input) of matter and energy; and the behaviour or trend of its entropy level.

The importance of the ecosystem concept, and the characteristic which makes it very different from the vague terms such as 'physical region' formerly used by geographers and others, is that it has four main properties which make it an ideal framework for research:

(1) It is *monistic*, that is, it brings together man and the plant and animal worlds in their environment so that the interactions amongst them can be analysed.

(2) It is *structured* in a rational way so that the pattern of the components can be determined (see Figure 39.iv, which shows a pattern of feeding relationships in a qualitative manner).

(3) Ecosystems *function*, that is, they involve the continuous throughput of matter and energy which can be measured, so that the efficiency of the ecosystem can be assessed.

(4) They have the characteristics of all *open systems* in that they always tend towards a steady state, and any interruption of this tendency will cause a reaction in the components of the system such that there will be a return to equilibrium conditions in which the interrupting component will be involved.

An ecosystem can be of any size, from a goldfish bowl to the Amazon rain forest. What is common to such varied systems is a relationship between every part of the system and all of the other components. Thus solar radiation provides the energy of wind and rain. It gives the energy used by plants in photosynthesis and the created plant tissue is the basic food for the animal population. The decaying plant and animal remains are incorporated into the soil and the nutrients they release are taken up again by plants. This transmission of energy through the ecosystem can be measured.

Energy

Every organism can be regarded as having a place in a hierarchy of energy exchange which is called a trophic level. The first level is that of the primary producers, the green plants which, by photosynthesis, use the sun's energy to create protoplasm; the second, third and fourth

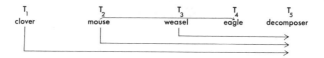

Figure 39.i. A food chain
showing 5 trophic levels. Any
of the intermediate links
may be by-passed.

levels are those of the consumers, the herbivorous and carnivorous
animals; the fifth level is that of the saprovores or decomposers. At the
first trophic level (T_1), part of the received energy is converted into
plant tissue and part is lost in respiration, thus at the T_2 level there is
less energy available for the creation of animal tissue than was originally
received in the system. The loss of energy at the first trophic level is
very high and may be as much as 99% but at succeeding levels it is
usually about 90% (this is another way of saying that ecological
efficiency is about 10%), so the length of any conversion system or
food chain is limited to about 5 links or levels (Figure 39.i). If, for
example, 1500 kcal (6300 kilojoules) of light energy per square metre
per day were absorbed by green plants, about 15 kcal (63 kJ) would end
up as net plant production. The animal that ate the plants—a herbivore

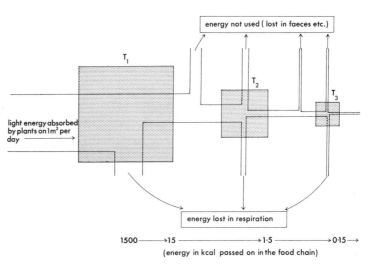

Figure 39.ii. An energy flow
diagram. The boxes represent
the energy absorbed by
organisms at each trophic
level and the figures and
paths show the energy losses
and the direction of energy
flow. The loss between the
first and second trophic
levels is usually higher than
losses between other levels.

—would produce tissue storing only 1·5 kcal (6·3 kJ), and the carnivore
which ate the herbivore would convert only 0·15 kcal (0·63 kJ) (Figure
39.ii). Furthermore, this would only be true if there were organisms
present which could fully convert the available energy; if there were
not, then the saprovores would take off energy at any place in the chain
(Figure 39.i).

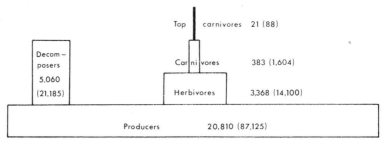

Figure 39.iii. An energy
pyramid for Silver Springs,
Florida.
After Odum

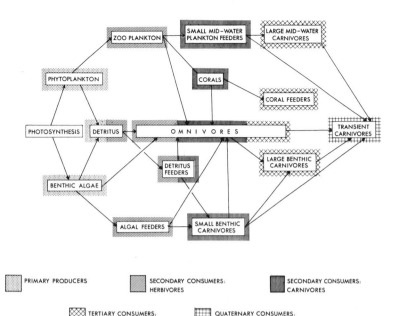

Figure 39.iv. A food web of coral reefs in the Marshall Islands, showing the trophic structure in a qualitative manner.
After Hiatt and Strasburg

A food chain is seldom as simple as it seems, for not only can steps such as T_3 be omitted from the sequence but the decomposers are active at every level. If the micro-organisms are taken into account as well as the great variety of possible energy exchange in any system, a food web rather than a food chain becomes a more useful idea and the concept of trophic levels is best illustrated by a biomass or energy pyramid. Such a pyramid has a broad base representing high energy at the consumer level, narrowing to an apex as energy is lost during respiration (Figure 39.iii). A food web of coral reefs in the Marshall Islands is shown in Figure 39.iv.

For study purposes small ecosystems such as those of ponds or small islands are often chosen because the number of components is relatively small. Even an area of under 3 square kilometres may have

Table 39.1. *Populations of Breeding Animals on 2·6 Square Kilometres, Santa Rit Range Reserve, Arizona (After Leopold)*

Number	Species
1	Coyote
2	Horned owl ⎱ predators
2	Redtail hawk ⎰
10	Blacktail jack rabbit
15	Hognosed and spotted skunks
20	Roadrunner
25	Cattle (over 1 year old)
25	Scaled quail
25	Cottontail rabbits
45	Allen's jack rabbits
75	Gambel quail
1,280	Kangaroo rat (*Dipodomys*)
6,400	Wood rat (*Neotoma*)
17,948	Mice, spermophiles and other rodents

far too many components for study—as is shown by Table 39.1, which disregards the numerous producers and decomposers.

Biogeochemical Cycles

The energy within an ecosystem is renewable but the plant nutrients are not. These nutrients are derived from the weathering of rock and the creation of soil by the action of organisms. Such transformations are exceedingly slow, with the result that unless nutrients are constantly recycled within the system by the return of decaying plant and animal remains to the soil, there is a decline in fertility. Under natural conditions there are losses of nutrients by leaching and erosion, but in a mature or climax community this is balanced by release of fresh materials through weathering. When man interferes with an ecosystem and removes a crop, he takes away a large proportion of the nutrients, which have to be replaced by fertilisers if productivity is to be maintained.

The main nutrients which account for 95% of the mass of all protoplasm are oxygen, carbon, hydrogen, nitrogen and phosphorus. Of these oxygen is readily available, but nitrogen is not readily available and has to be converted into usable forms by organisms which are important agents in the cycling of nitrogen (see Chapters 22 and 24). Carbon also has to be cycled, as it is absorbed from the atmosphere by plants and released by animals in respiration. There is not likely to be a shortage of these nutrients, but there could well become an alarming shortage of phosphorus. This element accounts for only about 0·1% of terrestrial matter, but it accounts for more than 1% of the weight of the human body and it is an essential component of all known forms of life. Each year about 3·6 million metric tonnes of phosphorus are washed from the land and precipitated in the seas, but unlike carbon, nitrogen and oxygen it is not recycled into the atmosphere but is deposited in marine sediments. A small proportion—possibly 3%—is recovered in the guano deposited by seabirds and a little more in fish, but the vast majority of phosphorus is lost to the land. Agriculturalists make up the deficiency by applying phosphorus fertilisers manufactured from phosphate rock, but the supply of this is shrinking. It may well be that phosphorus is one of the ultimate limiting factors in the world ecosystem.

Ecosystem Productivity and Cropping

The amount of living matter within an ecosystem is known as the standing crop, whether it be in animal or vegetable form. The amount of the standing crop can be expressed in terms of the biomass or organism mass per unit area. The units used are variable but the value of the concept is that it gives a basis of comparison between the

Table 39.2. *Production in Various Systems (After Simmons)*

Crop	Place	Production in grammes/m²/yr
Shrubs and grasses	Nevada Desert	40
Marine organisms	Sargasso Sea	92·5
Maize	Canada	790
Pine forest	England	3180
Algal culture	Tokyo	4530
Sugar cane	Hawaii	6700

Table 39.3. World Distribution of Primary Production (After Odum)

Type of vegetation or land use	Production in grammes/m²/yr
Oceans and deserts	150
Dry grasslands, mountain forests, continental shelf waters, poor agriculture	200–1000
Humid forests, grasslands and most agriculture	1000–3500
Best intense agriculture	3500–7000

productivity of one area and that of another. From Tables 39.2 and 39.3 it can be seen that productivity varies greatly. In the deserts and in cultivated areas the space between plants decreases productivity. It is estimated that as little as 0·2% of the light of suitable wavelength received on the earth is actually utilised in the production of a standing crop.

On a world basis there is an enormous disparity in the production rates of different ecosystems. The open oceans and deserts of the continents, which make up about 80% of the earth's surface, are extremely poor producers. The drier grasslands and poorer agricultural systems such as dry farming and extensive grazing areas are also poor producers; only the humid grasslands, forests and better agricultural lands are good producers, but these latter types are limited to relatively small areas of the earth's surface.

Traditional methods of producing animal protein are often highly inefficient; thus in Europe open range beef cattle consume only about 1/7 of the total primary production of grassland pastures and the remaining 6/7 are lost to other herbivores and decomposers. The methods commonly called factory farming seek to reduce such losses. The maximum use of food energy available to livestock is one way of raising production of the animal protein traditionally eaten by man.

A second principle in raising output is to take the crop not at maturity but at the age of maximum growth efficiency. Young chickens and beef cattle, for example, have a gross growth efficiency of the order of 35% compared with a mean value of 4–5% for their lives as a whole. To raise livestock beyond the age of maximum growth efficiency, which usually occurs prior to the age of reproduction, is wasteful in terms of food production for the use of man. Hens, for example, do not lay eggs until they are 5–6 months old but their gross growth efficiency declines markedly after an age of 3–4 months, so to raise hens for egg production is wasteful. Similarly, from a given amount of hay rabbits produce the same quantity of meat as beef cattle but they do so 4 times as quickly and are therefore more efficient producers of meat than are cattle.

The African Savannas

It is commonly assumed that European-style agriculture is the most productive way of using land for food production. This assumption was not tested until quantitative ecological studies became possible. Table 39.4 shows the wild mammalian biomass (mass of animals per unit area) for a number of contrasting environments. It is clear from this and other evidence that the semiarid marginal lands of east Africa can produce a high biomass per unit area and that the wild animal population not only has a higher rate of animal increase and output than domestic cattle, but that the financial return by way of meat, skins, and other animal products is much higher than if the areas were opened to domestic cattle. Controlled game-cropping also has the advantage that it prevents soil erosion and vegetation depletion and provides a valuable basis for a tourist industry. In some parts of the South African Republic, selected species of wild animals have been reintroduced to some of the drier areas, where they provide a higher income than that derived from domestic cattle ranching.

Table 39.4. Wild Mammalian Biomass (After Huxley)

Area	Biomass (*in metric tonnes/km²*)
Ruanda-Ruchuru Plain, Congo	24·4
Queen Elizabeth National Park, Uganda	19·5
Nairobi National Park, Kenya	15·6
Serengeti Plains, Tanganyika	5·2
Henderson Ranch, Rhodesia	4·9
Kruger National Park, Transvaal	1·8
Scottish deer forest (red deer)	1·0
Canadian barrens (caribou)	0·8
Southern Russia (saiga antelope)	0·35
Rain forest, Ghana	0·1

The reason for the superior biological efficiency of wild game over domestic cattle, in some circumstances, is that the wild species are part of an ecological community that has reached a high pitch of efficient adaptation through natural selection over a long period of time. In the typical east African savanna the ecological community is highly specialised and differentiated. The animal sector may contain as many as twenty species of herbivore, each occupying a slightly different niche, for it is an ecological axiom that two species in the same habitat do not occupy the same niche.

The largest of the herbivores is the elephant, which fills many roles. It is a path-maker in the forests; a plough in the scrub, where it pushes over trees and exposes the soil and so promotes aeration; a digger of ponds in dry water-courses and a browser which keeps trees and shrubs

65. White-bearded gnu and zebra in the Ngorongoro Crater occupy different niches in the environment. *Photo by Noel Simon, the World Wildlife Fund*

in condition for many species of antelope and buffalo. The ungulates range from the large eland down in a size scale through the buffalo, wildebeest, zebra, impala and the rest to the tiny dik-dik antelope. Not only does each species have its own niche, with giraffes feeding off the tallest trees and the dik-dik off low grasses, but the many operations of path-making, pan-making, soil aeration, seed dispersal, keeping water-channels open (the hippopotamus) and the fertilisation of waters (hippopotamus, waterbuck and lechwe antelope), are also completed. The more complex the niche structure in a biological community, the more efficient the conversion cycle, as a larger proportion of the incoming energy is used.

In contrast, domestic cattle occupy only one niche and therefore leave large sections of the plant communities ungrazed while destroying their preferred food plants by overgrazing. The result is not only low production, but the accumulation of litter of ungrazed plants and hence a greater danger of fire; serious loss of balance in the soil flora and fauna; slow recycling of nutrients; a liability to erosion; and a greater chance of disease in the domestic cattle, which lack the immunity of the wild game species.

The Future

It has been estimated that the amount of all protoplasm that can be produced on earth each year amounts to about 417,000 million metric tonnes, of which 295,000 million tonnes represent plant growth and the remaining 122,000 million tonnes represent all of the consumer organisms. This sets a limit to the productivity of the biosphere and the amount of life on earth. Whilst it is true that such production is far greater than present needs, the human population explosion is so great that man must rapidly develop an understanding of terrestrial ecosystems if he is to prevent destruction of food-producing resources. The aim must be to establish and maintain a balanced and productive ecosystem in which there is a cycling of resources to produce a constant high yield rather than the robber or destructive economy of many cash cropping systems.

The best hope for man is to manipulate ecosystems so that they use a higher proportion of sunlight, and produce a higher proportion of edible food. To do this he must reduce physical limiting factors, increase the season of growth to make use of a higher proportion of the light energy available during the year, prevent loss of nutrients from the biogeochemical cycles, and shorten the food chains so that there is less loss of energy within them before man himself takes his share.

40 The Dispersal and Migration of Plants

Plants tend to form associations or 'communities' which can advance or retreat as wholes, but individual plants and animals can also move independently. Furthermore, once established in a new place, plants can behave as though they have no ability to migrate. On Hawaii, for example, the endemic flora and fauna are of a kind which has come from many different sources across considerable expanses of ocean—there are, for instance, no flightless vertebrates. Yet many plants have radiated amongst the islands and evolved different species on adjacent islands which are separated only by narrow water gaps.

The effectiveness of dispersal across wide ocean gaps is obvious, for endemic land birds, land molluscs, insects and plants are found on every oceanic island capable of supporting them, so that unless fantastic and geologically impossible land connections are to be imagined to every oceanic island, dispersal across oceans must be continuously effective. It appears that ocean gaps are most effective in stopping dispersal of terrestrial mammals, even though other barriers like mountains, deserts and cold climates are passed by such mammals. This barrier is not absolute, however, for a few terrestrial mammals have reached islands—the Galapagos Islands, for example—presumably by rafting on floating trees, etc. The effectiveness of a gap seems to be directly related to its width.

Other effective controls on dispersal as well as on evolution and the number of species in a given place are the area and climate of that place. The number of species decreases in relation to a decrease in area. Thus in the West Indies division of the area of an island by 10 divides the number of species of amphibia and reptiles by 2, with the larger species being eliminated first. The observed tendency for animals and plants to disperse from large to small areas, more than the reverse, is shown by the general trend to move from the northern to the southern continents and from continents to islands. It seems probable that the greater number of individuals in larger areas allows advancement, through natural selection and competition, to a higher stage of perfection. This seems largely to account for the general trend to disperse from Eurasia to North America and from North America to South America, with little reverse traffic.

A favourable climate also encourages the development of a large number of species. For example, in the northern hemisphere there is a reduction in the number of species of trees towards the northern limit of forest, and in the southern hemisphere the Amazonian rain forest probably includes at least 2500 species of trees, but there are only 6 in Tierra del Fuego and none on Antarctica. Dominant forms seem to develop in favourable, usually tropical, climates and to disperse from there. That is, they develop where competition is greatest and disperse to areas of less competition.

The Ability of Plants to Disperse and Migrate

Dispersal of a plant involves dissemination from the parent and distribution to a new place. Migration involves successful dispersal and establishment (that is, ecesis). The distinction is important because only a small proportion of the disseminules of most plants actually establish themselves. Nearly all plants manufacture far more disseminules than will survive to produce mature plants. Some plants have two or more means of dispersal: thus seeds and suckers are commonly produced by many trees. The possession of what appear to be most effective means of dispersal does not, however, mean that a plant is widely distributed, or that a plant with poor means of dispersal is of limited distribution; nor does it mean that a plant will always be dispersed by the use of its special mechanisms. The most significant agents of dispersal are discussed in the following sections.

Figure 40.i. Adaptations for dispersal: (A) the parachute of a dandelion (*Taraxacum*); (B) the flattened wings of the maple (*Acer*); (C) the fibrous outer husk of the coconut (*Cocos nucifera*) encloses air and enables it to float; (D) the water hyacinth (*Eichhornia crassipes*) whose bulbous leaf-stalks keep it afloat; (E) the germinating seedling of a mangrove (*Rhizophora*) projecting from a fruit on the tree—detached seeds can float for many weeks; (F) fruit of the strawberry (*Fragaria*) showing resistant pips (seeds); (G) hooks on the fruit of *Triumfetta*; (H) rhizome of marram grass (*Ammophila arenaria*); (I) strawberry (*Fragaria*) plant developing on a runner.
After Polunin

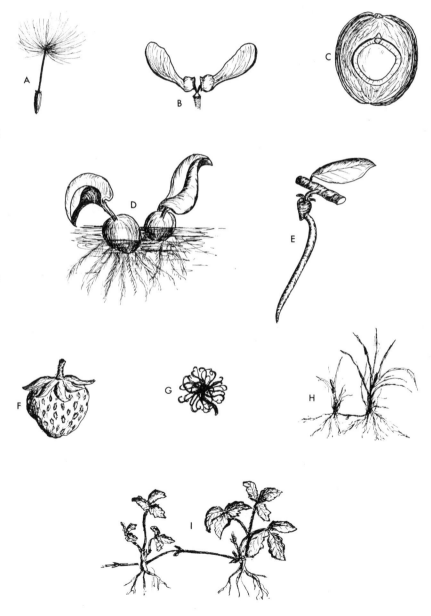

Wind

Many plants are adapted to use the wind as an agent. The common dandelion (*Taraxacum spp.*), with its fruits suspended from a parachute of fine fibres, is a good example. Other plants, like the sycamore, maple and many conifers, have winged seeds which are liberated well above the ground and can be widely dispersed in high winds. Pieces of the plant may also be dispersed by the wind: many epiphytes like Spanish-moss and the desert tumbleweed spread in this manner (Figure 40.i).

Many light seeds, and more particularly spores, can be carried to great heights by rising air currents and distributed over enormous distances. Thus spores have been found near the North Pole and elsewhere, far remote from their place of origin. Dispersal by such means is extremely wasteful and to compensate for this the production of disseminules is prolific. The giant puffballs, for example, can produce 10,000,000 spores and many mushrooms produce over 1,000,000. Another safeguard is the prolonged life of some seeds and spores, so that they may remain inert for many years before germinating.

Water

Almost any disseminule which has sufficient impermeability and buoyancy to remain on the water surface without being damaged can be dispersed by moving water. Some waterweeds and seaweeds can drift submerged and others like the water hyacinth float on the surface. The water hyacinth can reproduce so rapidly that it has choked formerly navigable waters in Africa and North America and has become a pest which is expensive to remove. Mangroves (species of both *Rhizophora* and *Avicennia*) with seedlings which float easily, and coconuts which have fibrous outer husks with trapped air giving buoyancy, are two coastal plants which are easily dispersed by ocean currents over great distances. The fringing of many tropical coasts indicates the ease with which the coconut palm is dispersed.

Streams and floods frequently carry disseminules, and it is noticeable that migration often takes place rapidly along a stream whilst plants spreading by other means are much slower. Uprooted trees may carry not only their own disseminules, but, in the soil trapped in their roots, those of other plants. Similarly, seeds in the crops of dead floating birds can survive long journeys.

Animals and Man

Animals and birds carry disseminules either adhering to their fur, hair, skin or feathers by means of hooks or sticky surfaces—such as the hooked burrs of the burdock (*Arctium spp.*)—or internally. Internally carried types are usually fruits like wild strawberries and peaches with a pleasant fleshy exterior and seeds resistant to enzyme attack in the interior of the animal.

Birds are also notable carriers of seeds on their feet. Darwin described how he removed 180 grammes of clay from the foot of one partridge and grew 82 plants from it. Macquarie Island, 800 km south of New Zealand, has 35 species of vascular plants all of which are thought to have been carried there by birds since the last glaciation.

Man is, of course, the most effective agent of dispersal, either deliberately by introducing new species, or unwittingly by carrying them in his ships, baggage, garbage, clothing and shoes, etc. In a few cases the indigenous vegetation has been largely replaced by the new arrivals—as in Ceylon and Hawaii. More commonly, new arrivals only colonise ground left bare after cultivation or fire. An interesting example is that quoted by Polunin of the Oxford ragwort. This weed was introduced into the University Botanic Gardens in the late seventeenth century but scarcely spread at all until the nineteenth century, when it migrated along the new railways. It is now found in many parts of Britain and even in western Europe. A new and potentially potent agent of dispersal is the aeroplane which can easily introduce spores and seeds into a new area. For this reason many countries insist on spraying aircraft arriving from countries with infectious diseases.

Mechanical Dispersal

This includes a number of powerful means of spreading. Some plants have seed pods which open explosively to scatter their contents in the wind—like some mistletoes and the rubber tree (*Hevea brasiliensis*). Other plants have overground runners and underground rhizomes or suckers. The elm tree has been known to put up suckers 46 metres from the parent tree.

Barriers to Dispersal

The barriers to dispersal are sheets of water, deserts, mountains, forests and the numerous environmental factors—mainly climatic, edaphic and biotic—which may prevent successful growth. In brief, any large area of uniform characteristics inimical to the plant can be a barrier, but the prolific production of disseminules and the numerous means of effective dispersal ensure that few areas of the earth are entirely free of plant life.

Good examples of the effect of distance are afforded by Easter Island and Krakatoa. Easter Island is about 3200 km west of South America in an isolated part of the Pacific Ocean. When Captain Cook visited it in 1774 it had only 50 species of flowering plants, ferns and mosses, 4 insects and 1 snail, and no other animals. Krakatoa, on the other hand, is a volcanic island in the Sunda Strait between Java and Sumatra, only 40 km from the nearest island. It was sterilised by the eruption of 1883, but within 3 years 11 ferns and 15 flowering plants had established themselves; after 13 years coconut trees, sugar cane and other shrubs covered the ground; after 25 years there were 263 species of animals—including 16 birds, 2 reptiles, and 4 land snails. In 50 years the whole island was again covered with dense forest containing 47 vertebrates—of these 60% drifted by sea, 32% came on the wind, and 8% on birds. If Krakatoa had been a remote island such a plant cover would have taken millions of years to establish itself. Krakatoa also provides a good example of the rate at which a primary sere develops in a humid tropical climate.

It is a characteristic of remoteness, therefore, for the flora and especially the fauna to be impoverished. For this reason the conditions in the southern hemisphere are of particular interest.

41 Biogeographical Relations between the Southern Continents

The distribution of many plants and animals in the southern extremities of South America, southeast Australia with Tasmania, and New Zealand is an old and outstanding problem. Darwin and Hooker argued about it and the discussion still continues. The problem is that there are close relationships between many of the plants and animals of these areas which imply that their ancestors must have been in genetic, if not geographical, contact yet there is an 8000 km ocean gap between Chile and New Zealand (Figure 41.i).

These southern lands have many features in common. On their southwest coasts there is a marked zonation of vegetation with belts of forest and mountain vegetation running north and south parallel to the coast. This is the result of the cool temperate oceanic climate with westerly winds blowing on to the mountain ranges. Another similarity is the narrowness of the land masses. In many respects, then, the environment of these areas is similar, and it may be reasonably assumed that before the Pleistocene glaciation parts of the Antarctic peninsula also had a similar climate.

Figure 41.i. Related vegetation in the southern hemisphere.
After Holdgate

On facing page :
Figure 41.ii. Distributions of (A) the northern or true beeches (*Fagus*) and southern beeches (*Nothofagus*); (B) carabid beetles of the tribe Migadopini; (C) bugs of the family Peloridiidae. These carabid beetles and bugs occur only in the southern hemisphere.
After P. J. Darlington

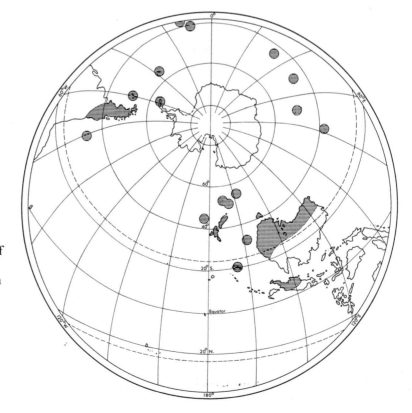

PRINCIPAL AREAS WITH PRESENT OR FOSSIL-RELATED
'SOUTHERN TEMPERATE' VEGETATION SHOWN THUS:

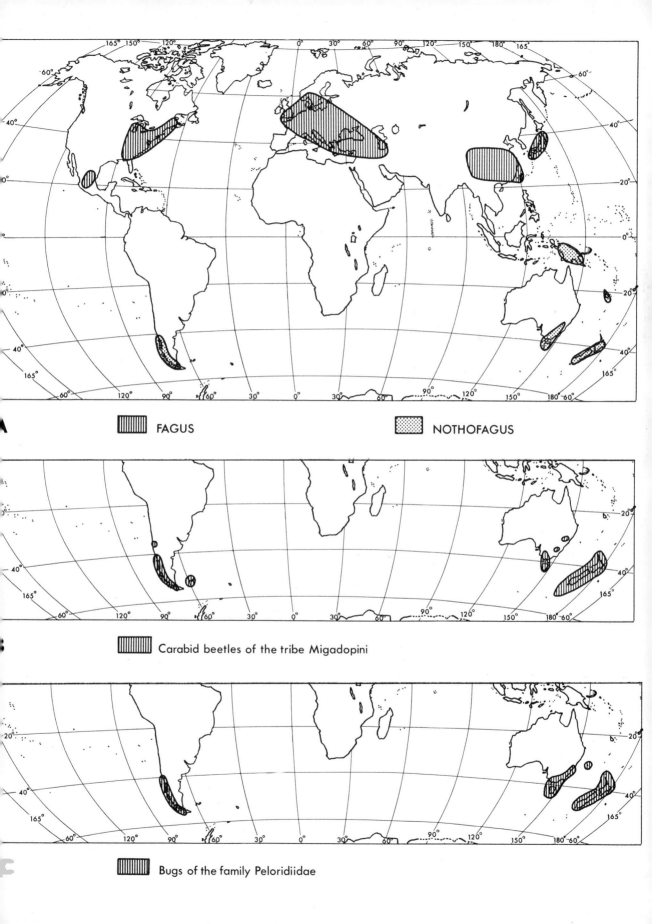

A   ▥ FAGUS   ▦ NOTHOFAGUS

B   ▥ Carabid beetles of the tribe Migadopini

C   ▥ Bugs of the family Peloridiidae

Many spiders, flies, small fish, seaweeds and coastal shellfish and plants have a wide distribution in these southern lands, but the outstanding examples are the southern beech trees of the genus *Nothofagus*. *Nothofagus* is related to the beeches of the northern hemisphere (*Fagus*), but it is only known in the north from its fossil pollen which occurs in Upper Cretaceous and early Tertiary strata in Kazakhstan and western Siberia. In the southern hemisphere its tropical montane form, *Nothofagus brassii*, occurs in New Guinea and New Caledonia, while *Nothofagus menziesii* and *Nothofagus fusca* now occur in all three cool temperate land areas of the south (Figure 41.iiA).

Two insects—carabid beetles and peloridiid bugs—might be taken as further examples (Figure 41.iiB and C). The Migadopini tribe of carabid beetles is distributed in southern Chile, Tasmania, New Zealand, small areas of southeast Australia, the Falkland Islands and the Auckland Islands. The peloridiid bugs have a similar distribution although they also occur on Lord Howe Island. These insects are associated with *Nothofagus* forest but are not dependent on it. Like *Nothofagus*, carabid beetles have northern hemisphere relations, but the Migadopini do not.

Darlington suggests that the *Nothofagus* and Migadopini type of distribution can be explained by migration from the north to the southern continents across the tropics. A hypothetical pattern of evolution and dispersal suggested by carabid beetles is shown in Figure 41.iv. Trans-tropical dispersal, however, explains only the relationships between northern and southern hemisphere groups; it does not explain the distribution of related genera at the southern end of each land mass. The theories which attempt to account for such distributions fall into three classes:

(1) Dispersal was across land connections in the far south.
(2) Dispersal was across wide ocean gaps.
(3) Dispersal was by northern routes, followed by a dying back of the migrating forms leaving relicts in discontinuous southern areas.

Dispersal across land connections implies either union of the continental land masses or the existence of isthmian links between them. Apart from the geological impossibility of juxtaposition of the land masses during the later Mesozoic and Tertiary eras, it is evident from the fossil record that there has been no interchange of terrestrial vertebrates between Australia-Tasmania, South America and New Zealand since at least the late Cretaceous, although pre-Cretaceous links are geologically possible. This therefore disposes of the idea of land links between all of the land masses. It is possible, however, that Antarctica and South America were linked either by an isthmus or by an island chain during the Tertiary and that migrating plants and some animals could move around the coasts of an ice-free Antarctica and radiate from there. Such a link is not essential to explain present distributions, but if the edges of Antarctica were habitable—and fossil trees and coal measures suggest they were—then the chances of successful dispersal of plants and animals round the southern hemisphere would be thousands of times greater than at the present.

The simplest explanation of the present distribution is that of *dispersal by water, wind and birds across the ocean gaps*. All these regions

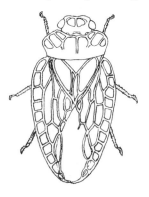

Figure 41.iii. A carabid beetle (*Migadops latus*) **of southern South America (above), and (below) a peloridiid bug** (*Peloridium hammoniorum*) **of the same area.** *After P. J. Darlington*

Figure 41.iv. Diagram of the hypothetical pattern of evolution and dispersal suggested by carabid beetles. (A) An initial stock evolves in the north and disperses southward to Australia, New Zealand and perhaps South America. (B) A second stock evolves in the north and disperses southward to Australia and South America, but not New Zealand. (C) The first stock dies out in Australia and (if it existed there) in South America, and the second stock dies out in the north except for local relicts. (D) Both stocks disappear completely in the north, leaving relicts in the three principal areas in the southern cold temperate zone. Many variations of this pattern are possible.
After P. J. Darlington

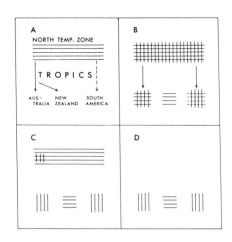

lie within the belt of westerly winds and the circumpolar current—the West Wind Drift—so in many cases this must certainly have happened. The islands of the southern Atlantic and southern Indian Oceans offer interesting evidence which any theory must take into account. They are volcanic, and many are geologically young and are thought never to have had land connections with continental areas, yet the Falkland Islands, the Auckland and the Campbell groups support vegetation obviously comparable with that of Tierra del Fuego and New Zealand respectively. Cushion bogs dominated by the same genera occur in Tierra del Fuego and on Gough Island. Related upland heath vegetation dominated by crowberry occurs on Tierra del Fuego, Tristan da Cunha, Gough and the Falklands, while *Azorella* occurs in southern South America, Kerguelen, the Crozets, the Falklands and Macquarie Island. The late Tertiary and recent volcanoes cannot have been stocked any other way but by wind, water and birds, and the subpolar islands like Macquarie Island have been stocked in the 8000 years or so since the last glacial period. Many of the ferns and small plants have spores easily carried great distances by the wind, or hooked seeds which may be carried by birds, or like the small tree *Sophora* can be dispersed by water. *Sophora* is found in Chile, several Pacific islands, Gough and New Zealand. It is a riverside plant with seeds which are known to be washed down to the coast and which can survive in sea-water for at least two years.

The third possibility—that of *dispersal by northern routes*—can occur. This is shown by the contemporary example of the oak (genus *Quercus*). It ranges from temperate Eurasia southward across the tropical Malay peninsula to the mountains of New Guinea, and from North America through Central America to the northern end of the Andes in South America. If it had already reached the south temperate zone, botanists might now be arguing about the direction of its dispersal.

The northern route hypothesis cannot explain the distribution of related genera at the southern end of each land mass. It can explain the stock at the south of one of the land masses but not the close similarities between them. Additional objections are that it cannot explain the unique fauna of New Zealand, which appears to have been isolated during the Mesozoic and since. New Zealand's only endemic land mammals are two species of bats, but it does have a large bird fauna which includes the extinct moas (*Diornis*) and the kiwis (*Apteryx*).

66. *Nothofagus* **forest in the Dart Valley, New Zealand.**
Photo by J. H. Johns,
NZ Forest Service.
Crown Copyright reserved.

These birds are flightless but are probably derived from flying birds which could have crossed oceanic gaps. Florin has also demonstrated that the northern routes have been ineffective for much of the Mesozoic and Tertiary; since the Permian, the conifer floras of southern lands have been very markedly distinguished from the contemporaneous northern floras—a feature that is most improbable if connections had been continuously open for dispersals.

Conclusions

It seems that the most probable dispersal pattern of southern flora and fauna is one of southward movement through Eurasia to Australasia, and from Eurasia to North America and then South America. From any one of the southern land masses, elements of the flora—but not the fauna—could disperse either through the oceans, possibly using the islands as staging posts, or around the shores of Antarctica in the pre-Quaternary eras of more equable climate. *Nothofagus*, for example, seems to have taken the Eurasia-Australasian route and left *Nothofagus brassii* in the montane forests of New Caledonia and New Guinea. Its pollen is very resistant to destruction and could be carried by wind and birds round the southern oceans. Dispersal is no doubt complex and no traces of the route may be left by many plants, for they would die out along part of the route.

In the Palaeozoic and early Mesozoic eras, terrestrial connections may have existed and allowed reptiles like the dinosaurs to occupy each of the land masses except New Zealand, which has always been isolated. Since the later Mesozoic, however, the continents have been separate and mammals have not been able to disperse once they reached the south; hence related forms on different continents are derived from invasions from the north by parallel stocks. The marsupials of South America and Australia seem to have reached their present areas independently from the north. Some forms have converged but others are sufficiently divergent to eliminate the possibility of southern land connections.

It seems, then, that the distribution of mammals is explained by the third theory—that of dispersal by northern routes followed by a dying back of migrating forms leaving relicts in the south. The same may have happened to some plants, but others could have migrated in the pre-Quaternary, using the shores of Antarctica and islands as staging posts. Since the last glacial period, and probably before it, many plants have been dispersed across ocean gaps, and this last explanation is the only one which will account for the revegetation of many of the remote islands of the southern seas in the last few thousand years.

Distributions of the flora and fauna of the southern islands seem to illustrate the basic principles of dispersal from large to small land areas and from favourable to less favourable climates. The climate has permitted only cold-tolerant plants and animals to enter the cool temperate areas of the southern hemisphere, although among birds and mammals the retarding effect is less than that among the plants.

The table showing the distribution of plant groups in geological time (Figure 42.i) indicates that many of the former groups of plants have died out leaving the modern plants belonging to relatively few groups. The most widespread and highly evolved are the flowering plants or angiosperms. They are relatively recent and presumably have not yet reached their fullest development. The second largest group, the gymnosperms, includes all seeding plants which do not flower. The best known of these plants are the conifers, which have a long history, and appear to have reached their greatest spread in the Mesozoic era. Included within this group is the ginkgo tree, which is often called a 'living fossil' because of its primitive leaf form. The ferns, or filicales, have an even longer history. They have a very wide distribution at present, ranging from the tundra to the equator, and present a great variety of forms from tree ferns to small epiphytes. The filicales are particularly well represented in the fossils found in Carboniferous coal beds, and they probably dominated the vegetation of that period. The bryophytes, or mosses and liverworts, are also widespread and appear to have originated in the Carboniferous period. The bacteria (schizophytes), algae and fungi are the most ancient types of plant. Their functions in breaking down organic remains and in forming soil are so important that it is difficult to imagine how higher organisms could have evolved, or still exist, without them.

Figure 42.i. Distribution of plant groups in geological time.

The distribution of contemporary plants depends upon three factors: (1) the adaptation of the plant to its environment; (2) the history of the plant type in geological time; (3) the migrational ability of the plant.

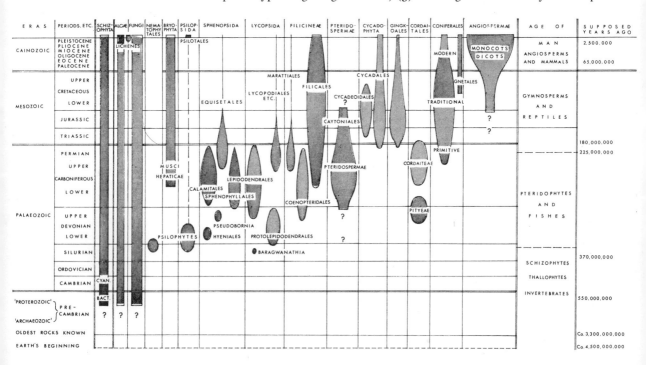

The first of these factors has been discussed in Chapter 37, and the third in Chapter 40. The other factor will be discussed here.

History of Plants in Geological Time

During the second half of the Mesozoic era and for much of the Tertiary, the earth's climates were different from those of the present (see Figure 12.ii): a humid tropical environment was more widespread than now and luxuriant vegetation grew near the poles. There appears to have been a much greater spread of forest than at present, with fewer climatic extremes and generally warmer conditions. Towards the

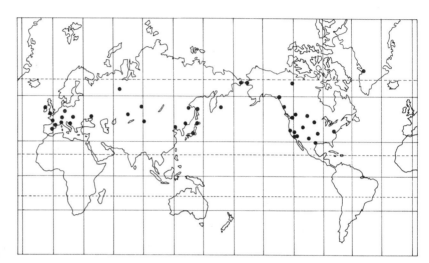

Figure 42.iia. Preglacial distribution of the redwoods.

close of the Tertiary there began a revolutionary period of earth history called the Quaternary, which has continued to the present, and in which the major fold mountain chains of the world were formed and the Pleistocene Ice Age occurred.

Figure 42.iib. Present-day distribution of (left) *Sequoia sempervirens* (redwood), and (right) *Sequoiadendron giganteum* (big tree).

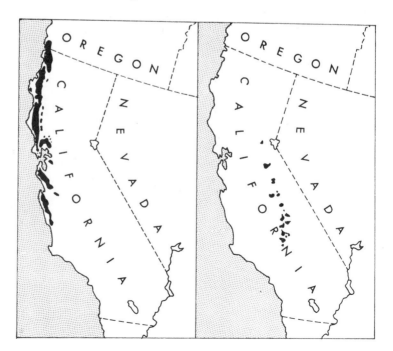

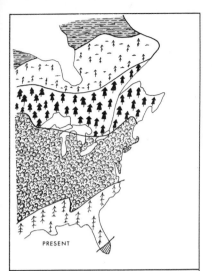

PRESENT

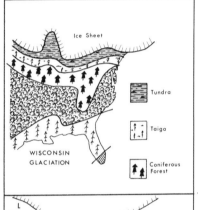

Ice Sheet

WISCONSIN
GLACIATION

Tundra

Taiga

Coniferous
Forest

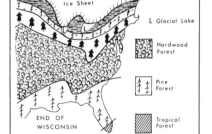

L Ice Sheet

END OF
WISCONSIN

L Glacial Lake

Hardwood
Forest

Pine
Forest

Tropical
Forest

Figure 42.iii. Vegetation zones in eastern North America at the present and during the last (Wisconsin) glacial advance.
After Hunt

67. Interior of a redwood forest, California.
US Dept of Agriculture

The changes of climate overwhelmed many plants, caused many to migrate, and induced adaptations in others. It appears that evolutionary changes were particularly rapid to produce the modern plants and animals. Many new cold-tolerant and other specialised species developed. The floras of small land areas to which access was difficult in the interglacial became impoverished. Thus Europe has relatively few species compared with North America or Asia, for restocking was partly prohibited by the Mediterranean and the mountain chains of the Alps and Pyrenees. Some plants like the giant redwoods, which were formerly widespread in Europe, Asia and North America and now survive as relicts only in California (Figure 42.iia,b), were confined to refuge areas which were not glaciated. Thus many species in central Asia survived in areas too arid or warm to be glaciated, or were able to migrate along mountain chains.

The effect of the advancing ice was to compress the climatic zones, but some of the zones may have been eliminated altogether. This is illustrated in Figure 42.iii, in which the type of ice advance which occurred in North America is represented. Such conditions may have been responsible for the local extermination of many plants.

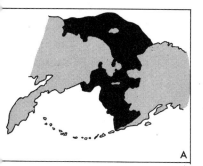

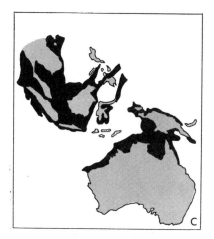

Figure 42.iv. Land bridges (solid black) which have existed in the geological past: (A) the Bering Straits link between Siberia and Alaska; (B) bridges between Europe, Britain and Scandinavia; (C) links between Australia and New Guinea and between southeast Asia and western Indonesia.

In the southern hemisphere isolation was even more pronounced than in the north, for Australia was cut off from Asia by the interglacial rises of sea level, and southern Africa was cut off from the north by the Sahara. The Ice Age therefore completed the process of separation which began when the drifting apart of the continents formed such large gaps between them that migration across the oceans became virtually impossible for many plants and animals—as it must have been during much of the Mesozoic and later.

The continental drift hypothesis (summarised in *The Surface of the Earth* I, Chapter 2) has had many supporters among biologists who found it hard to understand how striking similarities can occur between the floras of areas now widely separated, for the chance that evolution of genetically independent mutations will result in similar plants is incalculably low. This hypothesis now has much support from recent palaeomagnetic investigations, but the drift must have occurred much earlier in geological time than many biologists supposed. The other popular suggestion was that land bridges connected the continents. This theory has some support from the lowered sea level of glacial times, and Figure 42.iv shows these 'bridges'.

Postglacial changes of climate have also caused changes in the vegetation. In Europe and North America the ice gave way to climates that became progressively warmer until about 5000 years ago and then somewhat cooler. These changes resulted in modification of the dominant tree species.

The Geographic Distribution of Plants

There are no plants which are truly cosmopolitan—that is, found almost everywhere. There are a few grasses which are so widely distributed that they are almost cosmopolitan, but the vast majority of plants fall into one of three groups: (1) Arctic-alpine, (2) temperate, (3) pantropical. This separation into groups is mainly conditioned by the presence or absence of a winter. Within each group the actual distribution may be discontinuous or endemic.

Arctic-alpine plants belong to probably no more than 100 species. They are mostly perennial herbs, among which are grasses and sedges. The most extensive areas of such plants occur in the tundra around the Arctic Ocean but they also occur above the timber-line in the mountain

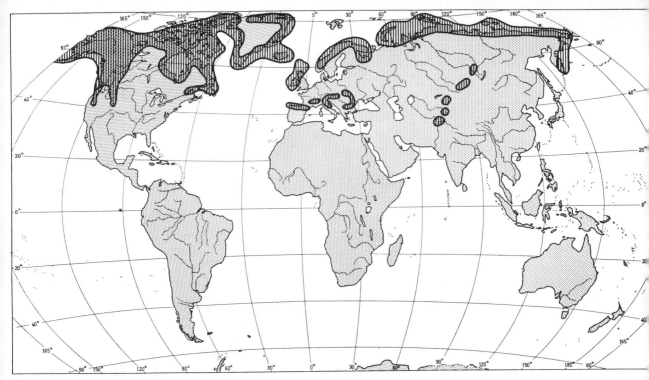

△ Figure 42.v. The Arctic-
alpine distribution of purple
saxifrage (*Saxifraga oppositifolia*).
After Hultén and Polunin

▽ Figure 42.vi. The temperate
distribution of poverty
grasses and wild-oat grasses
(*Danthonia*).
After Fernald

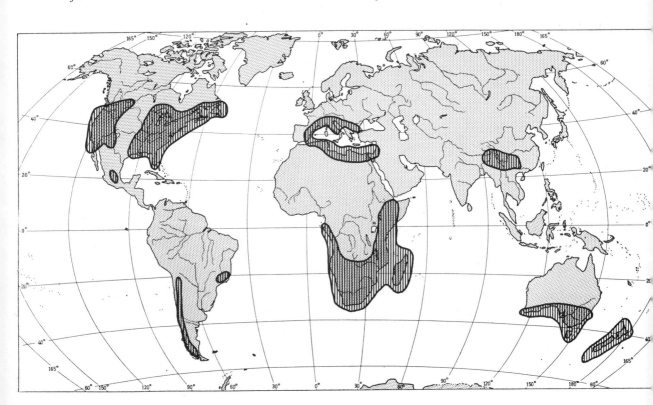

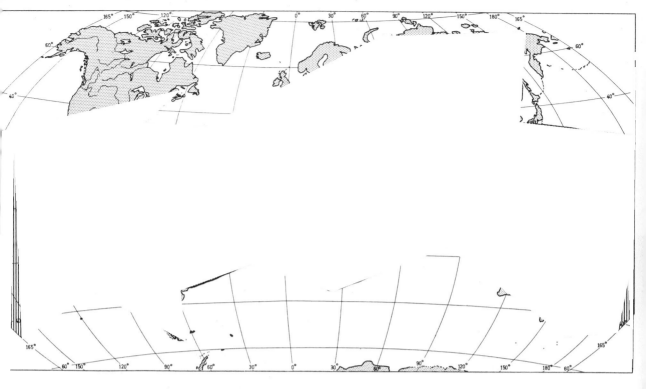

Figure 42.vii. A pantropical distribution of palms. These occur anywhere within the shaded area where environmental conditions are suitable. *After Good*

ranges to the south, and particularly in the great cordilleras of the Americas—the Rockies and Andes—and the east-west chains of mountains from the Pyrenees through southern Europe into the Himalayas. In their wide distribution in both hemispheres these plants come close to being cosmopolitan but because they are restricted to cold, snowy places they are not truly so.

Temperate plants are widely distributed through the moister parts of the temperate zone, but because of the configuration of the continents they are more widely spread in the northern than the southern hemisphere. Many of the temperate plants have been carried by men from the northern to the southern lands, so that in newly settled areas like Australia and New Zealand the northern deciduous trees like oaks, maples and lime trees, as well as the weeds like dandelions and thistles, now decorate or infest the parks, gardens and pastures created by European colonists in the south.

Pantropical plants are those which occur almost everywhere in the humid tropics except in the relatively undisturbed rain forests. The palms (*Palmae*) are among the most widely distributed family of plants (Figure 42.vii), and amongst them the coconut palm (*Cocos nucifera*) is found in coastal fringing groves wherever there are sandy shores and a sufficient rainfall. Its usefulness has led men to carry it across many ocean gaps, but the tolerance of the seed to lengthy immersions in salt water and its buoyancy have caused it to be widely distributed by the natural processes of ocean currents. Many other pantropical plants have been carried by men, so that useful foods like the yam have been brought to numerous Pacific islands by Polynesian voyagers, and many tropical grasses are found on each tropical continent.

A *discontinuous* distribution is common to most plants, mainly as a result of recent geological events. Many forest trees in the late Tertiary had a wide distribution across the northern temperate continental areas, but the southward advance of the Pleistocene ice sheets destroyed a number of forests and in Europe the climate was so severe that no breeding stock of many trees survived. In Asia, communities survived in the mountains of Japan and in Kwangsi and Szechwan in China. In North America, the southern Appalachians provided a refuge. With the retreat of the ice the forests once more spread northwards, but in Europe the flora was permanently impoverished by the destruction of many species. The result is that many trees like the *Liriodendron* (tulip tree) and *Liquidambar* (sweet gum) now occur only in eastern Asia and the USA, not in Europe.

An *endemic* is a plant which is restricted to one floristic region. Most vascular species belong to this group and a few, like the giant redwoods already discussed (Figure 42.ii), are extremely limited in their range either because they have been cut off by climatic changes or because they require a very specialised environment, such as those plants which only thrive in areas of serpentine bedrock. Many other endemics are restricted in their distribution by geographical barriers, so that both high mountains and oceanic islands often provide habitats for them.

43 Animals and the Environment

In order to live and breed successfully in a habitat, an animal must be adapted to it. There are three types of adaptation: structural, or morphological; physiological; and behavioural. Every animal might be described as a 'bundle of adaptations' to its environment. The different adaptations arose through the effect of the environment on organisms. In a herd of zebra or deer the animals will look alike, but in the struggle for existence those which vary too greatly from the norm will be eliminated—the very fast or very slow falling to predators. If the herd were to be moved to a new environment perhaps only the very fast deer would survive, and they by passing on through their progeny the genetic pattern which involves speed would give rise to a new stock of faster animals with that necessary beneficial adaptation. In this way the environment is acting as a genetic sieve removing harmful characteristics.

Structural Adaptations

The most obvious structural adaptations are those which fit an animal to live in a particular stratum, whether it be on the ground, in trees or in the air. The four-footed mammals can be simply classified on this primary basis into:

Aquatic (swimming)—seal, whale
Fossorial (burrowing)—mole, shrew
Cursorial (running)—deer, horse
Saltatorial (leaping)—rabbit, kangaroo
Arboreal (climbing)—monkey, squirrel
Aerial (flying)—bat

68. A hunting group of fleet-footed cheetahs with their kill on the Serengeti Plains.
Paul Popper Ltd

Figure 43.i. Peppered moths on dark and light backgrounds.

These adaptations are basically the same whatever general environment the animal lives in; thus the subterranean adaptations of tundra and hot desert animals are similar.

Some of the best examples of structural adaptations occur in the form of camouflage. Stick insects look like twigs; snow hares are white in winter and brown in summer. The peppered moth provides an example of the environment acting as a sieve. Until the industrial revolution of the mid-nineteenth century caused soot and grime to cover much of the landscape of central England, the moth was a pale speckled colour well camouflaged against a background of lichens and tree bark. Since the period of pollution and grime began, the speckled moths have been largely replaced by the dark ones which are camouflaged in their dirty environment (Figure 43.i). The speckled variety now only occurs in some rural districts. A similar example occurs in New Mexico, where on the white gypsum dunes of the Tularosa Basin a race of pocket mice is white, but on the nearby dark lava flows mice of a closely related species are black. If it were not for their protective colour, all the mice would soon be taken by owls and other predators.

Feeding adaptations are very common: thus carnivores, such as dogs and cats, have high-crowned, narrow pointed teeth suitable for tearing flesh; herbivorous ungulates and rodents have teeth that are flat-crowned and suitable for grinding harsh grasses and other vegetation; and omnivores have both sharp and flat teeth (Figure 43.ii). The bills of birds display great variety in shape and size, depending upon whether they are seed, insect, plant, fish or flesh-eating (Figure 43.iii).

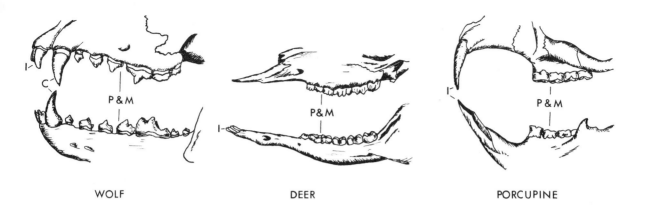

WOLF   DEER   PORCUPINE

Figure 43.ii. Large, sharp canines (C) of the carnivorous wolf are used for stabbing and tearing meat, and the premolars (P) and molars (M) for shearing. In deer grinding teeth predominate, with incisors (I) for cropping vegetation. The porcupine— a rodent—has long self-sharpening incisors for gnawing.
After Romer

69. Antarctic seals herd together for breeding.
Photo by J. Calvert, Antarctic Division, NZ Dept of Scientific and Industrial Research

Adaptations to climate seem to follow a number of rules, which are now named after distinguished ecologists:

Bergmann's Rule states that geographic races of a species possessing smaller body size are found in the warmer parts of the range, and races of larger body size in the cooler parts.

Allen's Rule states that tails, ears, bills and other extremities of animals are relatively shorter in the cooler parts of a species' range than in the warmer parts. This rule is illustrated very well by the ears of foxes, which tend to be small in the Arctic fox and large in the desert fox, as responses to heat preservation in the first case and heat loss in the second (Figure 43.iv).

Gloger's Rule states that in warm-blooded species black pigments increase in warm and humid habitats, reds and yellow-browns prevail in arid climates, and pigments are generally reduced in cold regions.

There are many exceptions to each of these rules, but their general applicability is shown by *Homo sapiens*, modern man. The pygmies and other dwarfed peoples live in the humid equatorial regions (Bergmann's Rule); the long thin arms and legs of some African peoples compare with the squat bodies of Eskimoes (Allen's Rule), and the coloration rule (Gloger's) applies to all races.

Physiological Adaptations

These are the internal adaptations which allow an animal to exploit a particular habitat. Wood-boring insects possess a special enzyme which digests the wood. The kangaroo rat of the southwestern deserts of the USA can live its entire life without drinking water, as it is able to extract all it needs from seeds. The north African jerboa is a similar animal with the same characteristic, which appears to be largely the result of its ability to conserve moisture by excreting a highly concentrated urine. The mammals of the Peruvian Andes—the llama, vicuña and viscacha—which live as high as 4600 metres, have haemoglobin with a much greater oxygen capacity than that of mammals living at lower altitudes where the oxygen content of the air is greater.

Behavioural Adaptations

A species of moth with banded wings which rests by day on the banded leaves of a lily plant has the structural adaptation which gives it camouflage protection, but the impulse to settle on the plant so that its wing stripes are always parallel with those of the lily is a behavioural response. Such adaptations are seen most readily in the animals of extreme climates. The jerboa re-uses some of its expired moisture by resting in a crouched position in which it breathes in the moisture again. The Tuareg and the Bedouin do much the same when they cover their entire heads with a cloth. Many animals hibernate for the cold, food-scarce winter, or avoid the heat of the midday sun in burrows or the shade, hunting and feeding at night. The caribou abandon the snow-covered tundra in early autumn and migrate southwards to the forest zone to seek sites where the snow is thin and easily removed to reveal their food supply. As the snow cover varies, so the caribou move to other sites.

Community behaviour is also seen in the defensive circle formed by musk oxen when they are attacked by wolves or the similar circle formed by resting North American quail. Predators like wolves also hunt in packs; lions hunt in pairs with the lion frightening his quarry towards the female lying in ambush. Community protection is also gained by the co-operative work in beaver communities and by the social organisation of baboons, who always move in a troop with adult males in the lead and in the rear, so that when attacked they can form a circle with

Figure 43.iii. Adaptations in the bills of birds: (a) a seed-eating sparrow; (b) an insect-eating warbler; (c) a plant-eating duck; (d) a fish-eating heron; (e) a predaceous hawk; (f) an aerial insect-eating whippoorwill.
After Kendeigh

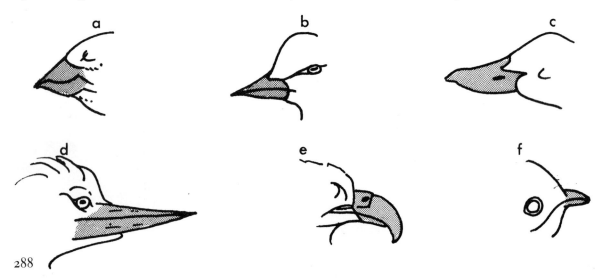

Table 43.1. Grassland Animals

	North America	South America	Asia	Africa	Australia
Leaping herbivores	Jack rabbit	—	Jerboa	Springhaas	Kangaroo
Burrowing mammals	Prairie dog	Cavy	Hamster	Ground squirrel	Wombat
Underground feeders	Pocket gopher	Tuco tuco	Mole rat	Golden mole	Marsupial mole
Running flightless birds	—	Rhea	—	Ostrich	Emu
Running herbivores	Pronghorn Bison	Guanaco Pampas deer	Saiga Wild horse	Springbok Zebra	—
Running carnivores	Coyote	Maned wolf	Pallas cat	Lion	Tasmanian wolf Dingo

females and young in the centre. A single baboon is weak but twenty will stop a leopard, a cheetah or even a lion.

Convergent Evolution

Animals with similar functions or niches tend to look alike even though they are unrelated; thus the Tasmanian 'wolf' of Australia, although a marsupial and not a wolf, behaves in much the same way when seeking its prey as the European wolf. The armadillo (a mammal), the wood louse (an isopod) and some beetles (insects) can roll themselves into a ball as protection against predators. The spiny anteater, the banded anteater, New World anteater, aardvark and pangolin are distantly related animals from four continents. All feed on ants and termites; all have cylindrical tongues, inconspicuous teeth and narrow, long faces—adaptations which enable them to seek their food. Table 43.1 shows grassland animals of the continents which fill similar niches and have similar forms, and are thus ecological equivalents. Such similarities have evolved because the adaptations are ideal for the particular niches which are available, and hence environmental conditions have influenced this convergence in evolution.

Niches

Table 43.1 indicates that in each continent's grassland areas there are similar 'professions' available which are filled by animals. This reaction

Figure 43.iv. Foxes of Arctic areas (left); of temperate areas (centre); and of desert areas (right).

to the physical and biotic environment, and the restriction of a species to a particular niche, depend upon the structural and physiological adaptations and developed behavioural patterns of the species. It has been observed that because of competition no two species can simultaneously and completely occupy one ecological niche in any area—a principle known as Gause's Rule, after the Russian investigator. There is a general tendency to avoid such intensive and continuous competition by behavioural patterns. Thus in an English winter it was noted that of birds feeding at a bird-table, the robin fed at sunrise and sunset, the blackbird around midday and the blue tit at mid-morning and mid-afternoon. In this way they reduced competition between the species. Similarly, in southern California the giant kangaroo rat is predominant in flat brush-covered country; on brushy slopes and rolling hilltops the Fresno kangaroo rat is predominant, and the Heermann's kangaroo rat is forced to live in the open plains because it cannot compete successfully with the other two rats in the brush areas.

An animal is not necessarily confined to one niche permanently, and many animals can rapidly adapt to changing environmental conditions by a change in their niches. In England there are ten species of rodents occupying various 'mouse niches'. They include the dormouse, the bank vole, the short-tailed vole, the long-tailed field mouse, the harvest mouse, and the house mouse. The last of these is the only species which has been introduced into New Zealand, where it has spread widely to occupy not only buildings—its English niche—but also the niches in forests, farmland and grasslands occupied by its competitors in England.

Those areas of the world with the greatest variety of available niches consequently have the greatest variety of species. Thus tropical forests, with their multiple synusia and great variety of plants, have a much greater number of species of animals than the impoverished and simple communities of the cold and hot deserts. Similarly, pioneer communities have fewer niches than the later seral and climax communities. Conversely, where the niches are discontinuous, as in a tropical forest, few communities can be continuous and populations of a particular species tend to be low; but where niches are continuous, populations of a species can be very high—as in the bison or caribou herds of North America before man reduced them and their habitat.

Animal Populations

As well as assisting the survival of the species by providing protection, group behavioural adaptations have other advantages. Flocking of birds at feeding grounds helps to ensure the full exploitation of the food supply, but this social behaviour also assists some birds to find their food in the first place. Some cormorants fly in formations which allow them to search large areas of sea for fish which, when sighted, will cause the lead bird to settle on the water. The other cormorants will then land and by steadily circling the fish will concentrate the shoal and so make the fish easier to catch. Cormorants have even been known to spread themselves across a bay or river and to drive fish into shallow water, where they are easily caught.

The advantages of such social behaviour keep animals in groups which are known as populations, when each group is composed of organisms of a single species concentrated in a given place at a given time. Within a population there are frequently regulatory mechanisms

which limit numbers in relations to the food supply or space requirements. Such mechanisms may be of the direct kind found among some rodents, which simply stop breeding if their living space becomes too crowded—as with many mice—or of the less direct kind found in many simple ecosystems of the tundra and northern coniferous forests. In such areas, the populations of carnivores like the fox and lynx fluctuate directly with those of their prey, the lemming and the snowshoe hare. These fluctuations appear to go in 3–4 and 9–10 year cycles. It seems that in a simple ecosystem, where the food supply is derived from only a few species of plants, the primary consumers can simply eat up their whole food supply and then either starve or fall easy prey to the secondary consumers. Alternatively, the primary consumers may stop breeding as the pressure on living space develops. In the snowshoe hares overcrowding produces stresses which cause their livers to degenerate, leaving an inadequate supply of glycogen available for emergencies, so that excitement or fear, instead of stimulating the hare to run, causes it to collapse and die. Whatever the cause, there is a sudden decline in numbers and the beginning of a new cycle which will gradually build up numbers again.

Such violent fluctuations do not occur in the complex ecosystems, where alternative food supplies are usually available and competitive and predatory conditions remain nearly constant. A change in external conditions can, however, cause animals of the humid tropics to respond in the same way as those of the cold zones. In some zoos it has been found that overcrowding causes many animals to develop heart diseases or to stop breeding. In producing changes in habitat, man can thus produce unexpected and quite disastrous effects upon animal populations—this is discussed further in Chapter 50.

70. **Barren ground caribou migrating in the interior of Alaska.**
Paul Popper Ltd

44 The Distribution of Animals

D. S. Jordan once stated that the general laws governing the distribution of animals can be reduced to three propositions. A species of animal will be found on any part of the earth with a suitable environment unless:

(1) its individuals have been unable to reach this region because of barriers;

(2) having reached it, the species has been unable to maintain itself because of an inability to adapt to the region or compete with other species;

(3) having arrived and survived, it has subsequently so evolved during adaptation that it has become a species distinct from the original type.

The nature of relations with the environment was discussed in the previous chapter; the nature of dispersal, barriers and establishment will be discussed in this.

Dispersal

Dispersal is the spread of animals away from their home sites. It is usually a slow process, being the cumulative result of short dispersions by successive generations, but over long periods of time it can produce considerable expansion in the range of an animal. Once a species has passed any barriers that may have hindered its spread, it can disperse very rapidly; the European starling, which was introduced into North America about 1890, spread at an accelerating pace until in 1940 it had become established over 6,475,000 square km. Birds have an ability to disperse very rapidly but other organisms, like the fresh water amphipod *Gammarus pulex*, which has taken the last 6000 years to disperse across 12 river systems from southern England into Scotland, can be very slow. Rapid dispersal is favoured not only by suitable physical environmental conditions, but by the absence of competitors and predators, which

Figure 44.i. The spread of the North American musk rat in central Europe between 1905, when it appeared south of Prague, and 1927. The lines with the year attached represent the outer limits of the invasion at that date. The area occupied more than doubled in every successive 5 years.
After Pearsall

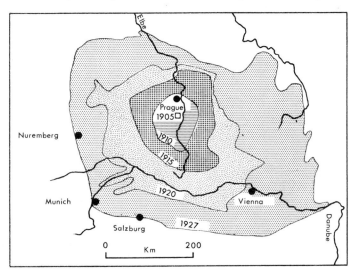

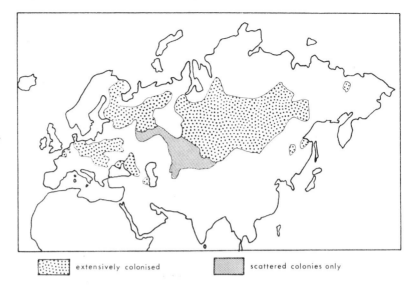

Figure 44.ii. Areas in Europe
and Asia invaded by the
North American musk rat by
1956.
After Pearsall

extensively colonised   scattered colonies only

71. A coypu (*Myocastor coypus*)
photographed at night on a
river bank in Norfolk,
England.
*Photo by John Markham, Nature
Conservancy*

accounts for the extremely rapid spread of the European rabbit in
Australia, and its multiplication to a population of about 500 million by
1950 when myxomatosis was introduced and killed off about 90% of
that population. Since this great epidemic both the rabbit and the virus
have undergone mutations, and the rabbit population is again on the
increase and may have been around 100 million in 1966. Since one
breeding pair, under moderately favourable conditions, can have 9
million descendants in three years, the situation is once more critical.

Ecological explosions of the rabbit-in-Australia type are not limited
to areas largely unoccupied by man. The spread of the North American
musk rat in Europe and Asia has been spectacular. This animal, the
'musquash' of the furrier, is one of the group of fur-bearing animals
which have from time to time escaped from fur farms. Five musk rats—
two males and three females—were liberated near Prague in 1905. By
1927 an estimated population of 100 million had spread into Austria,
Germany and Hungary (Figure 44.i), and since then musk rats have
spread widely in Europe and Asia (Figure 44.ii). The South American
coypu and North American mink are also spreading rapidly in Britain.
The mink endangers many small birds and animals, and the musk rat
and coypu both burrow into river and canal banks and so do much
damage.

Ecological explosions are seen at their extreme in this quotation from
S. D. Macdonald describing events on Sable Island, a small sand bank
of Newfoundland:

The English rabbit has at different periods been very numerous, and
threatened at one time to over-run the Island. But, to their misfortune, the
Norway rat landed from an old vessel, and in a short time became so
numerous that they nearly annihilated the rabbits, and then turned their
attention to the stores of the Island, so that during one winter the staff were
without bread for some months. In the spring the Government sent a
detachment of cats to look after the rats. The cats killed the rats and then
finished the remaining rabbits. In a short time the cats became so wild and
numerous as to be a source of danger, when dogs were sent to hunt the
cats, and with the assistance of shot guns in the hands of the staff, the cats
were finally extirpated.

But more serious, if not quite so dramatic are those ecological explosions which affect any oceanic island. Rats, goats and pigs landed from ships have largely denuded many islands of their vegetation and killed the bird life. New Zealand provides one of the best examples of the effect that man-carried organisms can have. Biological invasions began when the first Polynesians landed in New Zealand, but their main animal introduction, the 'bush' rat, did little harm. The main invasion came in 1773 when Captain Cook, on his second visit to the islands, introduced sheep, goats, pigs and fowl, and these were later supplemented by stock left by whalers and sealers. Rabbits had become a menace by 1870, and the weasels and ferrets introduced to prey on the rabbits turned their attentions to the wild ducks and other birds. The weasels and ferrets did not thrive in the open tussock grasslands where the rabbits multiplied so rapidly. Other introductions were 130 species of birds, of which 30 became established; moose, elk, white-tailed and mule deer from North America; red and fallow deer from Europe; sambar, rusa, and sika deer as well as thar from Asia. A gift of chamois from the Emperor Franz Josef of Austria produced more wild chamois

72. Two bull thar and groups of cow thar in the New Zealand Southern Alps.
Photo by J. H. Johns, NZ Forest Service. Crown Copyright reserved

in New Zealand than there are now in the European Alps. The opossum was introduced in 1858 from Australia, for its fur.

These invaders left their natural predators and parasites behind and entered an area where there were no indigenous land mammals and a vegetation which was not adapted to grazing or browsing. In the alpine zone deer, hares, chamois and the Himalayan thar have overgrazed the native herbaceous dominant flora, which is now either being replaced by grazing-resistant grasses or has been entirely destroyed so that accelerated erosion is a major problem in the high watersheds. In the lower tussock grasslands of the Canterbury and Otago hills, man's use of fire, and sheep have assisted the rabbit in opening up the tussock cover and allowing a new and more varied inter-tussock flora. In the forests the effect of mammals has been to reduce the density of the

73. The interior of a New Zealand forest in which all tree and shrub regeneration has been eaten off by deer, goats and pigs.
Photo by J. H. Johns, NZ Forest Service. Crown Copyright reserved

ground and shrub layers and to kill off saplings and regrowth so that some forests have a diminishing future.

The effect of an introduced mammal such as the red deer is in stages related to the size of its population. In stage one the maximum population density is built up, in a given area, over a period of 20–30 years. In stage two over-population occurs and the deer eat out their food supply so that the herds suffer from malnutrition and numbers decrease, and the vegetation is severely depleted. After a further 5–10 years, stage three begins, with the gradual change in the floral composition as the less palatable and browse-resistant plants replace those destroyed by the browsing. A new ecosystem with increasing stability is gradually introduced with a changed soil-plant-animal relationship, but in this process before a new equilibrium is reached there may be extensive destruction of soil and forests. Stage four is the final stable situation, with a new vegetation-soil complex and a deer population regulated by the new system.

Man can induce changes in animal populations and distributions not only by introductions into new areas but also by changing an environment. Thus many forest animals have had their range reduced by cultivation and clearances, but others are like the grey squirrel, *Sciurus carolinensis*, which was introduced into southern England in 1890 but spread rapidly only after the large scale cutting of trees during the First World War, when the open habitat favoured it over the native red squirrel, *Sciurus vulgaris*. Similarly the roe deer, *Capreolus capreolus*, has now spread south from Scotland and the Lake District into newly established forestry plantations.

Man has been by far the most successful and rapid agent of dispersal of animals in the last few hundred years, but in a historical sense his effect is only minor and is limited to a few thousand years. The major means of dispersal are similar to those of plants; thus animals spread most rapidly down wind, down water-courses, through areas of uniformly favourable environment, and small animals may travel on large birds or mammals. In such dispersals there is considerable trial and error, for few animals disperse deliberately in a particular direction; and barriers

74. Rata and kamahi forests on the west coast of the South Island of New Zealand, showing dead branches caused by over-browsing of the canopy by opossums.
Photo by J. H. Johns, NZ Forest Service. Crown Copyright reserved

can be of many kinds so that there may be an advance in one direction, a dying back and then another advance in a different direction, until a successful establishment or ecesis occurs.

Barriers to Dispersal

Barriers to animals, like those to plant dispersals, may be physiographic, climatic or biotic, but what is a barrier to one species may be a routeway to another. To fresh water organisms intervening land is a barrier, but to monkeys a river the size of the Amazon is a very effective barrier, and even to many fresh water animals rapids and waterfalls are barriers to dispersal upstream.

Terrestrial organisms are restricted by any extensive stretch of water, especially the oceans, and by physiographic breaks such as major gorges like the Grand Canyon of the Colorado river, which separates the ranges of the Kaibab and Abert squirrels. Mountain ranges may be a barrier if they are so extensive as to have climatic and vegetational zones on them which are unfavourable to the dispersing species.

Climatic boundaries are very effective, as many moist-skinned species cannot withstand exposure to low humidities, some animals cannot withstand extreme temperatures, and others may be limited by the length or absence of the daylight in polar latitudes.

Most animals have a great adaptability to food, so that vegetation may not be a barrier to some species, yet to others, like the moustached white monkey of Central America which relies on the fruits of the *Elaeis* palm, or the koala of Australia which can only survive on the leaves of certain species of eucalypts, or the panda which feeds on bamboo shoots, the range may be entirely limited by the distribution of the necessary food. The other biotic barriers are the presence of competitors which already fill all suitable niches, or of predators.

Dispersal Routes

For all but the most recent dispersals, and the latter are largely the work of human interference, dispersal routes have to be studied in relation to geological events and distributions. Most theories assume either that the continents occupied in the past much the same positions as now, or that they have drifted in the past. To explain most animal distributions it is not necessary to assume that the continents have drifted, and it is certain that if drift has occurred, then it took place before the time when the major predecessors of contemporary animals dispersed. The continents, however, have not been stable; sea level has fluctuated, drowning areas now land and at other times exposing parts of the continental peripheries, and climates have varied greatly (see Chapter 12). Of particular importance are the island chains linking continents, such as the Indonesian archipelago between Asia and Australia, and the land bridges which existed across the Bering Straits during the early Tertiary and parts of the Quaternary, and exist now across the Panama isthmus, even though this route was broken for much of the Tertiary. A typical dispersal pattern is shown in Figure 44.iii.

Land bridges may either allow most plants and animals to cross, in which case they are called corridors; or if they lack suitable niches they may cut out some organisms, in which case they are called filters; or they may only allow a few species to pass, and those largely by chance, when they are called sweepstakes routes.

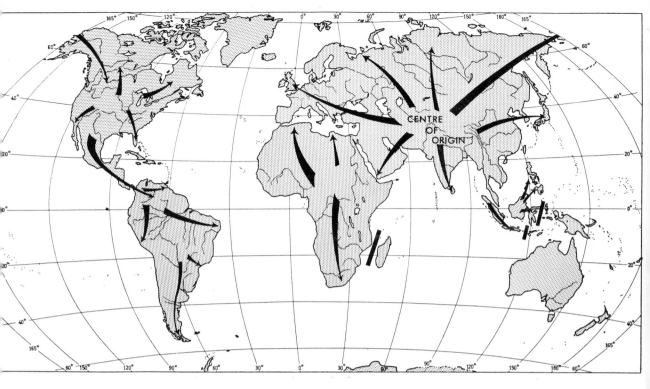

Figure 44.iii. The dispersal of toads throughout the world—except for Madagascar, part of Indonesia, New Guinea and Australasia.
After Darlington

The operation of corridors, filters and barriers has been such that many parts of the earth have distinctive fauna. Because only a few fish, some frogs, rodents, passerine birds and man can be considered cosmopolitan, the earth may be divided into distinctive geographical realms based upon their characteristic fauna.

Zoogeographical Realms

The world was first divided into zoogeographical regions, on the basis of its bird fauna, by Sclater in 1857. Sclater's regions were little modified when Wallace included all the land animals whose ranges were then known, in 1876. Wallace's realm correspond roughly with the continents, each realm being separated from its neighbours by either oceans, or deserts like the Sahara, or mountain ranges like the Himalayas and their east and west extensions. The 6 realms are shown in Figure 44.iv. These regions are generally accepted, although because the Australian and Neotropical are so very unlike any of the others, and there are similarities between the Nearctic and Palaearctic, some writers have proposed a threefold division of Neogea (i.e. South America), Notogea (i.e. Australia) and Arctogea (i.e. the rest of the world). Table 44.1 shows the distribution of many characteristic animals.

Nearctic

The Nearctic realm covers all of North America and Greenland and extends as far south as central Mexico. It was connected to the Palaearctic for much of the Tertiary across the Bering Straits and there was considerable interchange between the two areas, but virtually no exchange with the Neotropical realm to the south, from which it was

Table 44.1. Animal Distribution in Zoogeographical Realms

	Nearctic	*Palaearctic*	*Oriental*	*Neotropical*	*Australian*	*Ethiopian*
Tundra	Caribou, musk ox, lemmings, Arctic hare, wolf, Arctic fox, polar bear	Similar to Nearctic				
Coniferous forest	Moose, mule, deer, wolverine, lynx	Same genera, but different species from Nearctic				
Temperate grassland	Bison, pronghorn, jack rabbit, prairie dog, gopher, fox, coyote	Saiga, wild ass, horse, camel, jerboa, hamster, jackal		Guanaco, rhea, viscacha, cavy, fox, skunk		
Deciduous forest	Racoon, opossum, red fox, black bear	Similar forms to Nearctic but different species				
Chaparral Mediterranean woodland	This is an ecotone where animals of other areas mix	Ecotone				
Desert	Many lizards and snakes, kangaroo rat and cottontails	Jerboas, hamsters, hedgehog		Guanaco, rhea, armadillos, vulture	Marsupial mole, jerboa, parakeets, lizard	Springbok, porcupine, jerboa, rock hyrax
Savanna					Emu, red kangaroo, bandicoots, wombats, cockatoos, parrots	Zebra, eland, gemsbok hartebeest, gnus, giraffe, elephant, ostrich, lion, cheetah
Tropical forest			Gibbons, orangutan, monkeys, Indian elephant, sun bear, porcupine, tiger, many snakes and lizards	Monkeys, kinkajou, pygmy anteater, sloth, tree snakes, parrots, humming birds	Tree and musk kangaroos, wallabies, koala, opossums, cassowary	Okapi, gorilla, chimpanzees, monkeys, forest elephant

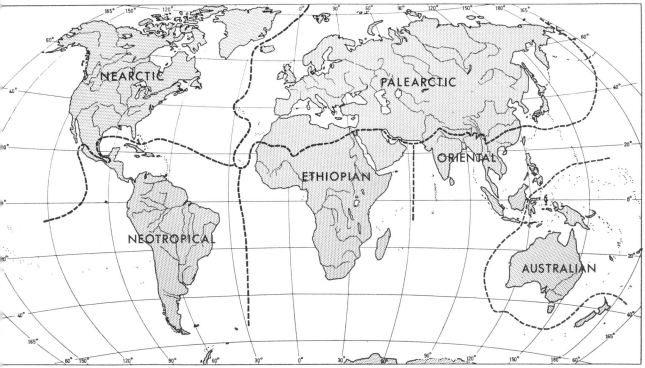

Figure 44.iv.
Zoogeographical realms.

separated by a water barrier. Because of the Bering connection and similar climatic and vegetational regimes, it has many faunistic groups in common with the Palaearctic although there are differences in species. It has no camels and the horse is a reintroduction from the Palaearctic although the horse is thought to have originated in the Nearctic. It shares the opossum and armadillo with the Neotropical. The pocket gophers, pocket mice and pronghorns are endemic to the region. Of the birds only the wild turkeys can be considered to belong exclusively to the Nearctic. It is also rich in reptiles and its fauna can be considered to be a mixture of Old World temperate and New World tropical.

Palaearctic

Like the Nearctic the Palaearctic has relatively few families. Nearly 1/3 of its 28 families are widely distributed outside the realm and the only two mammals exclusive to the region are rodents. Its birds are also widely distributed; there are very few reptiles, and none which are exclusive to it.

Ethiopian

This realm covers the whole of Africa south of the Sahara, and includes southern Arabia. It is largely a tropical realm with hot deserts, savannas and tropical forests, so it lacks the cold-adapted animals of the Nearctic and Palaearctic. It has the most varied mammalian fauna of all the realms and is second only to the Neotropical in the number of unique families. It is without moles, beavers, bears or camels (these latter were introduced from the Palaearctic), but shares many families with the other Old World tropical regions, so that species of elephants, rhino-ceroses, apes and pangolins occur also in the Oriental realm. Although the savannas had enormous populations of swift herbivores—zebras, giraffes, antelopes—and large carnivorous members of the cat family to feed on them, only the giraffe family is confined to the realm. Hippopotamuses, aardvarks and a number of families of rodents and insectivores are also unique. Of the 6 unique bird families the ostrich is the best known.

Oriental

This area includes all of southeast Asia and the large islands of Sumatra, Java, Borneo and the Philippines. The eastern boundary of the realm is usually taken as being Wallace's line. Like all other boundaries, except the edges of the continents the 'line' is really the centre of a zone of transition. To the east of it the number of Oriental species declines rapidly, and to the west of it the number of Australian species declines.

The Oriental realm has most in common with the Ethiopian largely because of similarities in climate. The northern boundary with the Palaearctic, being the Tethys geosyncline in the Tertiary and the Himalayas now, is a very effective one. Northern and southern India have distinctive faunas, probably because the extensive volcanism of central India in the Tertiary produced a barrier and gave rise to separate developments. An elephant, two rhinoceroses, several species of deer and antelopes, many pheasants and the jungle fowl—the source of the domestic chicken—tigers and a rich variety of lizards and snakes are found in the area. Tree shrews, the gibbon, orangutan and tarsier are exclusive to the region.

Australia

It is a matter of debate whether Australia was ever connected by a land bridge to the Oriental realm or whether the connection was by island chains. The realm is marked by an absence of most groups of placentals, although some have been introduced. The original mammalian fauna consisted of monotremes and a diversity of marsupials with rats, mice and bats. The marsupials belong to 6 families, none of which occurs in the New World, where the only other marsupials are found. The marsupials have taken the functions and often the form of placentals elsewhere, the mole and Tasmanian wolf for example. The uniqueness of this fauna makes up for a scarcity in types, for there are only 9 families of mammals and 8 of these are unique. The kangaroos and wallabies have no similarities of form with animals elsewhere, although they occupy the niches of fleet herbivores. The duckbilled platypus and the spiny anteater are the only living monotremes and occur nowhere but in Australia. The Australian region has little in common with any other—a fact suggesting long geological separation.

Neotropical

Large parts of this realm, which extends from central Mexico to the extreme south of South America, are in tropical areas with a small tip in the southern temperate zone and a large area of high mountains. The fauna is both distinctive and varied: excluding bats, 32 families of mammals are represented, 7 being of wide distribution and 16 being unique. This is the highest number of endemic families for any region. Some camelids—the guanaco and vicuña—live on the Andean plateaux but most of the unique animals are found in the forests. Horses are absent, although they existed until a few thousand years ago. It has been suggested that rabies-carrying bats were responsible for spreading the diseases which eliminated them. Two families of marsupials occur in the region but only one, the opossum, is shared and that occurs in the Nearctic. There are many monkeys, some with prehensile tails, large numbers of birds and many rodents and snakes.

Oceanic Islands

Islands like Britain can be regarded as continental because they are so close to large land masses. Britain's fauna is impoverished because recent glaciation drove out many animals and the post-glacial rise of sea level cut it off from Europe, but the situation of oceanic islands like New Zealand, 1000 miles from Australia and isolated since at least the end of the Mesozoic, is quite different. Because of its long standing isolation, New Zealand's fauna is impoverished: the only mammals are two families of bats although, in part-compensation, there is a unique array of flightless birds, half of which are extinct. The giant moa may have been 3·7 metres tall, but the modern flightless birds like the kiwi, flightless rails and kakapo are small. These birds are probably descended from flighted birds but in the absence of predators they took to the ground. There are geckos and skinks but no turtles or snakes. An extraordinary survivor is the tuatara, *Sphenodon*, the last surviving

75. The flightless kiwi.
Photo by P. Morrison, NZ Dept of Internal Affairs, Wildlife Branch

member of an order that disappeared from the fossil record a hundred million years ago in the Cretaceous period. This animal is now confined to a few offshore islands and does not exist in large numbers. There are no strictly fresh water fish. The vertebrate fauna is one which has arrived from across the sea—although the tuatara's route is unknown—and is now highly endemic. This isolation has given New Zealand a zoological uniqueness for an area of its size.

Animal Migrations

Migration, like dispersal, involves animal movements, but it also involves movement back to an area and usually temporary and repetitive occupation of areas. Birds are the best known migrators. Some like the Arctic tern breed in the Arctic and migrate south to spend the northern winter in the Antarctic summer. Such migrations are necessary because of the seasonal disappearance of their food supply. Some animals migrate to escape the cold, others to escape the heat.

Annual migrations are not limited to birds. Bison regularly moved south on the Great Plains of North America, during winter, distances of 320–640 km; fur seals breed in the Pribilof Islands in the Bering Sea during the summer and winter off California 4800 km to the south. Other animals, like some deer and birds, move altitudinally to seek food on lowlands during winter and return upwards during spring. The tendency to migrate presumably developed because survival was greater amongst those who did so than among those who did not.

76. A male tuatara.
Photo by C. Roderick, NZ Dept of Internal Affairs, Wildlife Branch

45 The Study of Plants in the Field

It has already been suggested in Chapter 38 that the stratification and life-forms in a community can be studied without a detailed knowledge of the taxonomy of the vegetation. Two other important aspects of field study are dealt with here—mapping and sampling.

Mapping Vegetation

Workers in many fields—forestry, conservation, planning, agriculture, military strategy, etc.—require to be able to map vegetation at a great variety of scales, and with various degrees of accuracy and speed. A number of methods have been suggested and tried. The system used and described here was proposed by Küchler in 1966. It has been refined from a number of earlier methods and successfully applied in a number of climatic regions, and may be presumed to have wide application. The system relies on symbols and in many ways it is like that proposed by Köppen for recording climatic types. Capital letters are used to indicate the basic life-form characteristics of the plant community; lower case letters indicate special features; and numerals indicate the height of each vegetation class.

The major subdivision is between woody and herbaceous plants. The woody plants fall into five groups distinguished on the basis of leaf characteristics, and the herbaceous plants fall into three groups.

Basic Woody Vegetation Categories

B. *Broadleaf evergreen:* These trees are never bare and have leaves broader than the needle type of leaf. Most trees of the tropical rain forest and many of the temperate rain forests belong to this group. Mangroves and eucalypts are examples.

D. *Broadleaf deciduous:* These trees defoliate periodically. Examples are the oaks and maples. In some tropical areas deciduous trees do not all lose their leaves at the same time, nor does any one tree necessarily become entirely bare. The mapper has then to be guided by the general appearance of the community. The eastern USA and western Europe have many trees of the broadleaf deciduous type.

E. *Needle-leaf evergreen:* The pines, hemlocks, and cedars belong to this category. The western USA has many of these trees.

N. *Needle-leaf deciduous:* Typical are the larches. Eastern Siberia has extensive forests of such trees.

O. *Aphyllous—leaves absent or nearly so:* These trees occur mostly in arid and semiarid areas. Photosynthesis is carried out by the green stems. The *Casuarina* forests of Australia are examples.

M. *Mixed:* Where E and D types occur together, so that each represents about 25% or more of the community, this symbol may be used.

S. *Semideciduous:* Where the broadleaf evergreen and broadleaf deciduous types mix, as in some tropical and subtropical areas, this symbol may be used.

Basic Herbaceous Categories

G. *Graminoids:* This term includes all of the grasses and the narrow leafed grass-like plants such as reeds and sedges. The bamboos are grasses but are excluded because they are woody. The pampa, steppes, prairies and many savannas are composed mainly of plants of this category.

H. *Forbs:* These are the broadleaf herbs. The category includes ferns (except tree ferns) and many flowering plants.

L. *Lichens and mosses:* Only ground plants are included (i.e. not epiphytes). They are important in the tundra and in the case of Ireland, for example, sphagnum moss covers many square kilometres.

Special Life-form Categories

Special categories are necessary because in some areas they give the vegetation a characteristic appearance.

X. *Epiphytes:* These range from mosses and lichens to many large tropical orchids. Their characteristic is that they grow on other plants.

C. *Climbers (lianas):* These are woody plants which climb trees and shrubs.

K. *Stem succulents:* These are characteristic of the warm dry climates. Cacti belong to this category.

P. *Palms:* These may give some areas a distinctive appearance.

V. *Bamboos:* These are usually in thickets. Being both woody and grasses, they do not fit into the main categories.

T. *Tuft plants:* These have trunks surmounted by tufts of leaves. Tree ferns are examples.

Leaf Characteristics

Lower case letters are used.

k : Succulent or fleshy leaves.

h : Hard, leather-like or sclerophyll leaves are common in the Mediterranean.

w : Soft leaves.

l : Large leaves more than 400 cm^2 in area are notable.

s : Small leaves with areas of less than 4 cm^2 also require special mention.

Stratification

The layers of the vegetation are often of great importance, and in many communities layers can be clearly distinguished. To make comparison between one community and another objective, Küchler divides the layers into 8 height classes, each occurring within indicated heights above the ground. The measurement is made from ground level to the average height of each layer at the period of maximum development of the plant in each year. Thus plants which die down after flowering are recorded at the time of flowering.

Table 45.1. Symbols for the Structural Analysis of Vegetation (After A. W. Küchler)

	LIFE–FORM CATEGORIES		STRUCTURAL CATEGORIES
Basic life-forms	*Special life-forms*	*Height stratification*	*Coverage*
Woody plants	Climbers (lianas) C	*Class height:*	*Class coverage:*
Broadleaf evergreen B	Palms P	8 = > 35 metres	c = continuous (> 76%)
Broadleaf deciduous D	Stem succulents K	7 = 20 to 35 metres	i = interrupted (51–75%)
Needle-leaf evergreen E	Tuft plants T	6 = 10 to 20 ,,	p = parklike, in patches
Needle-leaf deciduous N	Bamboos V	5 = 5 to 10 ,,	(26–50%)
Aphyllous O	Epiphytes X	4 = 2 to 5 ,,	r = rare (6–25%)
Semideciduous (B + D) S		3 = 0·5 to 2 ,,	b = barely present,
Mixed (D + E) M	*Leaf characteristics*	2 = 0·1 to 0·5 ,,	sporadic (1–5%)
	hard (sclerophyll) h	1 = <0·1 metres	a = almost absent,
Herbaceous plants	soft w		extremely scarce
Graminoids G	succulent k		(< 1%)
Forbs H	large (> 400 cm²) l		
Lichens, mosses L	small (< 4 cm²) s		

Coverage

The spacing of plants in the landscape is indicated by lower case letters which express the density by recording the percentage of the ground covered by the respective life-forms, assuming they are projected vertically to the ground. There are 6 coverage classes:

c. *Continuous:* The plants touch or nearly touch each other, giving a coverage of 76% and over.

i. *Interrupted:* The plants stand close together but give only 51–75% coverage. The trees seldom touch.

p. *Parklike, or in patches:* The plants grow singly or in groups with a coverage of 26–50%.

r. *Rare:* Coverage is 6–25%, indicating plants which are isolated in a community of different plants.

b. *Barely covering:* Few plants are to be seen. Coverage is 1–5%.

a. *Absent:* There are virtually no plants, as in a desert.

Recording Field Data

Using the symbols it is possible to record the vegetation in a particular place with almost any degree of detail, according to the magnitude of scale chosen. In developing a formula for expressing the nature of the vegetation the details of the major stratum are placed first: thus DG is a forest of broadleaf deciduous trees with a layer of grassland beneath a few of the trees. More completely it would be expressed as D5c G3r. Symbols for lianas and epiphytes are included in the formula for the highest life-forms in which they occur: B8cCX B6p3i P5r3p. Küchler gives two examples to clarify his method.

(1) S6i E8r B4p. This formula describes the physiognomy and structure of the following type of vegetation in California:

S6i: The major synusia consists of a mixture of broadleaf evergreen and broadleaf deciduous trees (S), 10–20 metres tall on average (6), and covering 51–75% of the ground (i).

E8r: Towering above this layer are needle-leaf evergreen trees (E) more than 35 metres tall (8), which cover 6–25% of the area (r).

B4p: Below the major layer is a patchy layer of broadleaf evergreen shrubs (B) 2–5 metres high (4), covering 26–50% of the ground (p).

(2) D7c4r H2p L1r. This formula describes the following vegetation in western New York. There are two layers of deciduous broadleaf plants; one layer 20–35 metres tall and covering more than 76% of the area (D7c), and one layer of shrubs 2–5 metres tall covering only 6–25% of the area (4r); under the shrub layer is a stratum of forbs averaging 10–50 cm in height, and covering 26–50% of the ground (H2p); the lowest stratum is of mosses and lichens less than 10 cm tall, covering 6–25% of the area (L1r). Depending upon the detail required, the last formula might become:

D7c4r H2p L1r	Detailed.
D7c4r H2p	Ignoring lowest stratum.
D7c4r	Concerned only with major features.
D7	Concerned only with main layer.
D	Concerned only with dominant trees of the formation.

In order to collect the information which is required for mapping, the observer should walk around in the vegetation to obtain understanding of its structure before proceeding to record details on his field record sheet. Table 45.2 is an example of a field record sheet upon which all the main details of the plants can be recorded by placing appropriate symbols in the relevant squares. Once sufficient observations have been recorded, the data can be used to compile maps of any scale or degree of detail.

Sampling

A study of vegetation, either to give detailed knowledge of particular communities or to map stages in plant successions, requires the plotting of each plant rather than the recording of layer details. Common methods involve the use of line and belt transects, and quadrats.

Detailed study of forests was begun by Davis and Richards (1933–4) who constructed to scale profile diagrams taken from narrow sample strips of tropical forests. They laid out narrow rectangular sample plots about 8 metres wide and seldom less than 60 metres long and then measured all of the plants within the plot. This often involved felling every tree—starting with the smallest—so that the method can only be fully applied in special circumstances. When all of the trees had been measured, a profile diagram was drawn (see Figure 46.i on page 313).

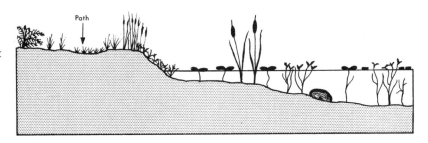

Figure 45.i. A profile transect to show the vegetation round the edge of a pond.

Table 45.2. Vegetation Record Sheet No. ... (After A. W. Küchler)

Location: . . . . Date: . . . .

Height above sea level:. . . . . Base map: . . . .

Slope and exposure:. . . . . Aerial photograph no.:. . . . .

Landscape:. . . . . Type (transect, quadrat, etc.) and size of stand

. . . . samples: . . . .

Structural Analysis

Life-forms: Height classes:	B Broadleaf evergreen	D Broadleaf deciduous	E Needle-leaf evergreen	N Needle-leaf deciduous	O Aphyllous	S B + D: Semideciduous	M D + E: Mixed	G Graminoids	H Forbs	L Lichens, mosses	C Climbers (lianas)	K Stem succulents	T Tuft plants	V Bamboos	X Epiphytes	P Palms
8 = > 35 metres																
7 = 20 to 35 metres																
6 = 10 to 20 ,,																
5 = 5 to 10 ,,																
4 = 2 to 5 ,,																
3 = 0·5 to 2 ,,																
2 = 0·1 to 0·5 ,,																
1 = < 0·1 metres																

Coverage: c = > 76%; i = 51–75%; p = 26–50%; r = 6–25%; b = 1–5%; a = <1%

Leaves: h = hard (sclerophyll); w = soft; k = succulent; l = large (> 400 cm²); s = small (< 4 cm²)

Figure 45.ii. The size of the quadrat will govern the number of individuals counted in each sample of the community.
After Kershaw

Quadrat A

Quadrat B

A line transect may be employed to record details of relationships between plants and soil or other environmental factors (Figure 45.i).

Detailed horizontal mapping is a laborious and very time-consuming operation and is usually reserved for study of small areas of vegetation made over a period of years. Permanent squares or rectangles—permanent quadrats—are marked out and pegs sunk into the ground to make relocation of the site possible. The quadrat may be subdivided by strings or tapes and the plants within each area recorded on squared paper.

The density of particular plants is measured by sampling the vegetation with a quadrat frame. A rectangular or square wooden frame with cross strings dividing it into small squares may be randomly thrown or placed, and the density of plants within it expressed as a percentage of the total area covered by all plants. The size of the quadrat has an effect on the results. Thus quadrat A in Figure 45.ii might record the plant represented by a dot in all of the areas it was used to sample, but quadrat B would indicate a much lower frequency of occurrence of the plant. In a similar way, the grouping of plants greatly affects the results from a sampling. Thus in Figure 45.iiiA sampling would always show the occurrence of the plant shown by a dot. In 45.iiiB the chances of recording the presence of the plant are greatly reduced and in 45.iiiC

Figure 45.iii. Using the same size quadrat, and with the same number of individuals in each case, very different counts will be obtained from each community.
After Kershaw

Quadrat size

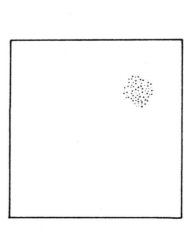

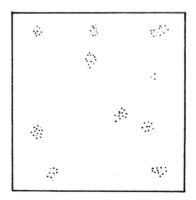

A   B   C

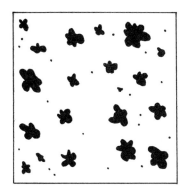

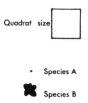

Quadrat size

• Species A

✖ Species B

Figure 45.iv. Using the quadrat size shown very different frequency values will be found although there are the same number of individuals of each species present.
After Kershaw

the chances are low. Under these conditions random sampling may be useless. A similar problem is shown in Figure 45.iv. It is clear, then, that the size of the quadrat and the number of samples taken have to be related to the circumstances of a particular area. Similarly, the intervals at which sampling is carried out along a transect must be related to the area concerned. Sampling on hummocky ground at metre intervals might record only the vegetation of the ridges or depressions, and would not give representative data.

For studying low, herbaceous vegetation, it has been shown that quadrats of $\frac{1}{4} \times 4$ metres are usually efficient. In a forest study, good estimates of tree population have been obtained by quadrats of 4×140 metres and 10×140 metres, with a 14% sample, when the long axis of the plots crossed the contours and vegetation banding.

There is no substitute for common sense in deciding how to select quadrat sizes and the number of samples required, but there are many statistical techniques available which allow the field worker to calculate the size of his quadrats and the number of his samples needed to include most of the variations occurring within the area being studied.

One very simple method is to plot on a graph the number of species accumulated in the sampling against the numbers or sizes of the samples. When the points are joined by a curve, it is usually found that the curve first rises steeply, because many species occurred in the first samples taken, and then it tends to level off as fewer species are added with increased sampling (Figure 45.v). The inflection in the curve represents the point beyond which sampling yields diminishing returns. In this example the community of grasses required a minimum quadrat of not less than $\frac{1}{8}$ square metres. If the number of quadrats used were plotted on the x axis instead of the quadrat size, then it could be discovered how many samples were needed.

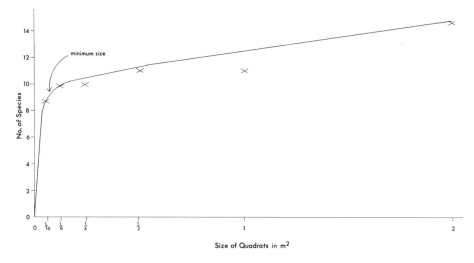

Figure 45.v. A plot of number of species against size of quadrats.

46 Biogeography of the Humid Tropics

The tropical zones encircle the earth in a belt extending to latitudes 23°27′, but the characteristic climatic and vegetational zones are not so rigidly or clearly definable. From a biogeographical point of view, a limit close to the 20°C annual isotherm, or the 18°C isotherm representing the mean temperature of the coldest month, might be a better representation; but there can be no exact or final limit. Away from the equator, vegetation of the humid tropics is gradually displaced by that of the dry tropics, so that an ecotone rather than a clearly defined boundary exists between them, and altitudinally temperature exerts a control so that there is a montane succession of vegetation types. Even the ecotones are not constant in position, as climate fluctuates or natural and man-induced changes in vegetation, soils and landforms occur.

In large areas occupied by the tropics there are enormous expanses of plain and plateau surfaces, especially in South America, Africa, Australia and India. These ancient continental shields have been subjected to extensive periods of successive erosion and uplift since the Jurassic period, so that large areas of their surfaces are occupied by geomorphologically similar features. The influence of slope and soil development on vegetation is very considerable in such conditions. The tropics of southeast Asia and the Cordilleras of the Americas, by contrast, are areas of recent tectonic activity, so that on their mountains temperate and cold climate floras spread into the tropics, and the recent soils of active volcanic areas and eroding slopes bear a vegetation which is not a climax. Man also has produced contrasts in these areas. South America and Australia have been less exploited than the old inhabited areas of Africa and southeast Asia; hence there exist not only cultural differences between types of vegetation use, but also differences in the degree of human interference with the vegetation.

Vegetation Classification

No complete classification of the vegetation of the humid tropics exists. In his *Plant Geography* Schimper spoke of tropical rain forest, monsoon forest, savanna forest, thorn forest and savanna grasslands. He assumed that each type of vegetation is a climatic climax. It has now been clearly established that savanna grasslands are not climatic climax formations; that there is no such thing as a tropical grassland climatic climax; and that tropical grasslands are always edaphic or biotic formations. Beard has given a very useful classification of tropical American vegetation which shows the influence of climate, altitude, soil, drainage and ground-water conditions. It is not relevant to all of the humid tropics but it is more complete than any other system. Most of his formations and their subdivisions do have similar correlative communities in the other tropical areas. Beard defines 28 formations:

A. *Optimum formation*   Climatic control
        Rain forest   ,,   ,,

B. *Seasonal formations*   Climatic control

    (1) Evergreen seasonal forest   ,,   ,,

    (2) Semi-evergreen seasonal forest   ,,   ,,

    (3) Deciduous seasonal forest   ,,   ,,

    (4) Thorn woodland   ,,   ,,

    (5) Cactus scrub   ,,   ,,

    (6) Desert   ,,   ,,

C. *Montane formation*   Altitudinal control

    (1) Lower montane rain forest   ,,   ,,

    (2) Montane rain forest or cloud forest   ,,   ,,

    (3) Montane thicket   ,,   ,,

    (4) Elfin woodland or mossy forest   ,,   ,,

    (5) Paramo   ,,   ,,

    (6) Tundra   ,,   ,,

D. *Dry evergreen formation*   Edaphic control

    (1) Dry rain forest   ,,   ,,

    (2) Dry evergreen forest   ,,   ,,

    (3) Dry evergreen woodland   ,,   ,,

    (4) Dry evergreen thicket   ,,   ,,

    (5) Evergreen bushland   ,,   ,,

    (6) Rock pavement vegetation   ,,   ,,

E. *Seasonal-swamp formation*   Fluctuating water table control

    (1) Seasonal-swamp forest   ,,   ,,   ,,   ,,

    (2) Seasonal-swamp woodland   ,,   ,,   ,,   ,,

    (3) Seasonal-swamp thicket   ,,   ,,   ,,   ,,

    (4) Savanna   ,,   ,,   ,,   ,,

F. *Swamp formation*   Permanent inundation

    (1) Swamp-forest and mangrove forest   ,,   ,,

    (2) Swamp-woodland   ,,   ,,

    (3) Swamp thicket   ,,   ,,

    (4) Herbaceous swamp   ,,   ,,

Tropical Rain Forest

Schimper diagnosed the tropical rain forest as being 'Evergreen hygro-philous in character, at least 100 ft [30 metres] high, but usually much taller, rich in thick-stemmed lianas and in woody as well as herbaceous epiphytes'. Such forests occur only in the non-seasonal tropics, with fairly uniform high rainfall seldom less than 150 cm and often in excess of 200 cm a year. The regularity of the rainfall is much more important than the quantity, and the dry season must not exceed 3 months. This dryness is relative, and those months with less than 5 cm of rain are regarded as dry. There are many areas of the tropics which receive an annual rainfall of more than 200 cm but have no rain forest because of the long dry season.

Mean annual temperatures vary according to locality between 24 and 26°C, but of greater significance is the very small diurnal range in the region of 5–9°C. In the shade of the forest, temperatures are usually between 22 and 32°C but invasions of cold air, as in the *friagems* of western Amazonia, can cause temperatures to fall as low as 10°C without damage to the vegetation. Within the forest relative humidities are constantly high, being at least 70% at 0900 hours; and wind velo-cities are so checked that the air is almost completely still at ground level—a factor which probably accounts for the lack of wind-pollinated plants in the lower strata of the forest. The very dense canopy of the forest excludes most of the sunlight so that, although flecks of direct light may reach the ground, light intensities are between 1/30 and 1/240 of those received by the canopy.

The areas covered by tropical rain forest correspond broadly with the climatic zones called Af and Am by Köppen. They mostly lie within the latitudes 10°N and 10°S, although in both central America and southeast Asia they extend into higher latitudes. These forests can be divided into three main geographical areas with a few outlying groups.

The largest single area of rain forest is the Amazon forest, which extends from the Andes to the Atlantic coast and covers about 3·2 million km². Separate strips of forest extend along the Atlantic coasts of South America as far south as 30°S, and on the Pacific coast it extends from northern Ecuador to Panama, where it covers the whole isthmus and extends north into the Caribbean area.

The African rain forest is the most restricted of the three. It is centred on the basin of the river Congo, from which there is an extension westwards along the south coast of west Africa with a gap (the Dahomey Gap) from eastern Ghana to Dahomey. Eastwards the forest extends to the high plateaux of central and east Africa.

The Indo-Malayan rain forest is widespread but fragmentary. It covers most of the Indonesian islands and many areas of the Melanesian islands. Southwards it occurs on the east coast of northern Australia and northwards it reaches 28°N in the Himalayas, Assam and southeast China, although the monsoon or semideciduous forests are more common near the northern limits. Outliers of the forest occur in Ceylon and western India. The tropical rain forests were once far more extensive than now and the Indo-Malayan and possibly the other two formations have been in existence for most of the Tertiary and Quaternary.

Horizontally the rain forests give way to seasonal forest or, where human interference has been effective, to various types of savanna. Vertically the lowering of temperature, rather than changes in precipitation regime, is effective and montane forest becomes dominant at altitudes varying between 600 and 1200 metres.

Characteristics of the Tropical Rain Forest

It is a popular misconception that the tropical rain forest is an impenetrable jungle through which the traveller has to hack his way with a machete. Along the banks of large rivers, beneath breaks in the canopy caused by falling trees or clearances for cultivation, this may be true, but in the primary forest the undergrowth may be easily passed and the major obstacles are fallen tree-trunks and masses of irregular roots. The low light intensities of the undisturbed forest do not favour vigorous growth of shrubs, so that although a march of 8–9 kilometres in a day may be a good day's travelling, the hindrances are the irregular thickets on swampy ground or at breaks in the canopy.

The forest traveller cannot get an accurate idea of the complex structure of the forest because of the tangle of slender saplings, lianas and hanging roots; serious study, therefore, is usually accompanied by clear-felling a strip of forest, starting with the smallest plants and ending with the tallest. Each tree is located on a map and measured after felling.

Figure 46.i. Profile diagram of tropical rain forest showing all trees over 4·7 metres high.
After Davis and Richards

Stratification

Figure 46.i shows a profile diagram taken from a pioneer study completed in 1933. This section of part of Guyana forest is an accurate drawing of the trees over 4·7 metres high in a strip of forest 41 metres long and 7·6 metres wide. It shows 3 layers of trees—the upper, middle and lower tree strata—each of which has a nearly continuous canopy. Beneath these 3 tree strata are shrub and ground layers, omitted from this diagram, making 5 strata for an optimum rain forest profile.

The uppermost stratum has rounded crowns reaching from 30 to 49 metres above the ground. At the optimum development these form a continuous canopy, but in areas where conditions are not optimum or where much of the forest is secondary—as it is in large areas of Africa and Asia—the upper stratum may be discontinuous and only represented by emergent specimens, or it may be absent. There are giant trees in the tropical forest, but none of them reach the dimensions of some temperate trees like the sequoias of California, over 91 metres high and 9 metres in diameter, or the similar-sized eucalyptuses of Australia. The highest trees of the tropical rain forest, generally a species of *Ceiba*, reach about 59 metres.

The second and third strata comprise both trees which do not grow beyond that level and also younger trees on the way to becoming uppermost stratum trees. As a result, the stratification is not always apparent. The third stratum trees frequently have elongated or conical crowns.

Tree Forms

A striking feature of the trees is the straightness of their trunks and lack of branches, which do not usually exist below about 18 metres. The foliage is not dense and is usually well spaced and able to use most of the incoming light. The bark is usually thin and smooth, although some species have spikes or ribs which make them look more like bundles of canes than single trunks. An unusual feature is the existence of cauliflory, or flowering on thick branches and trunks. On most temperate climate trees the flowers are usually produced only on the shoots of the current year or the previous year, but on many trees of the tropical rain forest the flowers are borne on the trunk or main branches and the whole crown remains purely vegetative. The cacao tree with its pods springing directly from the trunk is an example (Figure 46.ii).

The bases of many trees are also unusual. Most are only shallow-rooted, in response to the soil conditions. Because of the rapid decomposition of the litter layer and the rapid leaching of the soil the nutrients

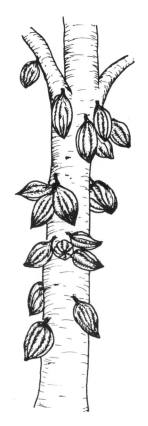

Figure 46.ii. An example of cauliflory: seed pods of cacao growing from the trunk.

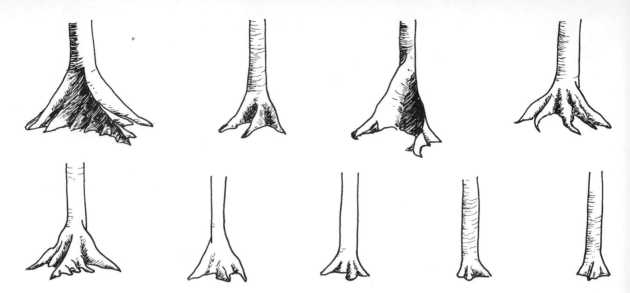

Figure 46.iii. Buttresses of tropical rain forest trees. *After Richards*

Figure 46.iv. Drip tips on the leaves of a fig tree.

are confined to the uppermost centimetres of the soil and there the roots concentrate. There is no need for long tap roots as the water supply is virtually permanent, but a few species do have them. Another, and largely unexplained, feature is the widespread occurrence of buttresses. The buttresses often start 3 or even 9 metres above the ground and are prolonged as vertical extensions of the great roots which spread out over the ground from the base of the tree. The buttresses often have a plank form and as such are frequently cut off by the natives for building purposes (Figure 46.iii). Buttresses occur on trees of many species, but not on all the trees of a particular species. Their function is not known. It has been suggested that they give structural support to the trees, but as only the emergent or upper story trees are affected by wind, this seems improbable. Buttresses no doubt give trees on swampy or shallow soils additional stability; they may assist the rapid movement of water and nutrients to the crown, and they may be related to the damp conditions of the forest floor which might favour the development of aerial roots. Plank buttresses can be formed only in trees with a superficial root system, but in trees with roots which are near the vertical, stilt systems may develop. Stilts are most common in damp areas but not restricted to them. It seems that there is no simple causal explanation for these features.

The leaf-buds of many trees and shrubs have no protective covering of scale leaves and, after bud break, the young leaves are often reddish in colour and hang down, apparently because of the late differentiation of mechanical tissue. The mature leaves of all forest plants are remarkably constant in size and shape, most of them being of intermediate sizes with areas of 13–180 cm^2. They are narrowly ovate to elliptic in shape and possess a drip tip (Figure 46.iv). Even the few forest grasses have similar shapes. The leaves are leathery dark green, with smooth surfaces—all features which assist in shedding water. If the leaf surfaces remained wet for long periods, transpiration, and therefore absorption of soil water by the roots, would be hindered.

Most of the plants of the tropical rain forests are woody and have the dimensions of trees. There are few herbaceous plants, and the members of families which are herbaceous in temperate climates are often woody

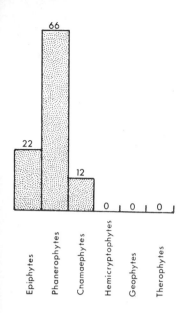

Figure 46.v. Raunkiaer biological spectrum of tropical rain forest flora at Morabilli Creek, Guyana.
After Richards

77. This garden in New Guinea has been abandoned for three years.
Photo by J. van Wessen

in the rain forests; thus there are violets the size of apple trees, and grasses like the bamboos 12–18 metres high. The life-form spectrum therefore indicates a preponderance of phanerophytes with large numbers of epiphytes, some of which are also woody, and a lesser number of chamaephytes (Figure 46.v). Because there is no season unfavourable to growth, plants like geophytes, which have a protected perennating bud, are rare and the palms are so lacking in protective adaptations that they have only one unprotected perennating bud in the crown of the tree.

Lianas, Epiphytes and Herbs

It has been estimated that over 90% of all species of climbing plants live in the tropics, and most of these are to be found in the forests. Ecologically the climbers belong to one of two main groups: either they are small, usually herbaceous plants which grow on the lower parts of trees in the deep shade, or they are large, woody lianas. Lianas are found throughout the forests but they are most abundant in the gaps where sunlight is greatest. Lianas grow by scrambling, twining or attaching themselves to trees by tendrils or roots. Most do not produce leaves until they reach the sunlight of the canopy. They are very rapid growers and able to take advantage of any break in the canopy, so that abandoned cultivation patches or riverside areas are choked by lianas; in the absence of suitable supports, some lianas assume a shrub life-form. Their main stems are seldom branched and produce no leaves or flowers, but the head can be so large as to shade out the support and kill it. Particularly effective killers are the stranglers such as the figs, *Ficus*. The strangling figs germinate epiphytically and their roots grow down and reach the soil so that they cease to be epiphytic. As the fig grows, more roots are produced and as these increase in girth and anastomose round the trunk of the host, the latter is restricted. This, together with the shading effect at the crown, kills the host which eventually rots away leaving a hollow fig tree (Figure 46.vi).

Epiphytes grow attached to the trunks and branches of rain forest trees and even to the surfaces of some leaves. In a closed forest the epiphytic habitat is the only niche available to small plants with high light requirements, although the better light supply is offset by precarious soil and water conditions.

There is seldom a closed layer of herbaceous plants in the tropical rain forest. The herbs are usually concentrated in small clearings by paths and streams where more light is available. Within the forest they are widely scattered rather than grouped, and even in the regions where they are greatest, their numbers are only a fraction of the number of trees and shrubs in the same area.

Floristic Composition

In no other type of vegetation are as many species of plants found as in the tropical rain forests. In west Africa there are about 6000 species of flowering plants, in Malaysia there are about 20,000 and in Brazil about 40,000. Because of this richness it is most unusual to find any stand dominated by one species of tree, although a few cases of concentration of one species, as in some teak forest areas of Burma, are known. Most commonly an enormous number of plants of many species exists in any area, but usually so scattered that concentrations of any one species are seldom more than 1–3 per 4000 square metres. The total number of plants in a small area can be very great as is shown by these figures for a plot of 1350 square metres of 30-year-old secondary forest on Singapore Island (after Burkill):

378 trees over 5·5 metres high
2,728 woody plants 0·6–5·5 metres high
<u>27,342</u> smaller plants, mostly woody seedlings
TOTAL: 30,448 plants on 1350 square metres

Such distributions seriously limit the usefulness of tropical rain forests for timber exploitation, compared with the single species forests of northern latitudes.

Cultivation and Regeneration

The traditional method of agriculture in the tropical rain forest is shifting cultivation. When an area of forest is prepared for cultivation, the shrubs and small trees are cut down, stacked against the bases of the tall trees and then burnt so that the large trees are also killed. Because of the very rapid leaching, the nutrients are exhausted in one or two years and the area has to be abandoned. Within a few months there is a tangle of saplings and climbers occupying the area, trees with wind-borne seeds dominating. The succession to primary forest again would take about 250 years but this seldom occurs, for in many areas the land will be cleared again within a period of from 5–20 years.

Within an undisturbed forest, tree saplings are rare because of the high mortality caused by low light intensities, high competition and attacks by animals. Any saplings which do survive are very slow growing, but when a gap occurs in the forest canopy, as when an old tree falls and breaks down smaller trees, climbers will develop rapidly in the fuller light so that the gap is of short duration (see below for discussion of human, edaphic, ground-water and montane influences on rain forests).

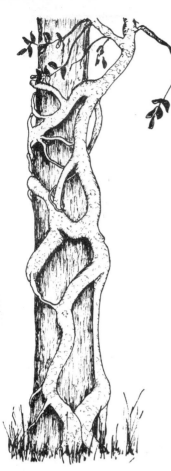

Figure 46.vi. A strangler fig growing round its host.

Tropical Forests with a Seasonal Rhythm

The optimum evergreen rain forest shows no seasonality of flowering or leaf fall and both occur throughout the year, but as the optimum conditions diminish, forests take on an increasingly seasonal appearance. Figure 46.vii shows the changing structure produced by the changing precipitation regimes.

In general there is a stepping-down of structure so that the evergreen seasonal forests have 3 tree layers but a discontinuous upper layer. The semi-evergreen seasonal forest has only 2 layers of which the uppermost is deciduous and the lower is still evergreen, and in the deciduous seasonal forest the upper layer becomes discontinuous. Thorn woodland has only 1 layer left and its drier margins are occupied by semidesert vegetation. This profile is highly schematic and in actual situations thorn woodland may occur at a sharp boundary with optimum rain forest, especially where human interference is strong.

Semi-evergreen Seasonal Forests

These forests are extensive in the West Indies and in a belt around the Amazon rain forest. In Asia, where they are often called monsoon

30 m

| Evergreen Seasonal Forest | Semi-Evergreen Seasonal Forest | Deciduous Seasonal Forest | Thorn Woodland | Cactus Scrub |

Figure 46.vii. The seasonal formation series of the South American tropics.
After Beard

forests, they are common in India, Burma, southeast Asia and Indonesia. They are rare in Africa. They are discontinuous and not easily mapped because of their patchy distribution and doubtful ecological status. In India and Burma, when protected from fire and human interference, they frequently take on the appearance of the true rain forests, so it can be assumed that the semideciduous forest is a response to interference. Typically these forests are found where the mean annual rainfall is 75–125 cm in America, but rather more in Asia. The dry season, or seasons, usually occupy 5 months of the year and each has a mean rainfall of less than 10 cm but more than 2·5 cm.

The structure of the semideciduous forests is simple, with an upper tree layer at about 21 metres composed of deciduous trees and a lower tree layer of small-leaved evergreen trees. The trees are more widely spaced, have thicker bark, fewer buttresses and lower branches on thicker trunks than the evergreen rain forest trees.

78. The patchwork appearance of shifting cultivation and rain forest in central Johore.
Photo by D. W. McKenzie

79. Thorn woodland in
northern Nigeria slightly
opened up by fire.
Shell Oil Co. Ltd

The shrub layer usually flowers in the dry season so the forest never has the true winter appearance of temperate deciduous forests. Grasses and ferns and lianas are common but epiphytes are smaller in number. In the often thick undergrowth, geophytes are fairly common. Stands of one species are still rare but trees like the teak (*Tectona grandis*) may represent 10% of a stand in parts of Burma.

Deciduous Seasonal Forest

Where there is a dry season of 5 consecutive months each with less than 10 cm of rain, the forest becomes more deciduous. It has 2 strata, an upper discontinuous one not exceeding 18 metres and a lower one between 3 and 9 metres with both deciduous and evergreen trees. The trees are gnarled, crooked, low-branched, often small-leaved, poor in lianas and almost devoid of epiphytes. The shrub layer and ground layer is thin, although in Burma dense thickets of bamboo are common. These forests occur in areas with as little as 100 cm of rain a year, but in regions of porous soils they may extend into areas with 190 cm a year.

Thorn Woodlands and Forests

No abrupt change in vegetation type occurs where the deciduous seasonal forests of America and Asia are flanked by even drier regions. As the dry season becomes longer and mean annual rainfall declines, the trees become lower and have a more gnarled and spreading habit. More and more species disappear, so that the communities consist of low-growing trees and shrubs, which are often thorny, while beneath them the ground may be bare, grassy or occupied by xerophytic plants like the cacti.

Thorn woodlands, thorn scrub and thorn forests, as they are variously called, occupy large areas in America, Africa, Asia and Australia. In America they occur in the northern coastal areas of Venezuela and Colombia, and are extensive in northeastern and southern Brazil. In Asia they occur in large areas of central India and in the dry zone of Burma. In Africa the various species of acacia, with trees of similar form, are almost universal in a belt south of the Sahara from Guinea to Kenya and through central and eastern Africa into the Union of South Africa. In Australia there are extensive areas in interior Queensland dominated by mulga and brigalow.

The Venezuelan thorn forests have a canopy at 3–9 metres. The trees are both deciduous and evergreen, but the evergreens have xeromorphic adaptations in their leaves, which are hard and small. Scattered amongst them are several species of cacti, some of which overtop the thorn trees. In the Brazilian 'caatingas' are bottle trees with swollen, fleshy, water-storing trunks up to 4·7 metres in diameter. The thorn forests occur typically in areas with a mean annual rainfall of 65 cm and a dry season of 6 months or more, but in areas of pervious rocks, such as many sandstones and limestones, they exist in areas of 165 cm of rain a year, especially where there are occasional severe droughts, as in northeastern Brazil, which prevent invasion by evergreen trees.

The Asian, African and Australian thorn forests are similar in appearance to those of America, but the species are different in each case and although bare ground is a feature in all areas, there also tends to be more grass cover in these continents than in America. The Australian bottle tree has similar form and functions to its Brazilian equivalent, although it is of a different family (Figure 46.viii).

Figure 46.viii. The bottle tree of Australia (*Brachychiton rupestris***) grows on the wooded savannas of Queensland.**

Edaphic and Hydrologic Modifications of Evergreen Rain Forests

Where ground-water conditions permit, evergreen rain forest may extend into the zones of the seasonal forests, so it is not unusual to find rain forest as a gallery forest along water-courses in areas with far too low a rainfall to maintain it. These forests have a profusion of lianas, palms, epiphytes and moisture-adapted plants. In the true rain forest, by contrast, climatic conditions may be suitable for the forest but soil and ground-water conditions can produce an environment which is unsuitable for it, being too dry in the first case and too wet in the second.

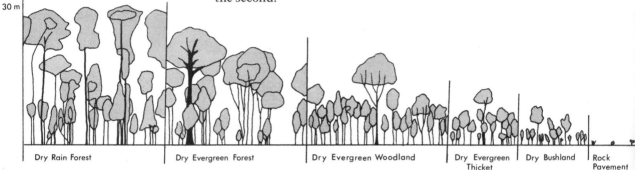

Figure 46.ix. The dry evergreen formation series.
After Beard

Within many rain forests there are outcrops of porous and permeable rocks like sandstone and limestone, and inselbergs, often of massive igneous rocks, through which the water rapidly seeps or runs off to leave an excessively drained soil, or on some inselbergs no soil at all. In some parts of the Guyana forests of South America, a soil with a massive upper horizon of very porous and infertile sand, which appears to be a tropical podzol, also gives excessively dry conditions. In such situations the rain forest becomes depleted, lower in stature and less complex in composition. In Figure 46.ix the types of resulting forest are shown. These vary from only slightly modified rain forest to the communities of lichens, mosses and grasses which colonise bare rocks, and the thin accumulations in the rock joints and cracks. Which type of community develops will depend upon the local circumstances.

In excessively wet conditions with a high water table, either of fresh or of saline water, a swamp vegetation replaces the rain forest. The main types of swamp are shown in Figure 46.x. The swamp forest is the fresh water equivalent of the salt water mangroves. Many swamp trees and shrubs have stilt and aerial roots; others have completely submerged root systems and vegetative parts above the water; and others like the water-lilies have vegetative parts which float on the water. Swamps are common in the humid tropics in shallow, depressed areas like Lake Chad, or around the edges of lakes or rivers. Many are dominated by one

Figure 46.x. The swamp formation series.
After Beard

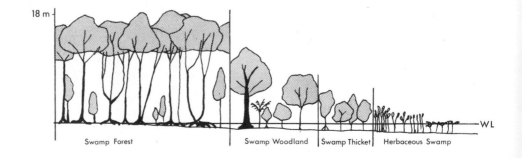

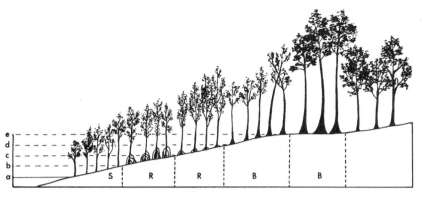

Figure 46.xi. Mangroves. The genera *Sonneratia* and *Bruguiera* have widespread root systems which support the trunk, and shoots and loops which breathe. *Sonneratia* is the pioneer genus (S) which colonises mud just above the level of the lowest tide (a). It is replaced by *Rhizophora* (R), the roots of which are regularly covered by tidal water. *Bruguiera* (B) flourishes on high banks washed only by spring tides (e).
After Di Palma

Sonneratia

Bruguiera

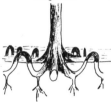

Rhizophora

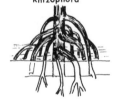

plant, like reed swamps of the Congo, or papyrus in the Chad and Nile regions.

Saline swamps are usually dominated by mangroves. Mangroves are not dependent upon rainfall and occur on many tropical coasts in Africa, Indonesia, and around many Pacific islands in Central and South America. They occur mostly within latitudes 25°S and 25°N but they extend, although reduced in size, outside these limits in warm seas, to southern Florida, the south of Japan and the north of New Zealand. Mangrove swamps everywhere have the same general appearance and occur on any low-lying, muddy coasts and around brackish lagoons and estuaries, where they may extend 95 km or more inland. As a pioneer the mangrove colonises mudflats and, by encouraging siltation, plays an important role in building up land. Mangroves have an economic use in some areas as providers of bark for the tanning industry, and of timber, and of fodder for camels in the Red Sea desert area.

Mangrove trees are of modest height, generally reaching 9–15 metres high in the tropics. They have tall, slender stilt roots and also long aerial roots hanging from their branches. The roots develop in completely oxygen-free mud and send up pneumatophores. The young plants develop on the parent tree and are able to take root as soon as they fall into the mud or are washed on to it by the sea. As a result they can colonise mudbanks very rapidly (Figure 46.xi).

80. Tree savanna in Australia, showing tufted grasses with bare soil between the tufts.
Photo by R. J. Blong

81. Grass savanna in
Australia.
Photo by R. J. Blong

The various species of mangrove, *Rhizophora*, *Bruguiera*, *Avicennia* and *Sonneratia*, are of similar appearance and may occur in mixed communities, but *Sonneratia* or *Rhizophora* is usually the pioneer, as it builds up a soil by encouraging silt deposition and accumulation of organic matter. *Avicennia* or *Bruguiera* tends to take over behind it, and then the ground is slowly prepared for invasions of the rain forest.

Savanna

The word savanna is derived from an Amerindian word used in Haiti and Cuba for treeless plains. Its use has been extended to tropical grasslands throughout the world and to include areas with a partial cover of trees over the grasses. The great zones of savanna are unevenly distributed. In Africa they cross the continent from east to west south of the Sahara, extending over much of the high plateau area of central and east Africa, and into southern Africa. In Australia they are largely confined to the north and east of the continent, where they are mixed with bush forest or open eucalyptus forest. There are small patches of savanna in Indonesia and Malaysia but cultivation has destroyed most of their former range. In Central America they occupy only small areas but in South America they are of great importance where they form the Llanos of the Orinoco Basin and the campos cerrados of central and southwest Brazil.

Figure 46.xii. Profile of
savanna grasses in Albert
National Park.
After Lebrun

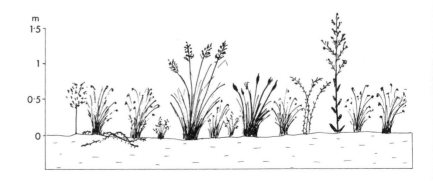

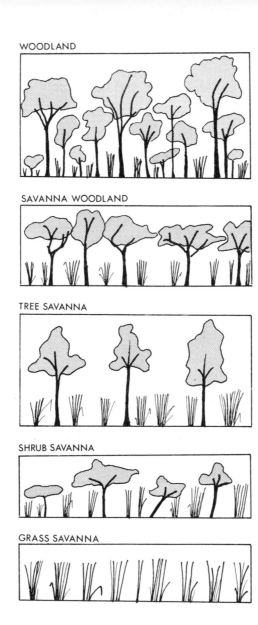

WOODLAND

SAVANNA WOODLAND

TREE SAVANNA

SHRUB SAVANNA

GRASS SAVANNA

Figure 46.xiii. Types of savanna in west Africa. *After Hopkins*

Composition and Structure of Savannas

Savanna grasses grow to a height of at least 80 cm and have flat leaves. Many of the grasses actually become 1·5–1·8 metres high in one season and some, like the elephant grasses (*Pennisetum spp.*), can form a dense growth 3–4·7 metres high. The grasses all have the tussock form and between them there are many herbs and forbs (Figure 46.xii). Most of the grasses and forbs have tubers, bulbs, rhizomes or rootstocks which are adaptations probably more effective in protecting them against fire than against drought.

In each major area where they occur there is a range of savanna types which can be classified according to the tree cover. In decreasing importance of woody plants they are: savanna woodland; tree savanna; shrub savanna; and grass savanna (Figure 46.xiii). A life-form diagram

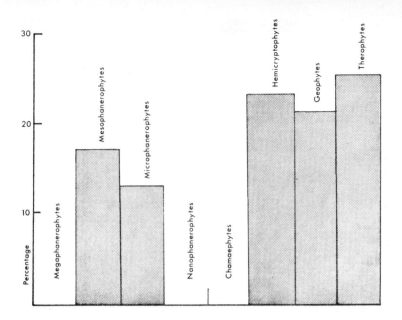

Figure 46.xiv. Savanna life-
form spectrum.
After Hopkins

will therefore reflect the structure of the area studied. Figure 46.xiv is
a spectrum diagram for a savanna woodland in Nigeria. About 1/3 of
the species are phanerophytes (trees), 1/4 are therophytes and 1/5
hemicrytophytes, and a further 1/5 geophytes. This contrasts with a
rain forest only 6·4 km away which had 90% phanerophytes and a grass
savanna which had no phanerophytes.

The trees in the savannas vary greatly according to climatic, soil and
fire factors, but in most savannas there is a marked poverty of species
and often a dominance by one species over very large areas. The
savanna woodlands of east Africa, for example, have acacias as the main
tree species for hundreds of square kilometres. Savanna trees are
predominantly deciduous and seldom more than 9–12 metres high,
except where they occur in groves or on the edges of true forest (Figure
46.xv). Many of the trees and shrubs are thorny with thick, fire-
resistant bark and small or medium-sized leaves. Adaptations to
withstand drought and fire are perhaps best seen in the giant baobab
(*Adansonia digitata*), with its thick bark and sponge-like, water-
saturated wood, which stands leafless for 8 months in the year amid
grass savannas in large areas of Africa. The arborescent euphorbias
(*Euphorbia abyssinica*) which look like American cacti are similarly
adapted (Figure 46.xvi).

Most of the savanna trees have a very large root system with long tap
roots able to use water at depth, and an extensive fan of shallow roots
which spread out horizontally some centimetres beneath the surface.
At the onset of each wet season, suckers spring up from these roots and
form the shrub layer.

Origin of Savannas

Schimper, writing at the beginning of this century, was convinced that
savanna grasslands are a climatic climax of the seasonally dry tropics.
Since his work it has become increasingly clear that this is not correct
and, with the possible exception of some of the Australian savannas, it

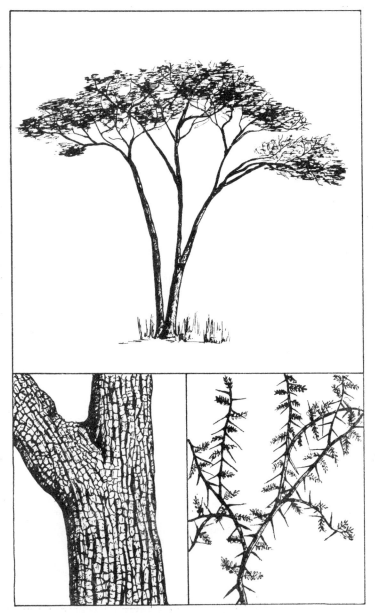

Figure 46.xvi. An arborescent euphorbia (*Euphorbia abyssinica*) typical of the savannas of Kenya.

is safe to say that all savannas are the result of either edaphic, or drainage, or fire factors, or of human interference, and that they occur in areas which would have a climatic climax vegetation of some type of forest or woodland.

It may seem odd that such enormous areas should have a vegetation that is deflected from its climatically controlled climax but three factors seem to be largely responsible for this.

(1) The savannas occur mostly on old erosion surfaces of the ancient continental shields with impeded drainage and other soil conditions unfavourable to trees.

(2) They occur in areas of ancient human occupance in Africa, with a long history of man-induced fires which are easily started in the long dry season.

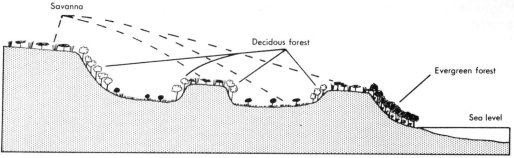

Figure 46.xvii. Diagrammatic section through eastern Brazil showing the relationship between plant communities and erosion surfaces.

(3) Quaternary climatic change probably started the retreat of the forests into their present diminished zones and so made restriction by fires and destruction of the forest equilibrium easy.

The topography of the African and Brazilian shield areas is composed of a series of uplifted erosion surfaces. The soils on these plateaux are either red and white impermeable clays or sandy topsoils, seldom more than 0·9 metre deep, overlying mottled clay or an ironpan. In either case, drainage is likely to be poor because water will only run off inclined surfaces and the clay is too impermeable to allow water to soak through it. Within the sand a perched water table exists, but for much of the wet season water occupies all depressions and stands to a depth of some centimetres over much of the savanna during and after heavy rains. Few forest trees are able to survive such conditions and hence only occur on the better drained sites. In eastern Brazil a section through the plateaux to the coast shows savanna grasslands and savanna woodlands (campos cerrados) on the flat erosion surfaces; thorny deciduous forest on the scarps between the surfaces (caatinga); and evergreen forest on the coastal zone slopes and along the river-courses (Figure 46.xvii).

It seems clear, then, that soil drainage conditions, possibly aided by the infertility of many of the leached soils, are responsible for large areas of savanna. The distribution of the various types of savanna also depends on local features. Figure 46.xviii shows this relationship quite clearly. In Venezuela the forest and woodlands show the usual response to increasing aridity, becoming lower and more open with decrease in rainfall. Locally, however, in an area of dominantly shrub savanna, trees will occupy well drained rock outcrops; shrub savanna the flat plains where there is a sandy topsoil overlying impermeable clay; and scattered bunch grasses only where the clay forms the surface (Figure 46.xix).

In Africa the situation is rather different from South America. Edaphic controls may still exist in some areas, but over most of the savannas of the continent fire is the main control. Many fires are

Figure 46.xviii. The relationship between vegetation and site in Trinidad. The evergreen seasonal forest (ESF) occurs on well drained sites, savanna on poorly drained plateaux, and secondary bush (SB) develops where drainage improves.
After Beard

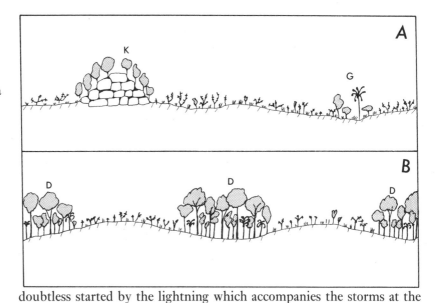

Figure 46.xix. The association of vegetation and site in the Venezuelan Guiana. [A] Forest on a granite knoll (K) and gallery forest (G) in a valley, savanna on flat areas. [B] Evergreen forest in moist depressions (D) and savanna on ridges. *After Beard*

doubtless started by the lightning which accompanies the storms at the beginning of the wet season, but for at least 10,000 years man has been occupying much of Africa and using fire to clear forests; for agriculture; to improve grassland for grazing domestic animals or game; to deprive game of cover or to drive game while hunting it; to repel attacks or drive enemies out of refuges; to protect settlements from larger fires by controlled burning of inflammable vegetation around them; and probably for the pleasure of seeing a good blaze.

82. A giant baobab in the Northern Transvaal. *South African Tourist Corporation*

From the descriptions already given it is clear that the savanna vegetation is adapted to withstand fire and that non-adapted plants are largely excluded. In most of Africa at the present, fires are used annually to burn off the dead grass of the previous season and encourage the young shoots to emerge from the centre of the clumps of grass. The burning probably also releases nutrients from the litter which, in the dry season, would only be released slowly, but there are harmful effects as well. The excessive leaching common to the soils of most of the humid tropics results in the continuous loss of plant nutrients and the burning prevents the accumulation of vegetable and animal organic matter which would help to retain them.

Where savannas are protected from burning, trees and shrubs multiply and a succession begins which would lead in most places to some kind of forest, except in those few places where the soil structure has become inimical to tree growth because of long periods of erosion and depletion, or where there are no trees surviving to provide the seed for regeneration. In those places where tree regeneration does not occur, it might well take thousands of years of weathering and gradual change before the effects of climatic fluctuations and burning are eradicated.

In conclusion, it seems that no savannas are climatic climax forms of vegetation. Over large areas of Africa, and maybe parts of South America, savanna is a response to fire. In much of South America and in parts of Africa soil, ground-water and topographic conditions, influenced by the climate, are responsible for the form of the vegetation, and the composition of the flora is adapted to fire and grazing animals which have destroyed many palatable species and left thorny species to dominate the trees.

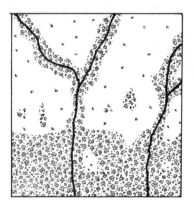

Figure 46.xx. Diagrammatic distribution of forest and savanna grassland around the forest-savanna boundary. Fringing forest is left along the rivers and relict patches in the grassland. Patches of savanna occur in the forest as a result of burning.

Savanna/Forest Boundary

Most savannas occur within areas with a rainfall of 50–100 cm per year, and a dry season of 5–7 months with less than 2·5 cm of rain a month, but some exist in areas of up to 150 cm a year and with a very short dry season.

In these wetter areas the forest/savanna boundary can be very narrow and the gradual modifications characteristic of broad ecotones may be replaced by an abrupt change from forest to grassland. In much of west Africa the boundary zone is as little as 30 metres wide. Each year the fires reach the edge of the forest and kill seedlings from it and so prevent it spreading on to the savanna. In an exceptionally dry year the edge of the forest itself may be burnt and so be forced to retreat a little. If the

Figure 46.xxi. The effect of fire on the climatic climax vegetation types in west Africa from north to south. Climatic climax types on the left, fire climax on the right. *After Hopkins*

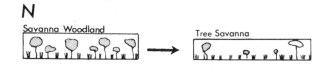

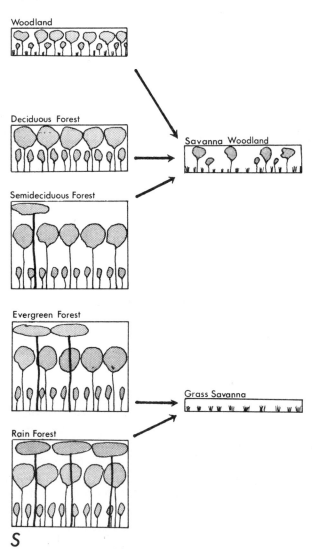

retreat is assisted by man, who cuts trees as well as fires the grass, it can be rapid, and only relict patches of forest on damper sites or fire-protected hills will indicate its former extent (Figure 46.xx).

Within the forest far from the savanna boundary, clearances for cultivation will rapidly be covered by forest regrowth, but near the boundary, wind-carried grass seeds from the savanna will introduce a grass vegetation which, because it is more liable to burning, will often be gradually extended until it becomes a grass savanna patch within the forest. The typical savanna trees do not easily enter this grassland because their seeds are not carried far by the wind. The effects of fire can then be summarised in a series of diagrams (Figure 46.xxi). The moist forests, when burnt, give way to grassland. The deciduous and semideciduous forests and woodland give way to savanna woodland, either because they contain some fire-resistant species or because they are near savanna woodlands, and the savanna woodlands on severe burning become tree savannas.

Animals in the Humid Tropics

Forests

In the tropical forests the animals are closely related to the various strata of the vegetation. Because it is the zone of light and most abundant food the uppermost tree stratum has the greatest wealth of animals. This is illustrated by species of birds in Costa Rica, where it was found that 14 species foraged on the ground, while 6 more drop to the ground regularly. There are 18 species in the shrub layer, 59 living in the lower tree stratum, 67 in the middle tree stratum and 69 in the upper canopy. Even mammals are primarily arboreal and in Guyana, for example, 31 species of forest mammals live in the trees but only 23 species are ground dwellers.

Animals have evolved a number of adaptations for living in the trees: there are suitable claws; feet that can grasp, as in monkeys and men; suction discs on toes, as in some tree frogs; adhesive slime like that of snails; and special organs like the prehensile tails of reptiles and mammals. In addition there are many adaptations for jumping, gliding and flying.

Prehensile tails help many animals to live in trees, like opossums, some anteaters, arboreal porcupines, as well as the New World monkeys (the African and Asian monkeys do not have prehensile tails). The great apes are also forest animals, but of the four kinds only the Asiatic gibbons are truly arboreal. The others, like the African gorillas and chimpanzees, are too heavy to be real acrobats, although they are still good climbers. It is probable that the opposable thumbs, binocular and colour vision, which give man and the apes so many advantages, originally evolved as adaptations to arboreal life.

On the forest floor, there are many animals but few large ones. Of the large African mammals, for example, only the elephant, buffalo, okapi, wild hog and leopard occur in the forest. In the American forest only the tapir and jaguar are large, and the deer are all dwarf species. The characteristic animals of the floor are the pigs of the Old World and the peccaries of the New. These omnivores move in bands and work co-operatively, moving rotting logs or ploughing up ground to get at their food. The other important animals are the rodents and insects, which are exceedingly numerous but often inconspicuous. The most numerous and notorious insects are the driver ants of Africa and the army ants of America. They forage in bands which may have 100,000 individuals and move along the forest floor and lower canopy killing and dissecting any animals which are too slow to escape.

The tropical forests are not the snake-infested jungles of popular imagination; and even an expert may go for days without seeing one, but snakes are there, often well camouflaged, and filling many niches in all strata of the forest.

83. Giraffes in Nairobi National Park. Kenya.
Paul Popper Ltd

Savanna Animals

Of the savanna grasslands only the African areas have a well developed population of large herbivores with attendant carnivores. In South America there are rodents, birds and small deer in the savannas, and in Australia kangaroos, but in Africa there is a varied and numerous fauna. The African herbivores are not all in competition with each other because they can occupy a great variety of niches. The giraffes, most obviously, are adapted to browsing low trees and shrubs; elephants frequent both forest and tree savannas, and in moving between the two they make paths for other animals by trampling vegetation. The hippopotamuses are always associated with water, coming out of the rivers and pools only at night except in the rainy season. In general the animals show marked preferences for wet or dry conditions and migrate with the seasons.

The balance between the herbivores and the pastures depends also on the carnivores, of which the largest are the lions. Lions, tigers (found only in Asia), cheetahs and hyenas are chiefly open country predators. Some, like the lions, work together as a hunting group, but others hunt singly like the tiger. Their prey also work in social groups, and in Africa it is a common sight to see large herds of gazelles, antelopes, zebras, giraffes and elephants; each species exploiting a different part of the environment and all depending upon co-operative warnings, and either speed or sheer size for safety from the predators.

If the savannas are largely fire-induced in Africa, and that by human interference, then the populations of large herbivores must have increased enormously and extended their range in the last few thousand years. Animals like the giraffe and zebra, by their coloration, seem adapted to open woodland and the forest edge, and animals like the gazelles were probably once inhabitants of the semideserts. Time has been too brief for evolution of adaptations in recent times, but those animals which were pre-adapted or could make behavioural adaptations could extend into the newly created open areas.

Tropical Mountains

It is often stated that on tropical mountains there are zones of soils and vegetation which correspond with the zones found near sea level varying with latitude. Thus at the base of tropical mountains rain forest will prevail, above it montane forest (which has no latitudinal equivalent), and then deciduous forests, coniferous forest, alpine meadows and tundra-type vegetation beneath the zone of perpetual snow. On many mountains there is such a general pattern, but the analogy with latitudinal zones obscures the real nature of zones on tropical mountains, and is most misleading.

Mountain ranges in the tropics have one similarity with latitudinal effects and that is a fall of mean temperature towards the summit, which has its analogy in mean temperature decrease towards the poles—but that is the only valid similarity and its effect is often obscured by other factors. Towards the poles length of night and day varies with the seasons and tundra plants are adapted to summer periods of 24-hour daylight and similar lengths of winter darkness. On tropical mountains night and day are about the same length throughout the year. The gradual fall of about 1°C for every rise of 100 metres may have a latitudinal counterpart in some hundreds of kilometres, but within the tropics there is little seasonal effect, so that at Quito in the Ecuadorian Andes, at 2850 metres and almost on the equator, the average daily temperature is about 13°C throughout the year and varies from the mean by only about half a degree. In the Rocky Mountains the seasonal difference at 2850 metres might be 17°C. Tropical mountains do not have the cyclonic storms and severe weather changes of temperate climates, and in every way their weather is more constant. The absence of seasonal effects and hence the lack of a resting period, caused by the high elevation of the midday sun throughout the year, together with the precipitation regime produced as moist air is forced to rise up the ranges, often to give increased precipitation and cloudiness with altitude, have produced a unique flora which is not comparable with those outside the tropics.

Figure 46.xxii. Schematic chart of the altitudinal distribution of a mountain plant. The density of large dots is proportional to the number of localities in which the plant occurs.
After Van Steenis

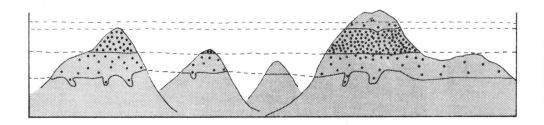

Plant formations have evolved which are adapted to the lack of seasons and to prevailing precipitation, so that there are some remarkable examples of convergent evolution to be found on widely separated, and often isolated, tropical mountains. There are differences between the various ranges, but enough similarities to make generalisations possible. The vertical zonation shown on any simple diagram is misleading. As can be seen from Figure 46.xxii, plants are not confined strictly within limits but an ecotone exists between the zones. In a valley one formation may tongue up into a higher zone because of the protection from adverse winds, while on an exposed ridge the plants typical of a higher zone will penetrate into a lower one.

Vegetation Zones

With altitude there is a general decline in the height and complexity of the vegetation of tropical mountains. This is shown in Figure 46.xxiii, which indicates the altitudinal sequence for tropical South America. Starting at sea level, a traveller would notice a gradual change of species composition of a tropical rain forest as he climbed upwards. The structure would remain the same for a couple of hundred metres but with altitude there would be an increase in precipitation as well as a slight cooling effect, so that a transition would occur into a montane forest with lower trees, more massive and lower-branching than the tropical rain forest, and without plank buttresses. Only 2 tree strata would be discernible and the foliage would be less dense; hence a thicker ground vegetation would develop.

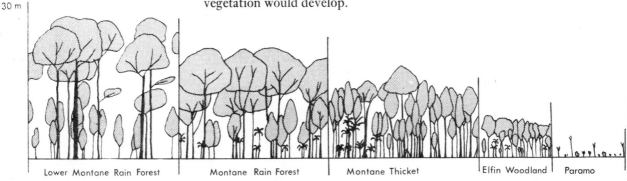

30 m

Lower Montane Rain Forest   Montane Rain Forest   Montane Thicket   Elfin Woodland   Paramo

Figure 46.xxiii. The montane formation series.
After Beard

With increasing height the total number of species declines but more temperate species enter the forests and the epiphytes are largely ferns, mosses and liverworts. In the high rainfalls and pervading cloud and mist, the upper parts of the montane forest have a mossy appearance which becomes one of its chief characteristics.

In the subalpine zone the trees become lower still and increasingly xeromorphic in structure, and sometimes deciduous, although this is not a response to seasonal effects. Temperate species may become increasingly common; thus evergreen oaks (*Quercus spp.*) become common in Malaya and Indonesia, and conifers, although usually of different species from those of higher latitudes, may become dominant. In New Guinea the montane conifers are close relatives of those which dominate in the forests of New Zealand and subtropical South America, so *Agathis alba* (a relative of New Zealand's kauri, *Agathis australis*) and *Araucaria cunninghamii* (of the same genus as Chile's monkey puzzle tree, or araucaria pine) are common. Several species of *Podocarpus*, a genus of southern hemisphere conifers, occur similarly. In Mexico the familiar pines (*Pinus spp.*), spruces (*Picea spp.*) and firs (*Abies spp.*) are common where, near the limits of the tropical zone, there is increasingly a seasonal regime on the mountains.

Maximum precipitation and cloudiness usually occur around 1830–3050 metres, and here the montane thickets or subalpine forests with

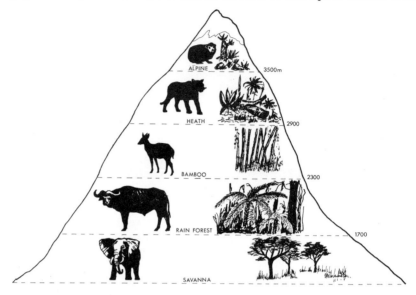

Figure 46.xxiv. Altitudinal zones on the Ruwenzori massif. Elephants are typical animals in the savanna, buffalo in the rain forest, duiker in the bamboo zone, leopards in the heath, and the rock hyrax in the alpine zone.
After Milne

deep coverings of moss are at a maximum. Where the forests extend above this zone they may become drier, less mossy and even taller, before the increasing exposure and cold cause a transition to the elfin woodland or low, ground-hugging forest. This can be merely a dense shrub forest, but frequently it is an extensive krummholz of horizontally growing trees.

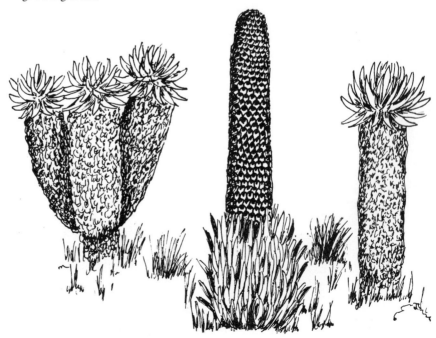

Figure 46.xxv. Giant lobelia (*Lobelia bequaerti*) in the centre, and tree groundsel (*Senecio johnstoni*). These plants are characteristic of the alpine zone of the high mountains of east Africa near the equator.

In the alpine zone above, of shrubs, heath and grasses, there frequently occur plants which are peculiar to their mountain range. These seem to have evolved in isolation, yet there are other species similar to those found in cool climates elsewhere. The east African Ruwenzori range has a number of peculiar features (Figure 46.xxiv). Not only does it have a zone dominated by bamboo, and a zone of giant mossy heath, related to the heaths of temperate climates but 3 or more metres higher, but in the alpine zone there are giant groundsels (*Senecio johnstonii*) and lobelias (*Lobelia spp.*) which may be 3–4·7 metres taller than their temperate relatives (Figure 46.xxv).

Mountains in the Dry Tropics

Where mountains rise from dry zones, the pattern already described may be broken and in the extreme case of the western slope of the Andes from northern Peru to north-central Chile the vegetation is desert or semidesert type to the tundra zone. This extensive area of *Stipa* grasslands is called the puna. It is caused by the extremely stable

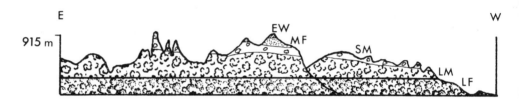

Figure 46.xxvi. East-west section through the Northern Range, Trinidad. The lowland forest gives way upwards to lower montane evergreen forest, and then in turn to semi-montane, montane rain forest and finally elfin woodland. The prevailing wind is from the west. *After Beard*

air over the ranges which prevent incursions of moist air from the Pacific.

Elsewhere, as in the Tibesti and Ahaggar ranges of the Sahara, sclerophyllous forest and scrub similar to that of the Mediterranean coasts is the most humid formation. Even in quite humid areas, local föhn effects or other climatic conditions can prevent the development of the full sequence of zones already described. The Ruwenzori range, for example, rises from a plateau with savanna vegetation and not from rain forest.

Altitude of Zones

It should already be evident that there is too great a variation in vegetation types and zones to lay down any useful guidelines on the altitudes at which one zone gives way to another throughout the world. The variations are caused by:

(1) uncertain upper limits of forests and trees where life-forms like the giant *Senecios* are arborescent but not comparable with normal trees;
(2) many trees being modified to shrubs by adverse climatic conditions;
(3) considerable local rises and falls of the timber-line, resulting from topographic and local effects of cloud, mist and exposure;
(4) limits controlled by seasonality in the outer zones of the tropics.

Number of humid months (*W. Lauer*)	*Tierra caliente* Lowland	*Tierra templada* Lower montane stage	*Tierra fria* Upper montane stage	*Tierra helada* High montane stage
12 11 10	Tropical lowland evergreen rain forest and tropical lowland semi-evergreen rain forest	Tropical lower montane forest	Tropical upper montane forest	Paramo (shrub and grassland)
9 8 7	Tropical moist deciduous forest and grassland	Tropical moist valley vegetation (forest and grassland)	Tropical moist sierra vegetation (moist sierra bush)	Moist puna (shrub and grassland)
6 5	Tropical dry deciduous forest and grassland	Tropical dry valley vegetation (forest and grassland)	Tropical moist sierra vegetation (dry sierra bush)	Dry puna (shrub and grassland)
4 3	Tropical thorn forest and grassland	Tropical thorn valley vegetation (forest and grassland)	Tropical thorn sierra vegetation (thorn sierra bush)	Thorn puna (shrub and grassland)
1	Tropical desert shrub	Tropical valley desert shrub	Desert sierra	Desert puna
0	Tropical desert	Tropical valley desert		

This last point is illustrated by Table 46.1 for the tropical Andes. Four vertical thermal belts are usually distinguished: *Tierra caliente*, *Tierra templada*, *Tierra fria*, and *Tierra helada* (or the hot, temperate, cool and frozen belts).

It is evident, therefore, that any figures given for limits can only be accurate for a particular locality. Some figures are given in Figure 46.xxiv for the Ruwenzori range.

47 Biogeography of the Dry Regions

With increasing dryness away from humid tropical regions, the thorn forests give way to semidesert scrub. The ecotone is often a broad one as the dominant plants become lower in height, more widely spaced and there is a gradual change in the dominant species. The plants of the thorn forests are characteristically deep-rooted, but those of the semi-deserts are shallow-rooted and able to use the moisture from light rain showers. The competition between plants is frequently intense and it is this which controls the spacing of the shrubs and low trees.

Semidesert Formations

Although there are similarities in structure, composition varies between the American, southeast Asian, Australian, Indo-Saharan, and southwest African formations.

The American formations are often referred to as 'cactus scrub', even though there are many plants present which are not cacti. In South America this type of scrub occurs along the fringes of the Atacama Desert, and in northwest Argentina. Isolated patches occur in Venezuela. In the north it covers much of Lower California, the northeast of the Mexican plateau and most of the Sonoran Desert. The density of the vegetation varies greatly with soil type and drainage conditions. There is a great variety of species but the same species appear in all similar habitats, so that there is a repetition of characteristic communities; this is illustrated in Figure 47.i, where the shallow, rocky, well drained soils of the mountain slopes support open communities of yuccas, agaves, ocotillo, some shrubs and other succulents. Near the base of the slopes on the upper bajada, the coarse but deeper soils will support taller succulents like the saguaro. Still lower down the bajada, where the soils are sandy and deep with a caliche pan, the creosote bush is dominant. On the bottom lands the soil is fine-textured and sub-surface water is readily available; alongside stream channels tall willows and cottonwoods, and away from them close stands of tall shrubs, make

Figure 47.i. Idealised profile of a range and basin in the Arizona Desert showing relationships between landforms, soils and vegetation. *After Benson and Darrow*

Vegetation	Yucca Agave Sotol	Palo Verde Saguaro Cactus	Creosote Bush	Mesquite	Willow Cotton-wood	Salt-bush
Soil	Mountain Slope	Upper Bajada	Lower Bajada	Bottom Land	Channel	Playa
	Shallow, rocky dry	Coarse well drained	Sandy & fine caliche pan no subsurface water	Fine poor drainage low salt	Fine poor drainage very salty	

84. Arid and spinifex-covered hills in the Kimberley region of Western Australia.
Australian News and Information Bureau

85. The Little Karroo showing thorny shrubs and tufted grass.
South African Tourist Corporation

the lowland areas appear quite fertile. A characteristic of all American semidesert formations is the absence of grass and the presence of xerophytic broadleaved plants forming the ground layer, while there are considerable areas of bare ground between them.

In southeast Asia euphorbias dominate parts of the dry belt of Burma, where there are areas of alkaline soils, but their ecological status is uncertain as they seem restricted by edaphic conditions.

The Australia formations are characterised by species of mallee (*Eucalyptus spp.*) and wattle (*Acacia spp.*), with extensive areas of porcupine grass or spinifex (*Triodia spp.*). Large areas of Australia with mean annual rainfalls of less than 25 cm are covered with this hummock grass, which may be regarded as a climax vegetation and hence quite different in form from the vegetation of other semidesert areas.

The Indo-Saharan formation extends between Mauritania and the Thar Desert. It is the most extensive semidesert formation in the world and, like that of the Americas, is dominated by thorny and succulent plants. Few plants exceed 1·8 metres in height.

In southwest Africa, from the coast of central Angola to the northern Karroo, there is a similar formation but with different species from those of north Africa, although once again the succulent euphorbias are prominent.

The Deserts

Desert plants are, of course, highly adapted to deficiencies of moisture. These adaptations are not always shown by the Raunkiaer type of classification, because it is not the position of the perennating buds which is significant but the nature of the transpiring organs, and the growth form. There are three main kinds of desert plants: ephemeral annuals, succulent perennials and non-succulent perennials.

Ephemeral annuals form 50–60% of the floras of some deserts. These herbs are capable of completing their life cycles within 6–8 weeks. Their growth is restricted to the period in which water is available and they then seed and die, so that they can be said to avoid rather than withstand drought. Their morphological attributes of small size and shallow roots, and their physiological adaptations of rapid growth, speed of germination, early flowering and maturity all enable them to complete the life cycle when water is available.

Some plants are so highly adapted that their seeds only germinate when there has been adequate rainfall. This feature is attributed to the washing of certain germination inhibitors off the seed coat. Plants also mature in accordance with the availability of soil moisture, producing a dwarf form with few flowers or large forms with many flowers, depending upon the conditions.

Succulent perennials have enlarged tissues which enable them to store water which can be used during drought. Some plants are extremely effective at retaining water, and a stem of *Ibervillea sonorae* stored dry in a museum formed new growth every summer for 8 years, decreasing in weight only from 7·5–3·5 kg. The cacti of American deserts and the euphorbias of Africa and Arabia close their stomata during the day and open them at night. Their transpiration rates are thus low during the day when evaporation stress is high.

The *non-succulent perennial* species have a variety of forms: woody herbs, grasses, shrubs and trees. Many have highly adapted seeds which may, for example, have to pass through the alimentary canal of an animal or be abraded in a mud flow before the coat is sufficiently weakened for germination to take place but, unlike the ephemerals, above-ground growth stops after the production of a few leaves until a large root system has been formed. Above-ground growth can then resume.

86. **Succulent aloes of South Africa.**
South African Tourist Corporation

Figure 47.ii. Part of Death Valley, California. Vegetation can only survive in valleys where moisture is available. Runoff on the slopes is too rapid and the soil too eroded for plants to survive. *After Hunt*

Figure 47.iii. (A) Transect across part of Death Valley showing xerophytes on gravel fans, phreatophytes at the foot of the mountains where ground water is shallow, salt-tolerant phreatophytes where the saltpan is close to the edge of the fan. There are no flowering plants on the salt-pan. The length of the section is 29 km. (B) Relationships of phreatophytes to water quality in part of Death Valley: creosote bush (c), desert holly (h), saltgrass (s) and pickleweed (p) occupy sites related to water-soluble soil salts in the soil water. Percentage salts are given by volume (figures). The broken line represents the position of the water table. Length of the section is 366 metres. *After Hunt*

The xerophytes—plants adapted to limited moisture—have a variety of morphological, anatomical and physiological features. A particularly common feature is the possession of a very large root system for the size of the shoots. The same species may produce deep tap roots in alluvial deposits to reach ground-water, or it may form a shallow fan in sandy or shallow soils. Alfalfa roots are recorded as reaching 39·3 metres beneath the soil surface and, during the building of the Suez Canal, tamarisk roots were found at a depth of 45·7 metres. A few plants can produce fine surface feeders, just beneath the soil surface, in response to light showers.

Reduction of the transpiring surface is effected by small leaves, no leaves, shedding of leaves or whole branches, and rolling of leaves. All of these features will reduce water loss during a dry period. Anatomical characteristics include heavy cuticularisation (i.e. the formation of a surface plaster-like layer of cuticle), cutinisation (i.e. the impregnation of the cell wall with cutin—a type of waterproofing device), wax, hair and resinous coverings, lignification, compactness, and recessed stomata. All of these features reduce the loss of water through the cells of the skin and lignification, or woodiness, prevents collapse of the plant tissue during wilting.

The outstanding features of the desert communities are the low but varied heights of the plants, the openness of the stand, the absence of decaying organic matter, which is rapidly consumed by termites or blown away by the wind, and the lack of competition between plants. Each plant contends with its environment and it is not possible to speak of a climax desert vegetation. Each type of environment produced by soil and landform features has its characteristic assemblage of plants. This is illustrated in Figure 47.ii, showing an area of volcanic lavas in Death Valley, California. In another part of Death Valley, ground-water and soil salt conditions control the plant distributions (Figure 47.iii).

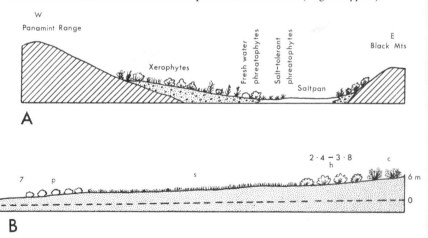

Oases

The usual image of oases is derived from those of the Sahara and Arabia (Figure 47.iv), but these are completely changed by man. The least modified are those of North America where, in New Mexico, Arizona and California, scrub willows, cottonwood, mesquite, Californian fan palms and many herbs form small communities. In the floors of some canyons, permanent streams support a riparian woodland of

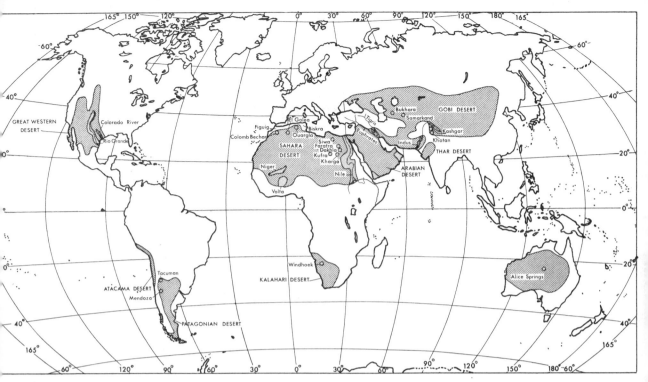

Figure 47.iv. Deserts and oases of the world.
After Cloudsley-Thompson

ash, sycamore, willow, cottonwood and walnut. Where they are large enough, these oases have their associated communities of animals which include mule deer and mountain sheep. In the deserts of temperate regions, as in the Gobi, the oases are characterised by willows and poplars. The other vegetation and crops are also temperate species. Polewards, these deserts are replaced by wide steppes.

The Asian and African oases probably had an original flora of tamarisk, oleander and other shrubs. These have long been replaced by date palms (*Phoenix dactylifera*), fruit trees and vegetables. The date palms are extremely versatile providers of fruit, fuel, thatch, fibre and building material. Even the stones of the fruit are crushed to feed camels. Only a few male trees are preserved, the female trees being pollinated by male flowers carried by hand.

Animals of the Desert

The limited resources of the desert do not support many species of animals, but those which do live there are often highly adapted to their environment. Desert animals have to adapt to shortages of water and temperature extremes—particularly to high temperatures in the tropical deserts. Most adaptations to the shortage of water consist of the conservation and maximum use of available supplies. The main losses of water from the body usually occur as sweating through the skin, exhalation with breath and loss in excreta. High temperatures may be overcome by dormancy during the heat of the day, by insulation, or by internal mechanisms which make high body temperatures harmless.

The Camel

Most desert animals are small, but the best known is the largest—the camel. The single humped dromedary is too large to avoid heat, but

87. A typical cactus range in the USA.
US Dept of Agriculture

being large it can afford to keep cool by the evaporation of sweat from the skin. Smaller animals cannot do this because they have a large surface area for their volume and would soon lose a high proportion of their body water. The camel has a coat which allows sweat to evaporate from the skin, where the maximum benefit is obtained. Sweat which soaked the hair would evaporate from the surface of the coat and allow the heat from the surrounding air to reach the site of evaporation, where the cooling effect would not be beneficial. The hair of the camel is able to act as a good insulator, and a camel shorn in one experiment produced 60% more sweat to control body temperature than unshorn animals. Moreover, the main fat of the camel's body occurs in the hump and not as a subcutaneous layer which would reduce loss of body heat. These advantages are, however, minor compared with those physiological adaptations which enable it to survive in the open desert for long periods. A camel can tolerate a loss of water equal to 25% or more of its body weight compared with the 12–15% for most other mammals. Losses greater than these cause death. In most mammals the water lost is taken from the blood, which becomes progressively more viscous until the heart can no longer transfer central body heat to the surface for cooling. In the camel most of the water is removed not from the blood, but from the tissues and gut. Camels have no ability to store water—certainly not in the hump, which is fat—but by losing water from the tissues and by being able to drink very large quantities in a short time—one camel drank 123 litres in 10 minutes—and to recharge the tissues, it can survive long periods without drinking. In addition, the camel has a variable body temperature which can fluctuate by as much as 6°C daily. These fluctuations are valuable because, as body temperature rises during the hot day, water otherwise used to keep the temperature down remains unexpended. The heat is stored in the body and dissipated to the cooler environment at night. A high body temperature also reduces the heat flow from the environment to the body, and therefore reduces the amount of water needed to prevent further temperature rises.

Small Mammals

Small animals—rodents, reptiles, birds and mammals—are the characteristic fauna of the desert. Even the largest predator in most areas is the size of a fox. Since they are small, most animals cannot reduce body heat by sweating and most have no sweat glands. They overcome this problem by avoiding heat, spending the hottest parts of the day in burrows and foraging for food at night. For many small animals the problem is often to keep warm during the cold nights and some, therefore, have well developed fur. The kangaroo rats, jerboas and many other small rodents are exceptionally efficient conservors of water. They will drink it when it is available but they can also survive without it. They achieve this by several mechanisms. They produce extremely concentrated urine and dry faecal pellets. They lose little water during respiration and by storing food in their burrows, where humidity is relatively high, they get that food to the most desirable condition. Other small animals obtain their moisture from succulent plants or from dew, and such behaviour probably also accounts for the survival of the large mammals like the oryx.

Reptiles and Insects

Reptiles have a very thin, non-fatty skin which allows them to lose heat rapidly, but in order to avoid heat, their chief adaptations are behavioural. Reptiles select their environment very carefully, basking in the sun or seeking the shade and coolness of rock crevices as necessary. Diurnally active lizards are compelled to move over hot surfaces, but their light-coloured, highly reflective bellies and the ability to run 'on their toes' help them to avoid excessive heating.

Many insects are nocturnal and cryptozoic (i.e. they seek shelter). The one desert arthropod about which a great deal is known is the desert locust—because of its potential threat to crops. Locusts are actually ordinary grasshoppers which have developed an ability to aggregate and migrate. The grasshoppers lay their eggs in bare, moist sand, and the young hoppers can only survive when there is a plentiful supply of fresh green vegetation. Because of their ability to migrate to areas which have these suitable conditions, they can concentrate into swarms. It appears that their direction of flight is largely controlled by the wind, so they are frequently blown into the convergence areas of air masses where rainfall provides suitable conditions for egg-laying and hopper development. The adults can go into a phase of arrested development until suitable breeding conditions occur; this, with their breeding habits, mobility and productivity, can produce very large swarms very easily. Irrigation projects have supplied suitable breeding grounds in some parts of the Red Sea area, and in these cases modern man has assisted one of the historic scourges of desert regions.

48 Biogeography of Temperate Regions

Between about 30° and 60° latitude are the areas with a relatively temperate climate. Within these latitudes there is a mixture of sclerophyllous formations, cool deserts and semideserts, extensive grasslands, deciduous summer-green forests, rain forests and mixed evergreen forests. The grasslands and forests were both far more extensive in the past than they are now, for they both occur in areas of old and densely populated industrial civilisations. The prairies of North America and the forests of North America and Europe are mere remnants from which, in Europe at least, it is often extremely difficult to reconstruct the former pattern of vegetation. In North America, where intensive occupation of the prairies has occurred only within the last hundred years, it is uncertain what part man-made and natural fires had in forming the boundaries of the grasslands.

The forests of contemporary temperate regions are greatly impoverished in species because of climatic and geologic events during the Quaternary period. During the Tertiary, a very extensive evergreen forest, similar in life-forms and appearance to that of the present North Island of New Zealand, stretched over the mid-latitude northern continents, while deciduous forests were confined within the Arctic Circle. At the onset of glaciation, some of the deciduous and evergreen species were able to migrate southwards, and some with an extensive north-south range also survived, but many species could not migrate, especially where they met the east-west barrier of the Alps. During the interglacial the deciduous trees had to migrate northwards, and again some species failed to do so and were eliminated in competition with the evergreens, which survived in areas like the Mediterranean basin. As there were 4 advances and retreats of the ice and hence 8 migrations, the forest species became greatly reduced. As a result, the boreal forests have very few species, the summer-green deciduous forests a few more, but both are very limited in variety compared with the present-day evergreen tropical forests. The greatest destruction of species occurred in Europe, and now both North America and eastern Asia have more species of deciduous trees.

In the southern hemisphere the land masses are very narrow and maritime influences on the continents are strong, so that both South America and New Zealand retained relics of their evergreen forests which were able to expand in the periods of climatic amelioration without great impoverishment of species.

In the following sections of this chapter, the main features of each of the temperate region vegetation types are described and the variations caused by altitudinal, hydrologic and edaphic conditions discussed in relation to two island groups—Britain and New Zealand—with greatly contrasting vegetation. They both show the great variation of vegetation types which can occur between two areas of similar size.

Sclerophyllous Formations

In areas which are said to have a 'Mediterranean' climate, the characteristic vegetation is a shrub or scrubland of woody plants, many of which have hard, leathery, evergreen leaves. Most of the annual rainfall of 50–100 cm occurs during the winter and the summer is hot and dry with cloudless skies. The natural vegetation over large areas of Mediterranean coastland is probably woodland but because of long occupance by man few remnants of this woodland still exist. In the other areas of 'Mediterranean'-type climate—South Africa, California, central Chile and southern Australia—both woodland and shrub vegetation occur.

Sclerophyll Shrub Vegetation

Small, thick, sclerophyllous leaves, with thick cuticles to reduce transpiration, are the main adaptation to summer drought. Most of the shrubs are evergreen and able to resume growth when moisture becomes available. In the Mediterranean region of Europe the two main types of scrubland are given many local names, but are most widely known by the French terms maquis and garrigue. The shrubs of the maquis all have the same life-form but the composition varies. Arbutus (*Arbutus unedo*), the olive (*Olea europaea*), cistus (*Cistus spp.*), myrtle (*Myrtus communis*), gorse (*Ulex spp.*) and broom (*Genista spp.*) are common shrubs, and between them geophytes, aromatic herbs and grasses cover the soil. The maquis usually occurs on siliceous soils as a fairly dense scrubland up to 3 metres high. The garrigue usually occurs on limestones or other pervious rocks with thin soils. It is composed of the same species as the maquis but the shrubs are more widely spaced and

88. Maquis vegetation, Cyprus.
Aerofilms Ltd

89. Chaparral vegetation, California.
US Dept of Agriculture

dwarfed. The general appearance of this vegetation is often rather drab in summer, as the leaves of the shrubs are grey-green in colour and most of the plants flower in the winter and spring, when they are bright and fragrant.

The climate and physiognomy of the plants of each of the sclerophyll formations are similar, but because they are isolated from one another each has a distinct flora. The chaparral of California contains many species, amongst them several evergreen oaks and some conifers. In the driest areas the scrub may be only 0·9–1·2 metres high, but in moister conditions the oaks can form a canopy 6 or more metres high. Many of the species are resistant to fire and can send up new shoots to replace burned stems. With increases in deliberate or accidental fire in recent years, the chaparral has been able to extend its range even into areas with 130 cm of rain a year where forest would normally occur.

The South African formation near the Cape of Good Hope has one of the most distinctive floras, with 2600 species in an area of only 518 km². A large proportion of these are endemic to the region. Many of these plants, like some heathers, ice plants and geraniums, have been introduced into other regions as ornamentals. The mallee scrub of Australia is also like the maquis in appearance but the composition is unique, being made up mainly of several species of the genus *Eucalyptus*. The shrubland was once so dense that it was often avoided by early travellers.

The outstanding feature of all of the Mediterranean formations is the remarkable similarity of life-form produced by convergent evolution. Such an occurrence in widely separated areas tempted early ecologists to make sweeping generalisations about the effect of climate on vegetation. Such climatic determinism may be justified in this type of climate, but in other areas the correlation may be lower or non-existent.

90. Native flowers in Cape Province.
South African Tourist Corporation

Mediterranean Woodlands

Although many of the drier coastal areas and islands of the Mediterranean probably had a climax vegetation of maquis, large areas would undoubtedly have been covered by forest in the absence of human interference. The coasts and islands of the Mediterranean, however, supported some of the earliest civilisations and the original vegetation has been replaced by arable farming, orchards and towns. Timber was required for shipbuilding, firewood and construction and no doubt accidental fires also took their toll of the trees. In many places the most damaging agent has been the goat, which has stripped the vegetation from large areas. With the destruction of the forest, the soil has also deteriorated or been eroded completely so that, even where the land is protected from fire and animals, the forest could not regenerate even if seed were available.

The original forest of the European area probably included several species of evergreen oak such as the holm oak (*Quercus ilex*) and the cork oak (*Q. coccifera*), as well as several species of pine such as stone pine (*Pinus pinea*), the maritime pine (*P. pinaster*) and the Aleppo pine (*P. halepensis*). The exact composition of the original communities cannot now be determined. These trees are adapted to the summer drought and have very deep rooting systems and small leaves, but the deep soils beneath a forest would no doubt have retained more moisture than the thin, humus-deficient soils of the maquis and garrigue. Elsewhere in areas of similar climate, forest is also the dominant. In Australia, for example, tall forests of eucalypts with open canopies, a shrub layer of acacias and other xeric shrubs, and a ground layer of grasses occur in areas with annual rainfalls of about 75 cm.

Animals

Because of the enormous changes induced by man it is uncertain what animals once occupied the sclerophyll ecosystem. This ecosystem is not very productive because of the long summer drought, so it is possible that the animal life was not particularly abundant. In California, herbivores such as the ground squirrel, deer and elk, and carnivores such as the mountain lion and grizzly bear were common until the mid-nineteenth century. In Australia, wallabies and other marsupials occupied similar niches.

Temperate Deserts and Semideserts

Large areas of central Asia east of the Caspian Sea, the intermontane plateaux of the western USA and parts of Patagonia all have semidesert formations. Their plants have adaptations to water deficit similar to those in the plants of the tropical deserts and sclerophyllous formations. These areas typically have a dull grey shrub vegetation 0·3–1·8 metres high. The shrubs are spaced to avoid root competition, although tussocks of wiry fescue grasses (*Festuca spp.*) and feather grasses (*Stipa spp.*) are common in Eurasia, where wormwoods (*Artemisia spp.*) and halophytic salt-bush (*Atriplex canum*) are characteristic. Sage-brush (*Artemisia tridentata*) is also the dominant in large areas of the USA, where the rainfall is 12–25 cm a year. In even drier zones, widely spaced creosote bushes (*Larrea tridentata*) are characteristic. The succulent cacti, euphorbias, aloes and agaves are also common in these climates.

Temperate Grasslands

Unlike the tropical savannas with their scattered trees, most temperate grasslands have no trees except along water-courses or on other humid sites, the northern ecotone areas of the North American prairies and Russian steppes with their stands of aspen (*Populus spp.*) providing minor exceptions. The temperate grasslands occupy large areas of the drier continental interiors and rain shadow areas, but because the rainfall varies from about 25 to 100 cm a year, and in places is sufficiently

Figure 48.i. The height of grass on the Great Plains of the USA as related to annual precipitation. The western plains are in the rain shadow of the Rocky Mountains, but the eastern plains receive moisture from the Gulf of Mexico.

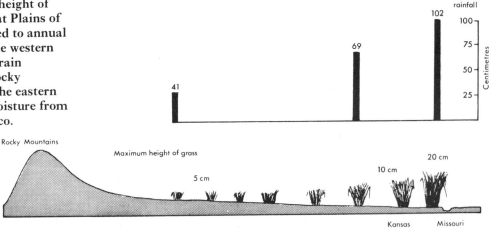

high for trees to grow, the ecological status of the edges of the grasslands is doubtful. In areas protected from fire, grazing and other interference, trees will often grow amid grasslands receiving only 50 cm of rain a year.

The main areas of temperate grassland are the great prairies of central North America and the Russian steppes from the Black Sea through the Ukraine and Kazakh regions to the Altai. In the southern hemisphere they include the Pampa of Argentina, the Veld of South Africa and the tussockland of the eastern South Island of New Zealand. Because most of these areas are relatively flat and easily converted into arable land, and because the soils and climate are particularly suitable for cereal-growing, vast areas of these grasslands have now been ploughed up or modified by grazing.

All temperate grasslands are dominated by grasses usually of various perennial species. The grasses are narrow-leafed hemicryptophytes with characteristically shallow, dense roots which form a resistant turf which prevents penetration of rainwater to deeper parts of the soil. Mature grasslands are thus secure against colonisation by trees even where precipitation is high enough for them. In addition, grasses are not damaged by fire and new growth springs from the clumps soon after a fire. Only trees like the aspen, which sends up suckers, can survive the natural and induced fires which are part of the summer regime of many grasslands.

In North America the appearance of the prairie depends upon the annual precipitation (Figure 48.i). Between the tuft grasses, forbs and turf grasses are common. Along the water-courses stands of trees—cottonwoods and willows—break the sweep of grasslands. The steppes

91. Tussock grasslands in the New Zealand Alps.
NZ National Publicity Studios

of Eurasia are essentially the same in form as the prairies but the genera of the grasses are different.

The Pampa of Argentina and Uruguay is the most extensive of the southern hemisphere temperate grasslands. Before European settlement, the area around the river Plate estuary was covered with tall bunch grasses with bare soil between them. Trees lined the rivers, and the floors of small depressions and scrub, called 'monte' by the Spaniards, occupied the low hills and the eastern parts of the plains. Because the western grasslands have a rainfall of over 76 cm a year some ecologists believe they were maintained by the fires of the Indians and that monte or even forest was once more extensive.

In South Africa also the ecological status of the red grass (*Anthistiria imberberis*) Veld, with more than 60 cm of rain a year, is doubtful. On the high plateaux west of the Drakensberg Mountains, between the Orange and the Limpopo rivers, the nutritious red grass sward has been debilitated by excessive firing by graziers and less valuable wire grass has invaded. Where the grasslands are protected from fire in the higher rainfall areas, scrub and forest gradually develop. At lower elevations and to the north, the temperate Veld is replaced by tropical tree-studded savanna.

In New Zealand around and over the Canterbury Plains, in the rain shadow of the Southern Alps and under the influence of the föhn winds (locally called the 'Nor'wester'), tussock grasses of uniform composition dominated by species of *Poa* and *Festuca* were seen by the European settlers who arrived around 1850. Between the tussocks there was a rich flora of herbs and low grasses, and above them an occasional cabbage tree (*Cordyline australis*). This vegetation has been largely replaced by cultivation and pastures of introduced grasses, and the remaining tussock grasslands of the hill country have been extended by firing the forests and by repeated burning and by sheep grazing.

The ecological status of most temperate grasslands on their wetter margins has frequently been questioned. It appears that burning, whether natural or by man, has been responsible for extending grass-lands into areas where trees will flourish and could dominate. In New

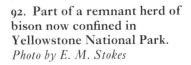

92. Part of a remnant herd of bison now confined in Yellowstone National Park.
Photo by E. M. Stokes

Zealand, the introduced *Pinus radiata*, willows and poplars thrive in areas with less than 50 cm of rain a year where tussock grasslands were once extensive. In North America the forest/grassland boundary was usually abrupt, with isolated patches of forest amid the grasses and patches of grassland in the forests, the boundary showing much inter-digitating of the two types: these are all signs of a fire boundary. The status of the drier grasslands, however, is not questioned, although it should not be assumed that there is a certain type of grassland climate, for in Australia a species of *Eucalyptus* has evolved which has produced a woodland in climatic conditions analagous to those of temperate grasslands elsewhere.

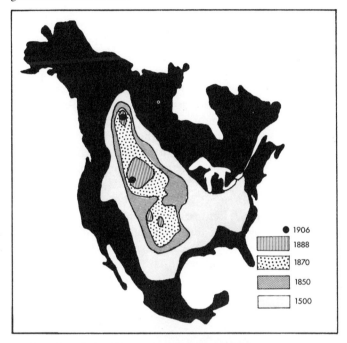

Figure 48.ii. The range of bison (i.e. buffalo) at various dates. The contraction of the range has been caused by the advance of European settlers. *After Farb*

Animals of the Temperate Grasslands

Grasslands have created conditions for animals which are very different from those of the forest. Forest animals can seek safety by hiding or climbing, but in the grasslands there is no cover and safety is to be found in speed or occasionally sheer size. Smaller animals may dig and burrow, but nearly all herbivores congregate into herds and many, such as the pronghorn, have a highly developed mutual warning system. Great herds are the chief characteristic of all grasslands—both temperate and tropical—but the bison and pronghorns of the early nineteenth-century prairies must have been one of the most spectacular sights ever available to any naturalist. Systematic destruction of the bison reduced the herds from an estimated 50–60 million individuals to a few hundred between 1800 and 1890. Today the bison is confined to two small herds—one in the USA and one in Canada (Figure 48.ii). The pronghorn antelope suffered a similar fate, although there are still about 350,000 left compared with the original 100 million. The ease with which grasslands can be modified by modern machinery, and the stored fertility of the best soils, means that the old ecosystem will never be re-established and that only a few remnants of the once enormous tracts of temperate grasslands will continue to exist.

Deciduous Forests

Summer-green deciduous forests, with the small exception of some southern beech forests (especially of *Nothofagus antarctica*) in southern Chile, all occur in the northern hemisphere. In areas of warm summers and cold winters the trees shed their leaves before the low temperatures of winter commence. During the cold period the tree roots can absorb little water and so winter is a period of physiological drought. The loss of leaves reduces transpiration and so prevents the death of the tree.

The forests occupied much of the eastern USA except for the southeast; Europe from England and France eastwards in a wedge to the Urals; and eastern Asia. The species are different in each of these 3 formations, although there are related species in each. All of these regions receive from 70–150 cm of rain a year, although the regime varies from a summer concentration in the continental interiors to a more even distribution in maritime western Europe.

Although there are considerable variations within each formation, most deciduous woodlands have 4 strata: the tree layer of the uppermost crowns forms an almost continuous canopy which varies in height from about 7.5 to 30 metres; the shrub layer may grow to a maximum height of about 4.5 metres; the field layer is composed of herbs and grasses, and the ground layer is of mosses and liverworts. The density of the lower strata depends upon the amount of light which filters through the canopy of broad leaves. These forests have a different aspect in each of the 4 seasons of the year, with spring-flowering perennial herbs and shrubs beneath still bare trees giving way to the sombre greens and deep shade of summer, to be followed by the vivid red and gold colours of autumn and then the bareness of winter.

The European formation has largely disappeared as cultivation and settlement have spread across it. In lowland Britain and northern France the dominant plant was frequently the pedunculate oak (*Quercus robur*)

93. A stand of mixed deciduous hardwoods, oak, poplar and hickory. Most of the trees are over 30 metres high.
US Dept of Agriculture

although beech (*Fagus sylvatica*) was more common on shallow and better drained soils, while the ash (*Fraxinus excelsior*) was often dominant on limestone. The sessile oak (*Q. petraea*) and the birch (*Betula spp.*) occur on shallow and more siliceous soils. Southwards in Europe other species of oak, the elm (*Ulmus spp.*), sycamore and chestnut became more important.

The North American and Asian formations are floristically much richer than the European formation. In addition to the many species of oak, and the beech, hickory and birch common in Europe, maples (*Acer spp.*), basswoods (*Tilia spp.*), tulip trees (*Liriodendron spp.*), chestnuts (*Castanea spp.*) and hornbeam (*Carpinus spp.*) occur. The composition of the forests of any particular area is often distinctive so that, for example, oak-hickory form the dominants in the south of the Appalachians and beech-maple in much of New England.

With the spread of agriculture both the forests and their animals have diminished, so that the once abundant deer and bears are now found in very small areas of their former range.

Rain Forests and Mixed Evergreen Forests

Temperate rain forests occur in areas with a high rainfall, usually in excess of 125 cm a year, with no season of drought. As a result of the abundance of water and relatively low evapotranspiration rate, the forests have many features in common with the tropical rain forests although in less abundance and size. The evergreen forests have a closed canopy and beneath it a second tree layer, frequently composed of tree ferns, nikau palms and smaller trees as in New Zealand, or of bamboos in Chile. A shrub layer is usually well developed and there is a ground layer of mosses, ferns and liverworts. Because of the slow rate of decomposition, fallen logs and litter rot slowly and, being covered with mosses, make a deep wet, spongy floor to the forest. Lianas are common and epiphytic bryophytes and ferns festoon the trees. Such forests can be extremely difficult to penetrate.

The North American rain forests of the west coast are dominated almost completely by large coniferous trees; in the south by the redwood (*Sequoia sempervirens*) (see Plate 67) which reaches a height of over 90 metres, and in the north by the Douglas fir (*Pseudotsuga menziesii*), giant arbor-vitae (*Thuja plicata*), Sitka spruce (*Picea sitchensis*) and western hemlock (*Tsuga heterophylla*). In southeastern USA the temperate rain forests (not those of southern Florida, which are more tropical) are dominated by evergreen oaks except where fire and ground-water conditions have favoured evergreen pine forest or savannas of loblolly (*Pinus taeda*), longleaf (*P. palustris*) and slash (*P. caribaea*) pines. All of these trees may be festooned with Spanish-moss (*Tillandsia usneoides*).

In the southern hemisphere the New Zealand rain forests are mainly dominated by the conifers of the family Podocarpaceae: rimu (*Dacrydium cupressinum*), kahikatea (*Podocarpus dacrydioides*) and others. In the cooler south the southern beeches (*Nothofagus spp.*) become more important. Podocarps and beeches of different species also compose the Chilean and Australian rain forests.

The available forage in these forests is not great so the biomass of the large animals is small, although deer, elk and mountain lions occur in the rain forests of western North America. In New Zealand isolation

353

94. A stand of loblolly pine, South Carolina.

Photo by B. W. Muir, US Dept of Agriculture

since the Cretaceous period has excluded mammals and reptiles from the original forests, and the ecological niches were filled by birds: ground-foraging ones such as the kiwi (*Apteryx*), weka (*Gallirallus*) and takahe (*Notornis*); kakapos, which are tree climbing; and the carnivorous parrots or keas (*Nestor*). The New Zealand forests have been greatly reduced by lumbering and by the introduction of many large mammals in the last hundred years, particularly several types of deer, pigs and small carnivores like the ferret, stoat, weasel and cat. This has not only caused destruction of much of the forest but also the virtual elimination of the flightless birds.

Mixed coniferous/broadleaf evergreen forests also occur in areas of lower rainfall in the Mediterranean basin and in South Africa, but in both areas they have been so affected by human interference that they are mere remnants.

Vegetation of New Zealand

The maps in Figure 48.iii illustrate the extent to which the vegetation of New Zealand has been altered by the European colonists since the first settlers arrived in 1840. The indigenous forests once covered about 18 million hectares or 75% of New Zealand's land surface but this has been reduced to about 6 million hectares. Even before the arrival of the Europeans, the Maoris had extended the area of scrubland and tussock grassland at the expense of the forest, but their impact was small compared with the destructive clear-felling and burning of the colonial settlers, who turned much of the formerly forested areas into pastures of European grasses, and who more recently have planted large areas with exotic pine trees—especially the Californian native *Pinus radiata*. Even the remnants of the forests have been so modified by introduced animals that it is probable that the only really natural communities surviving are to be found on offshore islands.

The vegetation of New Zealand is dynamic and constantly changing, not only because of human interference, but because of the climatic events of the Quaternary era which have affected all temperate zones of the world, and because of intense volcanic activity in the central North Island in the last 20,000 years, especially the vast rhyolitic pumice eruptions of the Lake Taupo district which occurred about 130 AD.

During the ice advances most of the South Island mountain valleys were occupied by valley glaciers and the mountain vegetation was forced to migrate to the lowland coastal areas. In the North Island the forests were forced northwards and to lower altitudes. New Zealand is too isolated to have received new stock in the post-glacial period and hence the post-glacial flora is impoverished because of the elimination of some species during the cold periods. During the climatic amelioration plants migrated southwards and upwards until, at the climatic optimum of warmer and wetter climate about 5000 years ago (see Table 48.1), much of New Zealand was covered with podocarp forest. Since the optimum the climate has deteriorated and *Nothofagus* forests have occupied larger areas in the south. Tussock grasslands have also extended, but to what extent this is the result of natural or induced fires is uncertain. Post-glacial changes can be summarised as in Table 48.1. Each of these stages can be seen in New Zealand today.

The volcanic eruptions of the central North Island destroyed all of the vegetation over a large area, and in the last 1800 years the pumice

Figure 48.iii. Pre-European and present vegetation of New Zealand.

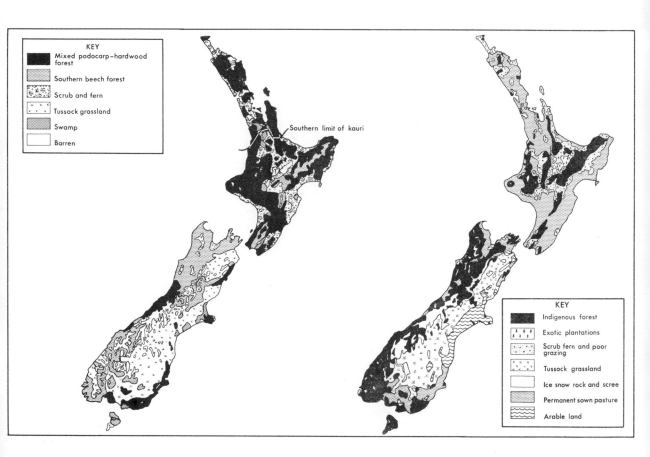

KEY

■ Mixed podocarp–hardwood forest

▦ Southern beech forest

▧ Scrub and fern

▨ Tussock grassland

▦ Swamp

□ Barren

Southern limit of kauri

KEY

■ Indigenous forest

▦ Exotic plantations

▧ Scrub fern and poor grazing

▨ Tussock grassland

□ Ice snow rock and scree

▦ Permanent sown pasture

▦ Arable land

Table 48.1. Post-glacial Changes in New Zealand Vegetation

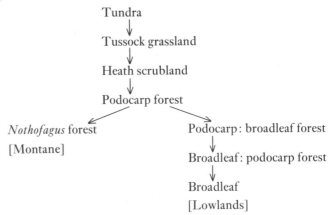

Tundra

↓

Tussock grassland

↓

Heath scrubland

↓

Podocarp forest

Nothofagus forest
[Montane]

Podocarp: broadleaf forest

↓

Broadleaf: podocarp forest

↓

Broadleaf
[Lowlands]

has been recolonised by grasslands→scrub→podocarp forest. The podocarps have spread back into the area from the surrounding ranges where they survived and appear to be preparing the site for eventual invasion by the broadleaf hardwood species. The more elevated and infertile areas are still occupied by tussock grasslands and scrub.

The vegetation of New Zealand can be classified into 6 main formations: (1) mixed podocarp-hardwood forest; (2) beech forest; (3) scrub; (4) tussock grassland; (5) swamp; (6) alpine formations.

Forest Formations

Although only 2 formations have been given, there are actually 4 elements in the New Zealand forests of which the first 3 often inter-

95. A grove of kauri trees. Note the size of the man for scale.
Photo by J. H. Johns, NZ Forest Service. Crown Copyright reserved

digite and occur in almost any degree of mixing: (1) kauri; (2) podo-carp; (3) broadleaf hardwood; (4) beech. All are of course evergreen.

Kauri forests are now limited to northern New Zealand and do not occur south of about 38°S latitude. The kauri tree (*Agathis australis*) forms a patchwork forest alongside broadleaf and broadleaf/podocarp forests. Most of the largest kauri stands and individual trees were cut during the early period of European settlement, for the tree with its massive trunk clean to about 18 metres above the ground provides excellent, easily worked timber for building, furniture or ship building. Much of the land it occupied was also dug over for the kauri gum (that is, fossil resin) which was an important ingredient of varnish.

Kauris form groves of almost uniform age with little or no regeneration of their own species beneath them. It is probable that the trees produce such an excessively acid mor humus and hence a very deep podzol that the shallow-rooting young kauris can only survive when broadleaf species have improved the soil and moisture conditions again. The result is a mosaic forest with groves of uniformly aged trees, each grove

Figure 48.iv. Profile through lowland podocarp-broadleaf forests showing a change from dominant podocarp with 2 tree layers (left), to 3-layer mixed (centre), to 2-strata broadleaf (right). Podocarps are stippled; all layers beneath 3 metres are excluded.
After Robbins

at a different stage of development. The kauri is not dying out, as is seen from the large number of buried podzols with forest maturing on the upper one; but it clearly does not dominate extensive areas.

Broadleaf and podocarp forest profiles are shown in Figure 48.iv. It will be seen that the structure ranges from podocarp forests, with a closed canopy and a lower tree stratum of broadleafs, to 3-storey forests with emergent podocarps over 2 broadleaf strata, and to 2-strata broadleaf forests. In all cases the shrub layer of broadleafs and tree ferns, and the ground layers, are excluded. These forests are all ever-green with virtually no bright colours. The trees are frequently fes-tooned with lianas and covered with a variety of epiphytes. Before introduced animals opened up the lower strata they were often difficult to penetrate, and even now the rotting logs, deep, raw humus and thick mats of moss, with the many lianas, make passage slow. Many of the large trees have buttresses, and were it not for the accumulations of organic matter and the coolness it might be imagined that the forests were tropical.

The *podocarps*—rimu (*Dacrydium cupressinum*), kahikatea (*Podo-carpus dacrydioides*), matai (*P. spicatus*), totara (*P. totara*) and miro (*P. ferrugineus*)—are gymnosperms and classed with the conifers although they bear berry fruit rather than cones. These trees are found throughout western New Zealand, especially on the lowlands, although in the far north they tend to become lower in stature and submontane. The most common tree is rimu except on gravel terraces and ridges,

96. A dense stand of podocarps, rimu, totara and
matai with an understory of
hardwood tawa, rewarewa and
hinau.
*Photo by W. Wilson, NZ Forest
Service*

where totara may be frequent, and on poorly drained areas, where
kahikatea may form stands. In general, however, podocarps do not
form single species stands. In some areas the northern rata (*Metro-
sideros robusta*), which begins life as an epiphyte and then sends down
aerial roots which surround and grow round the trunk of the host
podocarp, produces a distinct forest type. The rata eventually shades
out its host and is then left standing with a hollow interior as its host
rots away.

Of the *hardwood broadleaf* trees, tawa (*Beilschmiedia tawa*) is often
dominant in the North Island and kamahi (*Weinmannia racemosa*) in
the south, tairaire (*B. tairairi*), and puriri (*Vitex lucens*) being amongst
the other common species. Figure 48.iv shows that the podocarps and
hardwood forests are very mixed in composition. There is clear evidence
that the broadleaf hardwoods are invading and supplanting the podocarps and only in early seral communities as on the Volcanic Plateau
are podocarps still dominant and regenerating freely. It seems, then,
that there is a gradual progression from podocarp to mixed to broadleaf
forest. This process is very slow because many of the podocarps can

**Figure 48.v. Idealised pro-
files to show changes in
forest structure: (1) 2-layered
podocarp forest; (2) mixed
3-layer; (3) 2-layered broadleaf. Podocarps are stippled.**
After Robbins

30 m

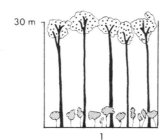
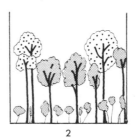
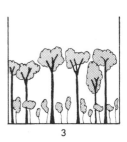

1   2   3

97. A rata vine on the trunk of a rimu tree.

live for a thousand years. The causes of this change are not entirely clear. It is not an altitudinal, hydrologic or edaphic control, for the mixed forests occur over a wide range of sites, nor is it entirely a climatic control, for podocarps like the rimu occur through the length of New Zealand. One hypothesis is that a type of fusion is occurring, with podocarp climax forest preparing the sites and giving way to broadleaf forests which will, at their climax, have a simpler structure and more limited composition than the present forests. Another possibility is that the forests are adjusting to late Quaternary climatic changes. Such a change accounts for the over-mature emergent podocarps which can be seen in many forests. These New Zealand forests are therefore in a state of flux and cannot be considered to be at the climatic climax state as a whole.

The southern *beech* forests are composed of the 5 New Zealand species of the genus *Nothofagus*. These forests have a simple structure and are more open than the mixed forests. They seldom have lianas and only a few large epiphytes. The beeches form the canopy and the sparse lower trees are mainly growing beeches with some shrubs (Figure 48.vi). Beech forests occur above the podocarp/hardwood forests on the ranges of the North Island and also on the western areas of the South Island, with some mixing in ecotone positions.

Scrub and Fern

In exposed localities along the coast or at high altitudes, where flooding prevented tree growth or on poor soils, a scrub or shrub vegetation was dominant. Even today, manuka (*Leptospermum spp.*) and bracken (*Pteridium*) with many introduced species like gorse and broom cover large areas. The original extent of this vegetation had been greatly

98. Red and silver beech forest, New Zealand.
Photo by J. H. Johns, NZ Forest Service. Crown Copyright reserved

Figure 48.vi. Profile of a lowland *Nothofagus* forest with an intrusive podocarp (stippled). The lower strata are very impoverished.
After Robbins

30 m

99. A stand of kahikatea behind a swamp bearing flax, Westland, New Zealand.
Photo by J. H. Johns, NZ Forest Service. Crown Copyright reserved

Figure 48.vii. Section across the central part of the South Island of New Zealand showing the variation of vegetation with altitude and climate. Compare this section with that in Figure 5.iii. which shows the mean annual rainfall and temperature.
After Taylor and Pohlen

increased by the Maoris through burning the forest. The forest advanced when free from burning and retreated when fired.

Tussock Grassland

In 1840 nearly 1/4 of New Zealand was covered with tussock, much of which owed its existence to Maori fires. There are 3 main types of tussock. On the east coast of the South Island up to about 900 metres, and in Hawke's Bay, short tussock dominated by *Festuca novae-zelandiae* and *Poa caespitosa* formed extensive areas of yellow-brown grassland. This community is adapted to the föhn winds, subhumid climate and gravelly soils. In wetter climates and boggy lands of Southland and in the central North Island, the tall red tussock (*Danthonia rigida*) grew in areas interspersed with scrub. The subalpine grasslands of snowgrass (*Danthonia flavescens*) occur on the mountains above about 900 metres. They do not form a continuous area but are interspersed with montane scrub, alpine plants, bare rock and scree.

Swampland

The most extensive swamps of New Zealand occur around the lower reaches of the Waikato river, in the Hauraki plains and in Southland. The great expanses of peat had a vegetation of rushes, moss, ferns and low manuka, but in the wetter areas bulrush, raupo and flax (*Phormium tenax*) were common. Large areas of the bogs have now been drained, but some of the small groups of kahikatea which occurred round their margins still stand amid pastures of introduced grasses.

The patchwork of vegetation shown by the maps in Figure 48.iii

W   Barren   Subalpine   Beech forest   Tussock   Beech   E

Rimu & rata forest   Tussock

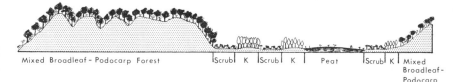

Figure 48.viii. Diagrammatic section from the west coast across the Waikato Basin, New Zealand showing the pre-European vegetation. On the lowlands kahikatea semi-swamp forest (K) grows on levees and slight rises; domed peat bogs occupy the former lake basins; and manuka and fern scrub covers large areas formerly forested.

Mixed Broadleaf - Podocarp Forest | Scrub| K |Scrub| K | Peat |Scrub| K | Mixed Broadleaf- Podocarp

gives little idea of the relationships between formations and relief. Figures 48.vii and 48.viii show cross-sections through the South Island and part of the North Island. In the South Island the west coast area has a mixed podocarp/hardwood forest which gives way upslope to beech forest or subalpine formations. In the valleys of the mountains the beech forests often form extensive uniform communities, but in the drier intermontane areas and the plains tussock is dominant. In the North Island the mixed forests occupy the ridges, but the floors of the valleys were often swampy and the Waikato Basin still has some shallow, open lakes. The peat swamps, with low domes, occupied large areas and on the levees of the rivers kahikatea stands were common.

Vegetation of Britain

The present vegetation of Britain is almost entirely the result of human interference. Large areas, especially of England, are occupied by crops, and much of the remainder is covered by various types of induced grassland. Between 1/2 and 2/3 of England, and about 3/4 of Wales, Scotland and Ireland are covered by grassland communities (Figure 48.ix). The climatic climax vegetation of Britain, however, would certainly be mainly forest. The actual nature of this forest can only be reconstructed from the few isolated remnants of woodland left after the

100. Scots pine (*Pinus sylvestris*) and undergrowth of heather, Inverness-shire. *British Forestry Commission*

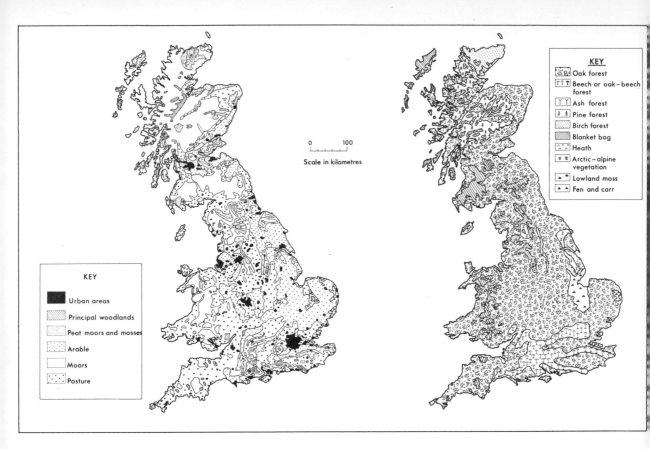

KEY

- Oak forest
- Beech or oak–beech forest
- Ash forest
- Pine forest
- Birch forest
- Blanket bog
- Heath
- Arctic–alpine vegetation
- Lowland moss
- Fen and carr

0   100
Scale in kilometres

KEY

- Urban areas
- Principal woodlands
- Peat moors and mosses
- Arable
- Moors
- Pasture

Figure 48.ix. Presumed original and present vegetation of Britain.
Original vegetation after Eyre

clearances which have been going on since about 3000 BC, when Neolithic men started their colonisation of the forest lands. A further problem is that the ecological status of the remnants is in question because of the shortness of post-glacial time available for the forest to spread and reach equilibrium.

During the maximum of the Pleistocene advance the ice reached an east-west line across southern England roughly corresponding with the course of the river Thames. The onset of the post-glacial occurred about 22,000 years ago, but the final withdrawal of glaciers from the Scottish highlands and southern Sweden took place only about 10,000 years ago, and a cold climate persisted over much of western Europe. The tundra and boreal vegetation which had occupied southern Britain, and possibly some nunataks further north, gradually spread northwards following the ice retreat, but the warm, dry climate of the Boreal period did not begin until about 7000 BC, when birch (*Betula spp.*) and pine (*Pinus sylvestris*) spread rapidly over the lowlands. About this time also Britain was cut off from the continent by the opening of the English Channel and spread of the North Sea, so that invasion by many plant species was reduced and Britain's flora remained impoverished. The amelioration of climate continued until about 500 BC, when a cooler, moister climate set in. Before this change the tree-line had been as high as 600 metres but the decline in summer temperatures and higher rainfall caused a lowering of the tree-line to a little over 300 metres and the spread of blanket peat bog on the uplands. The evidence for these changes is obtained from the pollen preserved in peat bogs and lake beds, and on the uplands remains of forest are found in the

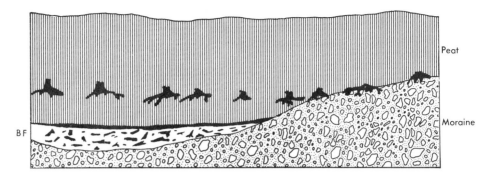

Figure 48.x. Layer of sub-boreal pine stumps in peat and beneath them remains of boreal forests (BF) on the moraine.
After Lewis and Tansley

blanket peat (Figure 48.x). It will be seen from Table 48.2 that the composition of the British forests has changed with climate and that at present beech and hornbeam forest would be spreading at the expense of the older mixed forest of oak, elm, birch, beech and alder, if human interference had not prevented it.

Table 48.2. Sequence of Late Glacial and Post-Glacial Changes (After Matthews)

	Approx. date	Period	Pollen zones	Vegetation	Climate
	BC 500	Sub-atlantic	VIII	Decline of mixed forest, beech, hornbeam	Cooler, more oceanic
Post-glacial	3,000	Sub-boreal		Mixed forest of alder, oak, elm, birch and beech	Warm, dry
	6,000	Atlantic	VII	Maximum alder, oak, elm, lime	Warm, oceanic
		Boreal	VI	Oak, elm, pine, hazel	Warm, dry
	8,000	Pre-boreal	V	Pine, birch	
			IV	(Hazel) birch, pine	Cold
Late glacial		Younger Dryas	III	Birch scrub tundra	Sub-arctic
	10,000	Alleröd	II	Birch tundra heath	Milder
	?13,000	Older Dryas	I	Open tundra	Sub-arctic
Full glacial		'Arctic plant beds'			

363

The map of the assumed or theoretical climatic climax vegetation of Britain in Figure 48.ix shows that vegetation which would occur if Britain were all naturally well drained, of gentle slope and undisturbed by man. These conditions do not apply, but the theoretical exercise is valuable because it helps to explain the origin of those relicts of the former cover which still survive. The existence of 5 distinct climatic climax formations can be postulated: (1) the deciduous summer forest; (2) boreal forest; (3) blanket bog; (4) heath; (5) Arctic-alpine.

The dominant species of each of the 5 formations have survived human interference but the structure of each has not. It is now uncertain what the profile of an original oak forest would be like, as most remnants are now more open than they would have been originally, a result of grazing, pollarding, wood collection and the prevention of regeneration by small animals like mice, rabbits and squirrels. Deliberate planting and selective cutting have also changed the species composition of many remnants.

The Deciduous Summer Forest Formation

Most of the well drained lowland areas of Britain were once covered with forest dominated by the pedunculate oak (*Quercus robur*). This tree sends down a long tap root to the permanent water table but it cannot survive waterlogging in the upper parts of the soil profile and then is replaced by the shallow-rooted alder (*Alnus glutinosa*) and ash

101. Pedunculate oakwood, New Forest, Hampshire.
British Foresty Commission

102. Natural ash woods on limestone in the Forest of Dean, Herefordshire.
Photo by Miss T. K. Wood, British Forestry Commission

(*Fraxinus excelsior*). On steeper areas with shallow soils the sessile or durmast oak (*Q. petraea*), which has a spreading root system, replaces the pedunculate oak. Sessile oaks occur up to heights of 465 metres in the Lake District and this seems to indicate that, if it had not been for interference and destruction, the tree-line could be higher than it actually is in many parts of Britain.

The original oak forests would probably have had a discontinuous shrub layer containing hazel (*Corylus avellana*) and hawthorn (*Crataegus oxyacantha*). In the spring small herbs of the ground layer like the bluebell (*Scilla non-scripta*) and dog's mercury (*Mercurialis perennis*) would flower before the shading effect became too great, but in areas of deep shade and much leaf litter there would be no ground layer. On siliceous rocks the upland forests would have more bracken (*Pteridium aquilinum*) and grasses in the lower layers and the sessile oaks would be mixed with silver birches.

The humus derived from oak leaves is rich in bases and forms a mull. On deep silts and clays, therefore, the soils beneath the forest are fertile brown earths, but on siliceous sands and shallow soils the canopy is more open, and the undergrowth is formed by heaths which do not demand a rich soil and so return to it leaves deficient in mineral bases. The litter is therefore acid and forms a mor humus with an underlying podzol. Some of the deepest podzols in Britain are found on the Bagshot Sands in the London Basin, beneath oak forest.

Oaks do not thrive on highly calcareous soils and in northern Britain, except where rainfall is very high and leaching rapid, the ash tends to replace the oak on Carboniferous limestone, the oolites and the chalk. In the south of England there are very extensive areas of chalk which was once occupied by beech forest (*Fagus sylvatica*). Beech grows on the chalk lands not because it requires a calcareous soil but because it requires good drainage. It actually develops best on the clay-with-flints overlying the chalk, which has an acid reaction, and it is widespread on

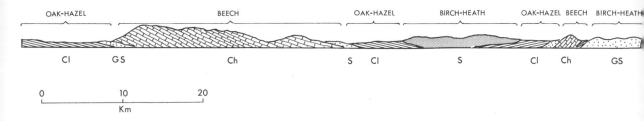

OAK-HAZEL   BEECH   OAK-HAZEL   BIRCH-HEATH   OAK-HAZEL BEECH BIRCH-HEATH

Cl   GS   Ch   S   Cl   S   Cl   Ch   GS

0   10   20
|_____|_____|
       Km

Figure 48.xi. Dominant vegetation on outcrops of clay (Cl), greensand (GS), chalk (Ch), and sands and gravels (S). The section is from the Midland Plain to the edge of the Weald across the London Basin.
After Tansley

the acid sands and gravels of the New Forest. It is not clear why the beech should have been restricted to southern England for in plantations it will grow near Aberdeen. It is possible that it could not compete with the oak in the cooler climate, but it is also possible that at the time of the great medieval forest clearances it had not reinvaded the northern areas from which it had been excluded during the colder and wetter Sub-atlantic period. The beeches do not take up much calcium and therefore produce an acid mor humus and soils in which podzolisation occurs. As a result the ground flora is very limited.

103. A sessile oak wood, Monmouthshire.
British Forestry Commission

A section through southern England (Figure 48.xi) shows the considerable variations in the forest formation which are the result of lithologic, soil and drainage variations.

Boreal Forest Formation

The remnants of pine forest still left in Scotland are at the narrow western extremity of the boreal forest zone. They only survive on a very few thousand hectares as the Ballochbuie and Rothiemurchus forests, with some smaller communities elsewhere. The pines (*Pinus sylvestris*) reach 610 metres but above this the birch dominates, and below about 185 metres the pedunculate oak is the dominant. Most of these forests were destroyed after the Forty-five rebellion, when forests which could have given cover to rebels were cleared and cutting for wood was increased. Grazing and invasion by the heath plants like ling (*Calluna vulgaris*) have prevented regeneration of most of the forests, and they appear to be doomed to extinction.

Blanket Bog Formation

In many parts of Britain the cool, wet climate prohibits the growth of forest trees and encourages herbaceous bog communities even on

104. Beech forest on the Chiltern chalk hills, Oxfordshire. Saplings are growing in the gap in the canopy, but elsewhere the ground layer is limited to mosses.
British Forestry Commission

105. This *Eriophorum* blanket bog has been drained. Young trees will be planted on the ridges of peat. The present vegetation is heather and deer grass, Sutherland.
British Forestry Commission

slopes of up to 20°. Bog communities cover much of western Ireland and the Western Isles, and in much of upland Britain they cover most gently sloping surfaces over 240–360 metres depending upon locality. This formation is dominated by bog moss (*Sphagnum spp.*) but there are usually other plants represented: cottongrass (*Eriophorum vaginatum*), ling, rushes (*Juncus spp.*) and sedges (*Carex spp.*) are all common.

The peat on which the bog plants grow has formed from the accumulation of the partly decayed bog plants. In the western Pennines above 600 metres it has reached a thickness of 3·7–4·7 metres. This thick peat is largely formed from bog moss but the thinner peats of the drier east have a large proportion of cottongrass and other grasses. The rain which falls on the peat contains almost no minerals, so that after a time the peat becomes extremely acid or oligotrophic, with a pH of about 4. Peats which form where ground-water with a higher mineral content circulates are said to be eutrophic.

Draining and burning has altered many of the blanket bogs, particularly in the last 100–150 years. With draining the sphagnum becomes less important and cottongrass and heather more important, to give the sequence:

Sphagnum bog ⟶ Mixed moor ⟶ Cottongrass or heather moor

Figure 48.xii shows the sequence along the edge of a drained and eroded peat bog.

Figure 48.xii. Distribution of peat plants on High Borrow Moss, Westmoreland.
After Pearsall

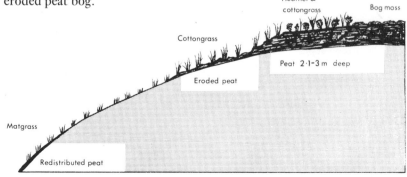

Heath Formation

In areas of extreme exposure to wind and very thin, poor soils, climatic climax heath forms. On uplands too steep for blanket bog formation the common heather or ling (*Calluna vulgaris*), heaths (*Erica spp.*) and heath grasses such as bent (*Agrostis spp.*) and mat grass (*Nardus stricta*) are co-dominants. All of these plants are very undemanding and produce a very acid litter, so that the soils beneath them are podzols although, as the heaths are frequently on steep slopes, the soil profiles may be truncated.

Composition of the heaths varies from one part of Britain to another, partly because of the wide range of climate and altitude at which they occur and partly because many of them are used as grouse-moors which are burnt at intervals to maintain the heather for grouse fodder. Burning probably maintains the heather and reduces competition from other plants. Drainage also causes the heather to spread on to former blanket bogs.

The Arctic-Alpine Formation

During the climatic amelioration of the post-glacial, the tundra plants which had occupied much of southern England became progressively less capable of maintaining themselves in competition with invading forests and hence became confined to areas unsuitable for trees, heaths and bogs. Such are the summit areas above about 600 metres in the exposed western areas of the rest of Britain. On the steep mountain slopes frost action is frequent, soil formation slow, erosion rapid, the growing season short and the period of snow cover long: the Arctic-alpine plants are adapted to such conditions. The most common and extensive plants are various kinds of grasses like bent and alpine sheep's fescue. These are probably the only fully natural grasslands in Britain. Most Arctic-alpine communities, however, do not form a complete ground cover, and individual plants like the dwarf birch (*Betula nana*),

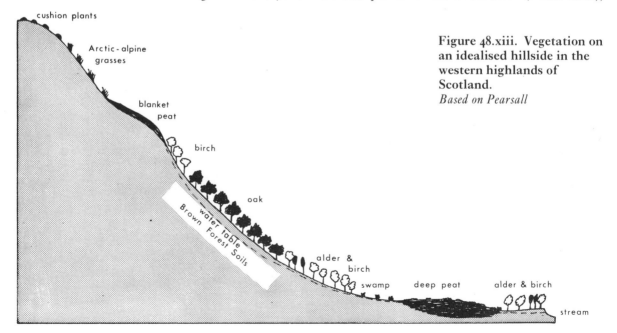

Figure 48.xiii. Vegetation on an idealised hillside in the western highlands of Scotland.
Based on Pearsall

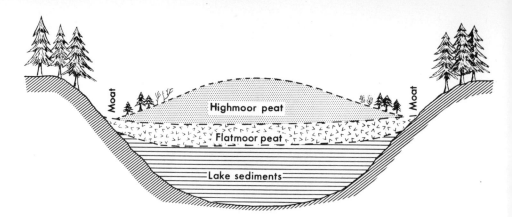

Figure 48.xiv. A lowland domed peat bog. The vertical scale is greatly exaggerated.

Arctic willow (*Salix herbacea*) and purple mountain saxifrage (*Saxifraga oppositifolia*) occur only where there are cracks in the rocks or gaps between stable boulders. Their ground-hugging, rosette, cushion or mat forms enable them to survive the severe winds and during the harshest winter they are covered by protecting snow.

In Figure 48.xiii an idealised section through part of highland Britain shows the relationships between slopes, altitude, water table and vegetation.

Hydroseres and Lowland Bogs

After the retreat of the ice, large areas of Britain were left with poor drainage and hence before artificial drainage was introduced hydroseres existed in various stages. Much of eastern England is low-lying and fens dominated by reeds (*Phragmites spp.*) and other plants produced a peat in the base-rich ground-water derived from the chalky boulder clay. On the higher areas were carr woodlands of willows and alders.

Peat also occupies much of the wetter lowlands in Ireland, Lancashire and the valleys of the uplands. The lowland acid peats are frequently domed, as the sphagnum grew from the centre and outwards (Figure 48.xiv) to form a raised bog. Many of the old bog areas still retain the name of 'moss'. Neither fens nor raised bogs can be regarded as climax communities because as drainage improves naturally both areas are colonised by forests.

49 Biogeography of Cold Regions

The cold regions are here taken as being the regions occupied by the boreal, or northern, forests, the taiga and the tundra. Their extent in the southern hemisphere is exceedingly small, being limited to the tundra of the southern extremity of South America, and parts of the Antarctic peninsula, and to the islands of the southern oceans. In the northern hemisphere the extent is vastly greater. The southern limit of the tundra approximately corresponds with the position of the 10°C isotherm for the warmest month, and the southern limit of the boreal forest with an isotherm representing a mean temperature of 10°C for the warmest 4 months of the year; in both cases the mean temperature of the coldest month is not higher than 0°C. The boreal forest has unbroken stretches of tall, mostly needle-leaved evergreen trees with scattered shrubs, herbs and a well developed moss layer. The taiga supports small, bunched or widely spaced trees with low shrubs, herbs and abundant lichens and mosses (it should be noted that some

Figure 49.i. Typical vegetation of tundra, taiga, boreal forest and deciduous forest areas. The central portion of well drained upland bears climax meso-phytic vegetation, the wet portion aquatic and marsh vegetation, and dunes and cliffs have drought-adapted (xerophytic) vegetation.
After Dansereau

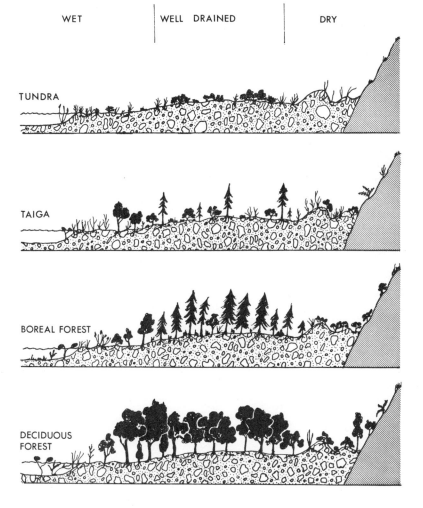

authorities use the term taiga as a synonym for 'boreal forest'). The tundra bears no trees except along some water-courses, and its vegetation consists of creeping shrubs, tufted grasses, lichens and mosses. Figure 49.i shows the features of these regions for the climax vegetation on flat and well drained sites, and for wet and dry sites.

The Boreal Forests

The southern boundary of the boreal forests is often quite sharp. In America there are few mature mixed needle-leaved evergreen and broad-leaved deciduous forests on well drained sites, and the transition from one formation to the other is well defined. The transition area between the short grass prairies of Manitoba, Alberta and Saskatchewan is often a parkland dominated by aspen, and in Siberia a poplar dominates a similar community, but it is possible that fire has been important in these areas.

Most coniferous trees have tapering single trunks with their longest branches near the base of the trunk to give a conical form to the tree. Most of the boreal forest conifers are evergreen, retaining their needle-like leaves throughout the winter. The leaves have thick cuticles and during the winter the stomata close, so that transpiration almost ceases—a necessary adaptation when the ground is frozen and no soil-water is available. The sugar concentration in the sap increases in winter and acts as a further protection against cold and water loss. The evergreen habit has the advantage that the tree can commence photosynthesis as soon as temperatures rise in the spring, and there is none of the annual wastage inherent in the deciduous habit. The northernmost sections of the Eurasian formations, however, have deciduous needle-leaved conifers, with larches (*Larix spp.*) as their dominants. In Siberia at about 72° 50′N and 105°E, the most northerly forest in the world has the Dahurian larch (*Larix dahurica*) as the dominant. Under such an extreme climate transpiration during the winter must cease altogether, and on the edge of the tundra the larches occur with the broad-leaved deciduous birches (*Betula spp.*), balsam poplars and aspens (*Populus spp.*) and the seral alders (*Alnus spp.*).

106. Boreal forest at Banff, Canada. Moose are grazing in the foreground.
Photo by E. M. Stokes

The forests have a simple structure with 1, 2 or 3 strata. Beneath very close stands the lack of light and the carpet of slowly decaying needle-leaves may inhibit growth, so that there are no ground or shrub layers, but in more open conditions small shrubs like the blueberry (*Vaccinium spp.*), crowberry, Labrador-tea and pale-laurel grow out of thick carpets of mosses and lichens. The dominant conifers are often 24–30 metres tall.

The main forests are composed of species of pine (*Pinus spp.*), spruce (*Picea spp.*) and fir (*Abies spp.*), but there are differences in composition between the North American and European formations and also within each formation. In Europe from western Norway to the Urals, the pine (*Pinus sylvestris*) and the spruce (*Picea excelsa*) are absolutely predominant, but eastwards into Russia and Siberia the species change and the larches become more common. In northern Siberia the active layer of the soil is so shallow that the trees have only some centimetres of soil above the permafrost. Their root system, therefore, has a spreading fan form with no tap roots. The trees are exceedingly slow-growing and are often only a very few metres in height even at an age of over a hundred years. In eastern Asia, Manchuria and Japan, the flora is much richer than in Europe and many species occur together in the forests.

In the richness of its flora the North American boreal forest resembles that of Asia. On the southern part of the Laurentian Shield the white spruce (*Picea glauca*) and the balsam fir (*Abies balsamea*) dominate on the better drained soils, with the black spruce (*Picea mariana*), the tamarack (*Larix laricina*) and in some areas willows (*Salix spp.*) on wetter sites. Westwards the dominants change until in British Columbia and Alaska they give way to the lodgepole pine (*Pinus contorta*) and the alpine fir (*Abies lasiocarpa*).

Edaphic and Hydrologic Conditions

The boreal forests have much the same appearance over enormous areas but edaphic and hydrologic conditions produce considerable variations. On sandy soils with low nutrient status pines are usually dominant. Their very acid litter assists in the production of podzol soils, but on better soils with higher nutrient status and higher clay contents the more demanding spruces dominate. They produce a less acid litter; because of the slower percolation podzolisation is not as rapid as on sandy soils, and the soils are therefore often grey-brown podzolics.

Large areas occupied by the North American and west European boreal forests have been subjected to continental glaciation. On the Laurentian Shield, for example, a mammilated topography with impeded drainage, scoured rock basins and moraine-dammed lakes has a large proportion of its area covered by water. Here bogs occupy a greater proportion of the land surface than in any other part of the world. It is probable that at least 60% of the world's peat occurs in these northern regions where decomposition is inhibited by low temperatures. Shallow depressions are filled in with decaying plants, and bog-mosses of the genus *Sphagnum* cover the peat and add to it, gradually extending the bog vertically and laterally. Figure 49.ii shows the development of a bog on the Laurentian Shield. In (1) the drainage is blocked and small shallow ponds and a larger deeper lake have been formed. In (2) rushes and sedges have occupied the shallow areas of

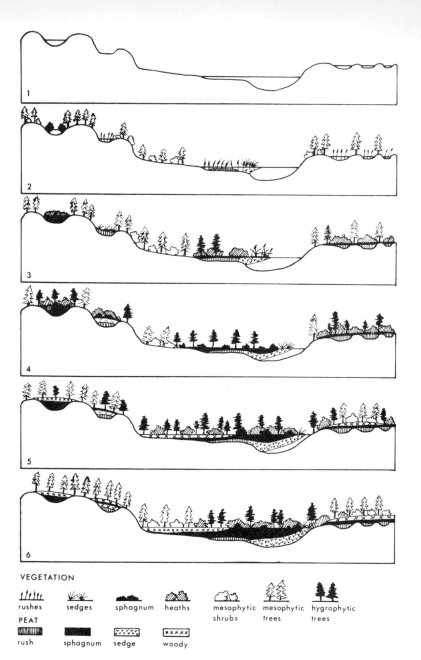

Figure 49.ii. Bog succession on the Laurentian Shield.
After Dansereau and Segadas-Vianna

VEGETATION

| rushes | sedges | sphagnum | heaths | mesophytic shrubs | mesophytic trees | hygrophytic trees |

PEAT

| rush | sphagnum | sedge | woody |

water, and conifers requiring well drained sites (i.e. mesophytic trees) have colonised the ridges. In (3) the small basins are completely filled with peat, and shrubs and heaths have covered the sphagnum mosses. A floating mat of marsh plants is beginning to extend across the open water and hygrophytic trees are beginning to colonise the edge of the bog. In (4) the bogs are extending laterally and mesophytic trees are being replaced by hygrophytic ones. In (5) the increase of woody plants on the bog has changed ground-water and drainage conditions so that mesophytic trees once more advance. By (6) all open water is closed in. In Canada the bogs are known as muskeg. They make cross-country travel very difficult and provide breeding grounds for the enormous numbers of mosquitoes which infest the northlands.

Figure 49.iii. Plant succession in the boreal forest. Lichens and sorrel are followed by goldenrod which is invaded by aspen or birch, which is replaced by fir and finally spruce and fir.
After Dansereau

Successions

Fire and lumbering have provided ample opportunity for studying plant successions in the boreal forest. On sites with thin rocky soils it will take thousands of years for slow weathering and lichens and herbs to form a soil suitable for trees, but on better soils regrowth will be more rapid. The first year after clearance will see a cover of the wind-dispersed purple fireweed, but in succeeding years this will be replaced by blueberries or bracken, brambles and goldenrod with patches of grass. Gradually the small and shrubby trees like hazel, aspen and birch will establish a forest, while the conifers grow beneath them at first and later firs and finally spruce and fir dominate. A simplified succession of this type is shown in Figure 49.iii. The more open conditions which follow burning and lumbering have favoured the spread of deer, but the drowning of shallow-water aquatic plants, or the alternate flooding and exposure produced during hydro-electric power production, have destroyed the shallow-water plants upon which ducks feed and the water-lilies which the moose eat in summer. Man, often unwittingly, has greatly modified the boreal forests in the last 50 years.

The Taiga

To the north of the boreal forest the trees become lower, generally not exceeding 12 metres, and the stands more open and parklike. The ground is covered with a thick carpet of lichens— the so-called 'reindeer moss'—and there are a few shrubs. The rate of growth of the trees may

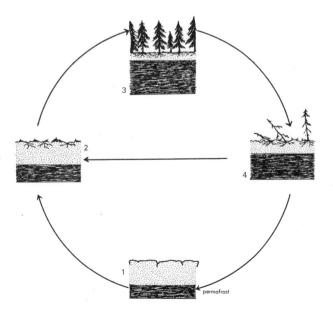

Figure 49.iv. The cycle of subarctic vegetation overlying permanently frozen ground. The annual thaw is deepest over bare soil (1), but herbs and low shrubs insulate it and allow a rise of the permafrost (2) which is highest under closed forest (3). As the permafrost becomes shallow, the trees fall over, inducing a recession of the permafrost and a return to a poorer vegetation cover.
After Dansereau

be very slow, so that a 100-year-old spruce may be only 1·5 metres high with a trunk barely 76 mm in diameter at the base. Successions are therefore also very slow. The typical taiga is confined to well drained upland sites, while in the valleys tongues of true boreal forest may extend northwards, or muskeg develop in areas of impeded drainage. In wet areas, disturbances by variations in the permafrost cause cyclic variations in the vegetation. Beneath the lichens, mosses and grasses the active layer may be some metres deep and so favours the advance of shrubs and eventually trees. The denser vegetation cover, however, insulates the soil so that the permafrost level rises, causing sagging and heaving of the surface with consequent tilting of the trees. This 'drunken forest' cannot maintain itself and is replaced by lichens and grasses beneath which the permafrost level again falls (Figure 49.iv).

The taiga is an important area for animal life. The caribou in North America and the related reindeer in Eurasia migrate into it during the winter, when they feed on the lichens.

The Tundra

North of the taiga, small patches of scrubby forest trees stand on favourable sites in areas dominated by heath, lichens and mosses of the true tundra; some authorities regard this as a separate zone called the forest-tundra, but others regard it as a mere transition to the true tundra, which is treeless.

In the original Finnish, tundra means barren ground, wasteland, hostile land. In normal usage it refers to the vegetation beyond or above the tree-line. The distribution of tundra seems to be conditioned not only by cold, although the 10°C isotherm for the warmest month is an approximate southern boundary in the Arctic, but also by the presence of continuous permafrost on the lowlands. Continuous permafrost is not a condition of tundra areas on high mountains, where wind and exposure are more significant. Wind is also important in limiting tree growth in the Arctic, where trees can often grow in sheltered locations while nearby exposed areas have no trees. The reduction of wind-speed near the ground is one reason for the success of dwarf vegetation in the tundra. Willows that grow no more than 0·3 metre high, and many herbs, whose relatives have an upright growth in less harsh climates, have a low cushion form which is an adaptation to wind and wind-chill. The cooling effect of wind is illustrated by measurements of temperature in and around a tuft of saxifrage: with an air temperature of − 12°C a temperature of + 3°C was recorded from the centre of the tuft and a nearby cushion of dark-coloured moss had an internal temperature of + 10°C. Growth in most cushion plants commences from within the cushion as soon as the plant is free of covering snow, so that frost-free periods are not significant in vegetation growth.

The growing season is always short, and often as little as 2 months, so most plants are perennials which are able to remain dormant for long periods and resume growth when conditions are favourable. There is an almost complete lack of annuals.

Arctic soils also provide limiting factors. Most are infertile because of the lack of nitrogen caused by the scarcity of leguminous plants and the lack of nitrofying soil bacteria. The vigorous growth of the orange nitrophilous lichen *Caloplaca elegans* on perching rocks used by birds,

107. The Siberian taiga in winter.
Society for Cultural Relations with the USSR

and the vigorous growth of grasses round human camps, indicate that growth could be much greater if nitrogen were more abundant.

Permafrost has similar effects in the tundra to those in the taiga. The annual rise and fall of the permafrost table, the frequent freeze-thaw producing expansion and contraction of ground ice and water, and the consequent movement of stones and rocks in the soil, cause frequent disruption of root systems. Cyclic growth and disruption of plants result from the increasing insulation beneath the plants and the consequent rise of the summer permafrost level to kill off the well developed vegetation and produce a return to barren conditions. The succession from bare soil or gravel to a covering by wood rushes, mountain-avens and purple saxifrage followed by willows, mosses, sedges and lichens can rapidly revert to bare gravels as a result of frost heaving. These cyclic successions are slow, probably taking at least 50 years for a well developed heath to form on initially bare gravel.

Because of these many severe limitations growth is very slow in the tundra, with 400-year-old junipers having a trunk diameter of no more than 25 mm and the Arctic willow (*Salix arctica*) having an annual increment of only 1/3 of the total plant weight—an increment which would take only one week in a temperate climate. Productivity is therefore very low and often no more than 1% of that of a temperate climate region of the same area. Tundra plants, however, have a high sugar content and therefore provide nutritious fodder although large areas are needed to support small herds of grazing animals.

The tundra plants belong to five main groups: (1) lichens; (2) mosses; (3) grasses and grass-like herbs; (4) cushion plants; (5) low shrubs. They are adapted to the environment by their forms—low-growing, cushion and tuft shapes, by their ability to become dormant, and by the presence of underground runners and rhizomes. The low shrubs are deciduous (*Salix*—willows, *Alnus*—alders, *Betula*—birches) and the heaths have small leaves with curled edges.

108. Tundra on the Arctic coast of Alaska.
Photo by F. C. Schrader, US Geological Survey

Close to the ice margins the vegetation becomes sparse, and nearer to the forest margins it is continuous, but over most of the tundra the type of community is largely controlled by topography and ground-water conditions. The main topographic contrasts are between the upland plateaux and mountains cut by valley glaciers and fiords in Norway, Greenland, Baffin, Alaska, Anadyr', Novaya Zemlya and Taymyr, while much of coastal Siberia and the lower Mackenzie Basin are low-lying marshy areas dominated by grasses, sedges and rushes.

The marshes of the wet tundra have a hummocky surface, with thick mosses forming the understory and sphagnum mosses producing hummocks, which are often ruptured by ice. Organic accumulations can be deep. Many birds nest in the tufted grasses and sedges. Willow and alder thickets spread along river banks throughout the tundra and form the only wood supply of the region.

In the upland areas on the fell-fields, scoured by glacial ice and strewn with boulders, cushion mosses, lichens and low cushion plants like the moss campion (*Silene acaulis*) and various saxifrages provide an incomplete cover. On areas with a better soil development, the shrubs and heaths form the most colourful area of the tundra.

Animals of the Cold Regions

The most important large mammals of the boreal forests are the numerous species of the deer family. Unfortunately their common names are confused. *Cervus* is the red deer of Europe and the elk of America; a subspecies in eastern Siberia is the wapiti, although this name is sometimes used to replace the elk in America. *Alces* is the genus of the elk of Europe called the moose in the New World—these are the largest surviving members of the deer family. *Rangifer* includes the caribou of North America, and the smaller and often domesticated reindeer of Eurasia. All of these animals belong to the forests, although the *Rangifer* spend the summer on the tundra grasslands and migrate to the taiga and forest during the winter to seek the 'reindeer moss' or

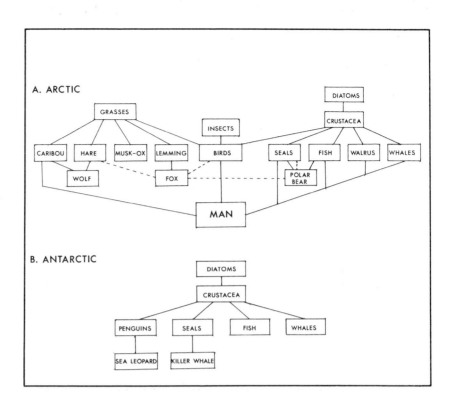

A. ARCTIC

B. ANTARCTIC

Figure 49.v. Polar food chains.
After Baird

lichens beneath the thin snow cover. The caribou of Greenland and some parts of northern Canada remain on the tundra all the year. These animals are preyed upon by wolves, which stay close to the herds taking the weak and young stragglers. Both the caribou and the musk-ox have declined greatly in numbers as a result of excessive hunting and although it appears that caribou numbers are still large, many herds are so seriously depleted that their survival is in doubt. The most obvious inhabitants of the north are the birds. Most of these migrate southwards in winter and return to nest and live on the prolific insects of the tundra in the summer.

Figure 49.v shows that in the Arctic the sea and the coasts have their own communities, with only man and to a lesser extent the birds, or the fox using the scraps left by the polar bear, sharing the resources of both land and sea. In the Antarctic the absence of higher plants from most of the land areas confines the larger animals to the sea, and even the penguins spend much of their time on floating ice or in the water.

Animals have a variety of adaptations to the cold environment. Most sea animals have layers of subcutaneous fat—blubber—which is an efficient insulator, and land animals have insulating fur or feathers. The ptarmigan even has feathers on the soles of its feet. The thick short underwool of the fox and musk-ox, and their cover of long guard hairs, are extremely good insulators. The hairs have large numbers of air cells in them, also increasing insulation, and this partly accounts for the white colourings, for the air replaces pigment. The caribou probably has the most efficient coat for its weight as a result of its tapering hairs, which are thickest at the tip and thus trap much air. These structural adaptations are reinforced by the behavioural adaptations of hibernation and migration. To fall through the roof of the snow cave of a hibernating polar bear with her young is one of the more perilous Arctic experiences.

110. A reindeer farm in
Siberia.
*Society for Cultural Relations
with the USSR*

The blue whale of the Antarctic is only one of the polar animals
which seems doomed to extinction by over-hunting; others like the
musk-ox and the Alaskan fur seal have only been saved by strong
conservation measures. Many more conservation schemes will be
needed if many species of whale, seal and the caribou are to be saved.
Productivity of the land is very limited even though large herds may
appear to deny this, and the small numbers of species but large numbers
of easily exploited animals are ready attractions to commercial enter-
prises concerned with quick profits rather than maintenance of resources.

Mountain Tundra

During the retreat of the Pleistocene ice sheets the general warming caused many tundra plants to migrate into mountain areas while others migrated polewards. There are now many related species of plants in the alpine and polar tundras and, because of the isolation of many mountains, relationships are often closer between polar regions and a mountain range than between different mountains.

The similarities between life-form and species content of alpine and polar tundra should not be allowed to obscure the differences. Alpine vegetation experiences high light intensities with daylight warming throughout the year; permafrost is usually absent and sites are usually well drained. As a result, physical drought is common. Adaptations to physical drought also provide protection against the physiological drought of Arctic winters and as both areas are subjected to high winds the vegetation of both areas has similar adaptations to the environment. Rosette and cushion plants, sedges and grasses are common mountain tundra plants, but the actual flora depends greatly upon precipitation, so that on wet mountains marsh plants and mosses may dominate, as on the flat moorlands of northern Scotland, but in the dry desert areas of the Pamirs desert plants occur. Similarly the altitude of the alpine zone varies greatly with climate and latitude—from 900 metres in the Cairngorms of Scotland to 3600 metres in the western Himalayas.

◁ 111. Alpine grasslands of green fescue at 2500 metres in the Wallowa Mountains of Oregon.
US Dept of Agriculture

50 Ecology and Man

During his growing dominance of the terrestrial environment during the last 4000 years, man has so modified the landscape and natural ecosystems that only the ocean has survived largely unscathed. The degree of change varies greatly from area to area, so that industrial and urban landscapes are almost entirely the product of man but remote mountains and forests are still essentially natural. Most of the changes are to man's advantage, but some are not, and since the beginning of the industrial revolution our ability to alter and destroy has exceeded our ability or desire to preserve and manage those parts of the environment on which we rely. This inability is partly caused by our lack of knowledge, for in an age when man has been able to travel to the moon he has not succeeded in ridding himself of such scourges as malaria or plague, and partly by our failure to respect the complexity of natural ecosystems.

The subject of man's relations with his environment is too large to be dealt with in detail here, and aspects of his relations with climate and soils have already been mentioned in Chapters 10 and 35, so the problems of pollution, pest control and conservation will be briefly discussed as examples. Productivity of ecosystems was discussed in Chapter 39.

Pollution

Man's destruction of his own habitat has reached a climax in this century as increasingly large amounts of toxic materials have been released from industrial processes and as toxins for pest control have

112. Spreading fertilisers by aircraft.
NZ Soil Conservation and Rivers Control Council

become more common. It might be thought that fertilisers could not be regarded as pollutants, but two examples will perhaps show the dangers of interfering with natural systems. In New Zealand fertilisers are often applied to pastures by low-flying aircraft. Around many of the lakes in the central North Island much land has recently been converted from forest to pasture and aerial top-dressing of the new grasslands has been frequent. Much of the dust so dropped drifts with the wind into streams and lakes so that there is a gradual buildup of the nutrients in the water. Most of these lakes are dammed by lava flows or occupy volcanic craters and are too young to have much organic matter in them. Their waters were originally clear, rich in oxygen and ideal for trout and so very popular with tourists and sportsmen. As a result of the higher nutrient levels weeds started to grow round the shores of some lakes and algae grew in the offshore waters. Animal life also increased as worms and small fish found a more plentiful food supply, and bacteria and fungi multiplied as more humus became available. By the activity of the bacteria the nutrients from humus were again released into the water. The higher populations used up the available oxygen more readily than in the past, so that fish like the trout, which require much oxygen, finding that they could only survive in the surface waters which were too warm for their health, began to decline compared with the coarser fish. The continual growth of weed crowding the shores and the smell produced as it decays makes the lakes less attractive to swimmers, fishermen and sailors; hence there is a distinct loss in amenity caused by enrichment of the water with nutrients. All lakes are temporary and are gradually filled with sediment, but the discharge of waste products into them is an irreversible process which accelerates their decline, as has been found in Switzerland, where the discharge of sewage into them has already destroyed many lake trout fisheries and made some lakes unfit for sports.

A second example is one common to many rivers which have sewage or other organic waste discharged into them. This will eventually be

Figure 50.i. Organic sewage is a rich resource for decomposers which thus use up the oxygen in the water near the sewage outfall. The life in the stream is thus controlled by the oxygen and light requirements of the fauna.

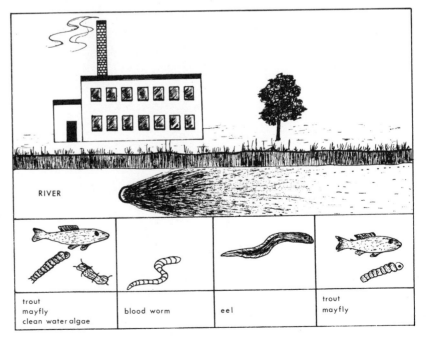

| trout
mayfly
clean water algae | blood worm | eel | trout
mayfly |

oxidised and end up in the sea, but at the point of discharge the oxygen may be so reduced as oxidation takes place that anaerobic conditions will occur and the water will be quite unfit for fish for someway downstream. On a river up which fish migrate, such as many salmon rivers, this can completely destroy the fisheries (Figure 50.i).

Radiation

A very modern type of pollution is that produced by radioactivity. So far relatively few people have died as a result of this, and most of these after the atomic bomb attacks on Hiroshima and Nagasaki by the US air force in 1945. Elsewhere a few scientists and some Pacific islanders have been affected. The dangers of radioactivity are, however, not confined to bomb explosions, for atomic power stations, X-ray machines and atomic waste products are becoming increasingly common. Not only death but illnesses such as radiation sickness, cancer and mutations can result from exposure to excessive radioactivity.

The area of the Bikini and Eniwetok atolls in the central Pacific Ocean has probably been subjected to higher levels of radiation than anywhere else in the world. In the 12 years from 1946 to the end of 1958, 59 nuclear bomb tests were made, and then in 1964 biological surveys of the islands were made to see what effect the nuclear explosions had had. The areas closest to the fireballs were still barren because of the destruction of vegetation and soil by blast and heat, but where the soil had not been seriously disturbed vegetation had reestablished and the fish, bird and invertebrate populations seemed to be thriving. Rats, many generations removed from the ancestors which had survived massive doses of radiation, seemed healthy and showed no

113. Destruction of forest by ionising radiation from a caesium-137 source housed in the metal pipe at the end of the path in the centre of the picture.
Brookhaven National Laboratory

physical abnormalities. In the explosion craters healthy fish and corals were growing. Some crabs and clams, however, had concentrated radionuclides into their tissue, hence they could be dangerous to eat. Bikini was clearly still not safe for permanent inhabitants because of the radioactive caesium, cobalt, iron and strontium concentrated in the organisms of the islands, but the waters were safe for swimming and the land safe to walk on after 12 years.

Studies elsewhere have indicated that coniferous and deciduous forests are far more susceptible to damage than the shrubs and herbs of the atolls. Studies at the Brookhaven National Laboratory on Long Island, in which a near-climax forest of oaks and pines was subjected to ionising radiation, revealed that pine trees were the most sensitive plants followed by oak trees and then herbaceous weeds. It seems that a complex ecosystem which has a sensitive dominant, such as pine trees, could be very rapidly changed by radiation whilst those with less sensitive dominants would be more resistant. In general it appears also that plants with buds near or in the ground have greater protection than those with buds above ground level, and that the plants of the early stages of a succession are more resistant than those characteristic of the climax.

Toxic Chemicals

In 1775 Sir Percival Pott declared that the scrotal cancer so common among chimney sweeps must be caused by the soot with which they came in contact. He could not 'prove' his statement but there was plenty of evidence that many industrial occupations had associated diseases. Many miners working with coal suffered from lung diseases, and those engaged in extracting lead, copper and other ores which contained impurities like arsenic were known for the occupational diseases which attacked them. The twentieth century has produced many more toxins than were known before and also the technology for their easy and widespread distribution.

Among the most common of toxins are pesticides, which Rachel Carson, author of *Silent Spring*, declared should be called *biocides* because they kill many organisms which are not pests. The early pesticides are represented by such inorganic chemicals as arsenic, which occurs in rocks in association with mineral ores. It is still used in many weed and insect killers and particularly against chewing insects. Nicotine has also been in use for some time against chewing insects.

The second generation of pesticides are the organic substances which are usually either chlorinated hydrocarbons like DDT or organic phosphorus compounds such as malathion and parathion. DDT (short for dichloro-diphenyl-trichloro-ethane) was first synthesised by a German chemist in 1874, but its properties as an insecticide were not discovered until 1939, when it was universally hailed as the protector against insect-borne disease and crop pests. It is not readily absorbed by the skin when it is in powder form and hence has been regarded as harmless to animals, but when dissolved in oil it is readily absorbed and concentrated and stored in the body fat. It is then toxic, and if because of exercise or illness the body fat is reduced, the stored DDT can be dangerous. Because it is stored in the body, DDT is passed down the

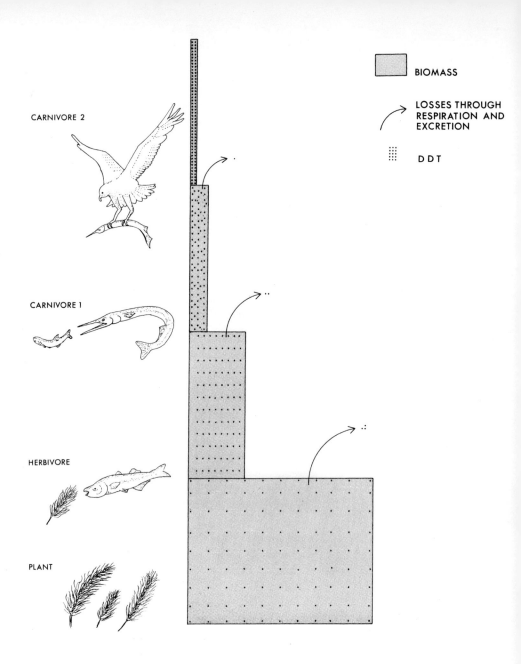

CARNIVORE 2

CARNIVORE 1

HERBIVORE

PLANT

BIOMASS

LOSSES THROUGH
RESPIRATION AND
EXCRETION

DDT

**Figure 50.ii. The trans-
mission of DDT through a
food chain.**
After Williams

food chain (Figure 50.ii) so that the final carnivore, which may be man,
can receive a very heavy, and possibly fatal dose, although the intermediate organisms may be unaffected.

Because of their initial effects, cheapness and the ease with which
they can be applied, the second generation pesticides have been liberally
and even wildly distributed over the developed countries and as a
special gift have been spread over some underdeveloped countries.
These chemicals are non-specific or broad spectrum killers which kill
useful as well as harmful organisms; hence in many areas, especially in
the USA, enormous numbers of useful as well as harmful insects,
many birds, fish and animals have been destroyed during crop spraying
programmes. They leave harmful residues in the soil and in water so
that some areas have been irreparably polluted, and in a few cases they

have been synthesised in the presence of sunlight and other chemicals into even more toxic substances. Because they cannot be washed off fruit and vegetables, and cannot be detected by ordinary procedures used to test water supplies for human consumption, many thousands of people must carry quantities of these chemicals in their bodies. Recent research in the Antarctic has revealed large doses of DDT in Antarctic fish, seals and birds which are far from any human habitation. This seems to prove quite conclusively that DDT is readily distributed by wind and water, so that we can reasonably assume that no part of the earth is uncontaminated.

Other problems are the rate at which many organisms have evolved mutations, which allow them to escape the effects of the insecticides, only to reproduce in increasing numbers because the original spraying has killed off some of their competitors. As a result the fight against the *Anopheles* malaria-carrying mosquito, which was once thought to be over, is again a tough battle. To overcome resistance, more and more powerful chemicals are being produced at ever-increasing cost and the dangers of death, sterility, mutations and diseases like cancer in both humans and beneficial organisms are also increasing.

The deaths of thousands of birds and fishes and the predators of the pests has given a boost to the research which may lead to the third generation of insecticides, although it will not solve the problems of massive doses of toxins already in the environment and in body tissues, nor the destructive attitude which leads people to spray whole communities with pollutants in order to attack just a few of the organisms in them. Agricultural crop systems are very simple and any monoculture is an easy target for a pest. A mixed community, like all natural systems, is much less vulnerable and less subject to the buildup of pest populations. Destruction of hedgerows, roadside mixed vegetation, natural ponds, woods and patches of vegetation is harmful because it destroys balance and the breeding areas of the enemies of the pests.

The third generation of insecticides which is under development at present will, if successfully produced, solve many of our problems. In most insects hormones are released into the blood for part of the life cycle only, and at other times a release would be harmful. For example, an immature larva has an absolute requirement for juvenile hormone if it is to progress through the usual larval stages, but before it can metamorphose into a sexually mature adult the flow of hormone must stop. After the adult is fully formed, juvenile hormone must again be secreted. Hence if insects come into contact with the hormone at certain critical times they will be so deranged that they will die or lay sterile eggs. It has already been possible to synthesise some insect hormones, so that there is a possibility of producing insecticides which are entirely specific to the pest species and to which the pest could not develop a resistance without committing suicide. It seems, then, that we are on the edge of another revolution in pesticides which should have a far less damaging effect on the environment than the earlier types.

The problem which remains is that of cost. Already over £20 million is spent in Britain each year on pesticides and over $500 million in the USA. There is thus a powerful vested interest in the form of manufacturers and retailers who do not want to see this industry decline, but if we are to avoid spiralling costs and further harm we must turn to biological controls as well as chemical methods.

Biological Controls

Biological control of pests is essentially the use of some form of life to overcome another form which is causing an economic loss to man. The opportunity to exert such controls arises because pests compete amongst themselves for the food supply which man wishes to crop. Biological pest control methods can be roughly classified as follows:

(1) *Against animal pests the use of:*
   (*a*) Parasites or predators
   (*b*) Lethal genes
   (*c*) Sterile matings
   (*d*) Resistant varieties
   (*e*) Competing species
   (*f*) Changes of the environment
(2) *Against plant diseases the use of:*
   (*a*) Parasitic fungi or bacteria
   (*b*) Swamping with other fungi harmless to the crop
   (*c*) Resistant varieties
   (*d*) Changes of the environment
(3) *Against weeds the use of:*
   Parasites or predators, which could be insects, fungi, viruses or any form of life.

Predation and Parasitism

The earliest forms of biological control all involved the use of predators. The use of predatory ants to control citrus pests in Asia predates written records but certainly was in use by 900 AD. In Arabia, growers also used ants from the mountains to control a pest ant which destroyed their date palms. The Indian mynah bird was introduced to Mauritius in 1762 and has successfully controlled the red locust, but introductions of birds have not always been successful as birds tend to behave like broad spectrum insecticides, killing useful as well as harmful insects.

The successful experiment which brought biological control before the public notice was the deliberate introduction into Californian orange groves of the lady-bird beetle, *Rodalia cardinalis*, to control the accidentally introduced cottony-cushion scale insect, *Icerya purchasi*, which could destroy the citrus trees (Plate 114). The cottony-cushion scale insect came originally from Australia so when it became a menace in California, Albert Koebele was sent to Australia to look for natural predators or parasites. With help from local entomologists he discovered that the lady-bird beetle lays its eggs in the scales of the cottony-cushion insect and that when they hatch the larvae feed on the eggs of the scale insect and so destroy it. In 1889 about 500 beetles were sent to California and liberated on the citrus trees. The success was spectacular and the scale only became significant again when a DDT spraying programme killed the beetle, which has since been re-established. The total cost of Koebele's expedition was less than $5000; it would now cost more, but certainly not as much as the $1,000,000 it often costs to put a new insecticide on the market.

A spectacular success of biological control methods occurred in Australia in 1925. A number of species of prickly pear, *Opuntia*, had been introduced from Mexico in the 1880's and 1890's for ornamental purposes, but these spread and became such pests that by 1925 24

114. Adult lady-bird beetles feeding on white cottony-cushion scale.
Photo reproduced with permission from: P. DeBach, Biological Control of Insect Pests and Weeds, *published by Chapman and Hall Ltd*

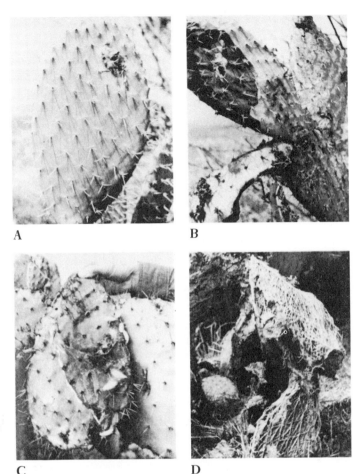

115. The destruction of *Opuntia* by *Cactoblastis*. **(A)** the appearance of the *Opuntia* soon after larvae have entered it. **(B)** the pad partly destroyed by larvae and wilt. **(C)** *Cactablastis* cocoons attached to *Opuntia* pad. *Photos reproduced with permission from P. DeBach*, Biological Control

A   B

C   D

million hectares were infested with them. Spraying was effective but very dangerous to animals and exceedingly expensive, and furthermore it did not eradicate the plant. Several insects were introduced in the hope that they would control the prickly pear, but all failed until the eggs of a moth from Argentina, *Cactoblastis cactorum*, were introduced in 1925. The eggs hatch to tiny caterpillars in 3–6 weeks, and the larvae immediately start tunnelling in the plant until only an empty shell is left. The success of *Cactoblastis* was instantaneous and complete.

Another successful example is the giant toad, *Bufo marinus*, which was introduced into Puerto Rico, where it successfully controlled the white grubs of sugar cane.

Use of Viruses

Perhaps the best known use of a virus for control purposes was the use of the virus of myxomatosis to reduce the rabbit population in Australia. Another example occurred during the 1950's in the south of France, where the pine forests were attacked by the pine processionary caterpillar, *Thaumetopoea pityocampa*. Control was achieved by infecting the caterpillars with a virus which killed them, and then grinding up their bodies in water to make a spray which could spread the infection through the forest. The great value of this method was that it used a virus which was specific to the harmful caterpillar.

Competition

A major problem of man-made simple ecosystems is that natural predation cannot always occur. This is particularly true when an animal or plant is introduced into a new area in which it has no natural enemies and may then so thrive that it becomes a pest. Man can sometimes introduce a competitor which will occupy the same ecological niche as the pest. This has been done in parts of Malaysia, where the *Anopheles* malaria-bearing mosquito can be partly combated by the introduction of the non-malaria-carrying *Culex* mosquito.

Resistant Varieties

Many cultivated plants are very susceptible to diseases like the rust which infects wheat. Breeding can produce rust-resistant strains of wheat but mutations in the rusts can produce virulent strains so quickly that the plant breeders have a constant battle to keep ahead of the mutants.

A long-term success was the use of grafting to overcome *Phylloxera*. *Phylloxera* is an insect of the aphid type which is native to North America, where it breeds and feeds on the leaves of native grape vines, to which it does no harm. During the 1860's it was accidentally introduced into Europe, where it attacked not only the leaves but also the roots of the vines and killed them. In a few years France lost a large proportion of its vineyards and became a wine importing country. Many thousands of people were ruined. The resistant American grapes produced but poor wine and so were of little value as substitutes, and the chemical methods of control which were tried were both expensive and dangerous. The successful solution was to graft the European vine on to the resistant American rootstock. In France alone ten thousand million vines were replaced by grafted ones. Grafting was a once and for all success against *Phylloxera* and hence relatively cheap.

Changes of the Environment

When a species becomes a pest it is often because there is a failure of its natural enemies to increase at the same rate as the pest, possibly because of some environmental factor such as water or food supply or shelter. Changing this factor can favour the enemies of the pest.

Migratory locusts (*Locusta migratoria*) do harm when they swarm, although they are seldom a pest when they are not in concentrations; removing the causes of swarming can thus prevent the locust from becoming a menace. The main causes of swarming are floods which make the locusts concentrate on high ground and droughts which make them concentrate around water and green plants. Flood control, drainage and irrigation have already prevented the deltas of the Danube and other Mediterranean and Black Sea rivers from becoming swarming grounds. Drainage of swampy breeding grounds has also gone far to eliminate the *Anopheles* mosquito from much of Europe and Malaysia.

Sterility

X-rays and gamma rays cause sterility in many insects. Sterilising and then releasing sexually active insects which will compete with normal insects will cause normal females to either not lay eggs or to lay sterile ones. This will reduce the insect population. The method has been so successful that some pests have been completely destroyed by this means and much more cheaply than could be done with chemicals. Furthermore, the method is entirely specific to the pest. An example is the campaign against the screw-worm (*Callitroga hominovorax*) which is a severe pest of cattle in parts of the Caribbean and America. The release of sterile male screw-worms has cleared Florida of the pest and the method is now being used in Texas, where 170 million sterile flies are produced each week for release.

This method has the advantage that it is very effective where pest populations are low. Chemical means, by contrast, are most efficient where populations are high, so there are often opportunities for integrated schemes in which pesticides can reduce the population and release of sterile insects can complete the eradication.

Conclusion

At present biological control methods have produced a few spectacular successes and many failures. In the USA up to 1950, 18 pests had been controlled out of the 91 against which biological methods had been tried. Many of the failures occur because we do not yet always understand sufficient of the ecology of the pest. It is obviously of little value if a parasite hatches out at a time when the pest is not at a suitable stage for attack. With increasing knowledge, however, more success may be expected although the future will probably see greatest success in the use of integrated schemes which will use several methods of control. Ultimately we should strive to find the most economic solution to our problems which will do least harm to the ecosystem.

Conservation and Preservation

In Neanderthal times there was a rough check kept on animal populations by competition between animals, and man had relatively little influence. He was one of the animals which either took other animals for food or ended up in the stomach of some large carnivore. With the development of technology and increase in human populations, man's impact increased. In the Mediterranean area destruction of forests for building materials and fuel, and the proliferation of herds of goats, caused extensive soil erosion even during Roman times, but the major

116. A barren area of Jordan which was well vegetated when the Romans built the forum.
Photo by E. M. Stokes

effect of man on animal populations has come only in the last 400 years, during which exploration and colonisation of the Americas, Africa and Australasia by Europeans armed with guns have exterminated many species and threaten many more.

The use of firearms is not the only cause of pressure on animals. Any change in the environment, and especially the clearance of forests, draining of swamps or ploughing of grasslands, can destroy the habitat and with it the means of existence. Political considerations have also been important as in the case of the North American bison, popularly called the buffalo. The plains Indians hunted the bison and from it obtained their food, clothing and tents, but did not deplete the herds. The colonists with their firearms did deplete the herds, for the profit in their skins, but the final near-extermination came at the urging of politicians and generals who wished to drive the Indians from the plains and could most easily do so by destroying their food supply. Francis Parkham described in his *Oregon Trail*, written in 1847, how he came across vast seas of bison corpses, shot down and left to rot, with only the most palatable and marketable part, the tongue, taken away. A few survivors have been gathered together to form a herd in Canada so at least the bison is not extinct. In North America also the pronghorn antelope and caribou were once present in herds of many millions and are now reduced to small relicts. The caribou is still threatened by Eskimo hunters and few really mature animals are to be seen. In South Africa mass slaughter of the Veld herbivores to make way for settlers, and in Australia the introduction of cattle, sheep and European grasses, have similarly reduced native herds. With an ever-increasing world human population such changes are inevitable and necessary, but many of the consequences are quite unnecessary, harmful and wasteful.

Unthinking destruction was seen at its worst in the Pribilof Islands in the Bering Sea, where Russian and Japanese sealers reduced the fur seal population from about $2\frac{1}{2}$ million to 200,000 by the time the USA took over and controlled the Islands just in time to save a breeding nucleus. In Antarctica the largest creature the world has *ever* known in its entire history, the blue whale (it can reach a length of 30 metres and

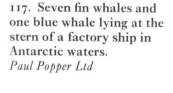

117. Seven fin whales and one blue whale lying at the stern of a factory ship in Antarctic waters.
Paul Popper Ltd

118. **An Arabian oryx trans-fered to Phoenix, USA, to form part of a breeding nucleus.**
Photo by G. Gerster, the World Wildlife Fund

a weight of 102 metric tonnes), is apparently doomed to extinction. The Norwegian, Russian and Japanese whalers have been warned and, although the Russians and Norwegians have shown a willingness to retire their ageing fleets, the Japanese with new ships and equipment have shown no willingness to reduce their profits and seem determined to continue hunting the blue and other whales as long as it is profitable to do so.

Because of the rate at which it was settled, North America has the blackest record in destroying animals. Since 1914 the following have become extinct: Labrador duck, great auk, passenger pigeon, heath hen, Carolina parakeet, Arizona elk, eastern elk, the California, Texas and plains grizzly bears, eastern forest bison, giant sea mink, plains grey wolf, eastern puma, Badland bighorn and many species of fish.

The passenger pigeon is an almost incredible example of wanton destruction. In early colonial times a single flock may have numbered 2000 million birds. When they flew south on migration in the autumn they passed continually in clouds some kilometres wide and filled the skies for 3 days on end. At these times they were slaughtered in such numbers that farmers drove their pigs up to 160 km to fatten them up on the carcases. Shooting and clubbing them out of trees alone made little difference to their numbers, but the destruction of their forest habitat with these other factors finally eliminated them, so that the last passenger pigeon died in Cincinnatti Zoo in 1914.

Conservation societies and governments have become more active and conscious of the problems in recent years but even their actions are not always successful. Many animals were rescued from the rising waters behind the Kariba Dam but then were released into the tsetse fly control area of Rhodesia, where all large game is shot in the belief that if the host of the fly is destroyed so is the fly. In fact the fly still survives on the small ground squirrels and rodents, so that destruction of the larger animals is pointless. Other efforts have been more success-ful. National parks and reserves have been established in all continents where breeding colonies of animals can survive, but money is needed to police the parks against poachers—as in the Everglade swamps of

Florida where the alligator (*Alligator mississippiensis*) is endangered by hunters after the skins, and in east Africa where poachers seek elephant ivory and meat. In the absence of a natural balance selective culling of herds is necessary. In Queen Elizabeth Park, Uganda, the herd of 15,000 hippos, each of which can eat 180 kg of grasses every 24 hours, has to be culled of 3000 annually to prevent overgrazing, erosion and sickness in the protected environment. In Murchison Falls Park and Tsavo Park in Kenya, herds of elephants cut off by roads and agricultural settlements from their traditional migration routes are destroying the landscape and will need culling.

A few simple species have been transplanted. A few Arabian oryx have been established in Arizona, where they are safe from Cadillac driving sheikhs with high-powered rifles, and the herd of Père David deer established at Woburn Park in Bedfordshire, England, after their rescue from the Royal Palace in Peking following the Boxer rebellion, have so built up their numbers that other herds have been started from this stock.

The advantages in preserving rare species are largely scientific, but conservation of many species is both ecological and financial good sense. Game ranching, which includes controlled cropping and culling, can assist in restoring and improving the habitat while obtaining maximum production. To come to this conclusion it was necessary to conduct extensive ecological studies under natural conditions. Hence it is essential to preserve some wilderness areas to act as 'controls' for comparison with man-made ecosystems, so that scientific research can indicate the most fruitful forms of land use. Such areas not only give pleasure to many people but are a source of enormous revenue from an ever-expanding tourist industry. Multiple use of resources, and preservation of controls, should help us to minimise the unfavourable consequences of transforming large areas of the earth into cities, artificial waterways, cultivated fields and one-species forests.

Appendices

Note: Find the known temperature to be converted in bold type. Then read the Celsius (Centigrade) conversion to left and Fahrenheit to right.

e.g. −2·8 **27** 80·6
27C = 80·6F
27F = −2·8C

C		F
−273	**−459**	
−268	**−450**	
−262	**−440**	
−257	**−430**	
−251	**−420**	
−246	**−410**	
−240	**−400**	
−234	**−390**	
−229	**−380**	
−223	**−370**	
−218	**−360**	
−212	**−350**	
−207	**−340**	
−201	**−330**	
−196	**−320**	
−190	**−310**	
−184	**−300**	
−179	**−290**	
−173	**−280**	
−169	**−273**	−459·4
−168	**−270**	−454
−162	**−260**	−436
−157	**−250**	−418
−151	**−240**	−400
−146	**−230**	−382
−140	**−220**	−364
−134	**−210**	−346
−129	**−200**	−328
−123	**−190**	−310
−118	**−180**	−292
−112	**−170**	−274
−107	**−160**	−256
−101	**−150**	−238
−96	**−140**	−220
−90	**−130**	−202
−84	**−120**	−184
−79	**−110**	−166
−73	**−100**	−148
−68	**−90**	−130
−62	**−80**	−112
−57	**−70**	−94
−51	**−60**	−76
−45·6	**−50**	−58·0
−45·0	**−49**	−56·2
−44·4	**−48**	−54·4
−43·9	**−47**	−52·6
−43·3	**−46**	−50·8
−42·8	**−45**	−49·0
−42·2	**−44**	−47·2
−41·7	**−43**	−45·4
−41·1	**−42**	−43·6
−40·6	**−41**	−41·8
−40·0	**−40**	−40·0
−39·4	**−39**	−38·2
−38·9	**−38**	−36·4
−38·3	**−37**	−34·6

C		F
−37·8	**−36**	−32·8
−37·2	**−35**	−31·0
−36·7	**−34**	−29·2
−36·1	**−33**	−27·4
−35·6	**−32**	−25·6
−35·0	**−31**	−23·8
−34·4	**−30**	−22·0
−33·9	**−29**	−20·2
−33·3	**−28**	−18·4
−32·8	**−27**	−16·6
−32·2	**−26**	−14·8
−31·7	**−25**	−13·0
−31·1	**−24**	−11·2
−30·6	**−23**	−9·4
−30·0	**−22**	−7·6
−29·4	**−21**	−5·8
−28·9	**−20**	−4·0
−28·3	**−19**	−2·2
−27·8	**−18**	−0·4
−27·2	**−17**	1·4
−26·7	**−16**	3·2
−26·1	**−15**	5·0
−25·6	**−14**	6·8
−25·0	**−13**	8·6
−24·4	**−12**	10·4
−23·9	**−11**	12·2
−23·3	**−10**	14·0
−22·8	**−9**	15·8
−22·2	**−8**	17·6
−21·7	**−7**	19·4
−21·1	**−6**	21·2
−20·6	**−5**	23·0
−20·0	**−4**	24·8
−19·4	**−3**	26·6
−18·9	**−2**	28·4
−18·3	**−1**	30·2
−17·8	**0**	32·0
−17·2	**1**	33·8
−16·7	**2**	35·6
−16·1	**3**	37·4
−15·6	**4**	39·2
−15·0	**5**	41·0
−14·4	**6**	42·8
−13·9	**7**	44·6
−13·3	**8**	46·4
−12·8	**9**	48·2
−12·2	**10**	50·0
−11·7	**11**	51·8
−11·1	**12**	53·6
−10·6	**13**	55·4
−10·0	**14**	57·2
−9·4	**15**	59·0
−8·9	**16**	60·8
−8·3	**17**	62·6
−7·8	**18**	64·4
−7·2	**19**	66·2

C		F
−6·7	**20**	68·0
−6·1	**21**	69·8
−5·6	**22**	71·6
−5·0	**23**	73·4
−4·4	**24**	75·2
−3·9	**25**	77·0
−3·3	**26**	78·8
−2·8	**27**	80·6
−2·2	**28**	82·4
−1·7	**29**	84·2
−1·1	**30**	86·0
−0·6	**31**	87·8
0	**32**	89·6
0·6	**33**	91·4
1·1	**34**	93·2
1·7	**35**	95·0
2·2	**36**	96·8
2·8	**37**	98·6
3·3	**38**	100·4
3·9	**39**	102·2
4·4	**40**	104·0
5·0	**41**	105·8
5·6	**42**	107·6
6·1	**43**	109·4
6·7	**44**	111·2
7·2	**45**	113·0
7·8	**46**	114·8
8·3	**47**	116·6
8·9	**48**	118·4
9·4	**49**	120·2
10·0	**50**	122·0
10·6	**51**	123·8
11·1	**52**	125·6
11·7	**53**	127·4
12·2	**54**	129·2
12·8	**55**	131·0
13·3	**56**	132·8
13·9	**57**	134·6
14·4	**58**	136·4
15·0	**59**	138·2
15·6	**60**	140·0
16·1	**61**	141·8
16·7	**62**	143·6
17·2	**63**	145·4
17·8	**64**	147·2
18·3	**65**	149·0
18·9	**66**	150·8
19·4	**67**	152·6
20·0	**68**	154·4
20·6	**69**	156·2
21·1	**70**	158·0
21·7	**71**	159·8
22·2	**72**	161·6
22·8	**73**	163·4
23·3	**74**	165·2
23·9	**75**	167·0

C		F	C		F	C		F
24·4	76	168·8	55·6	132	269·6	78·3	173	343·4
25·0	77	170·6	56·1	133	271·4	78·9	174	345·2
25·6	78	172·4	56·7	134	273·2	79·4	175	347·0
26·1	79	174·2	57·2	135	275·0	80·0	176	348·8
26·7	80	176·0	57·8	136	276·8	80·6	177	350·6
27·2	81	177·8	58·3	137	278·6	81·1	178	352·4
27·8	82	179·6	58·9	138	280·4	81·7	179	354·2
28·3	83	181·4	59·4	139	282·2	82·2	180	356·0
28·9	84	183·2	60·0	140	284·0	82·8	181	357·8
29·4	85	185·0	60·6	141	285·8	83·3	182	359·6
30·0	86	186·8	61·1	142	287·6	83·9	183	361·4
30·6	87	188·6	61·7	143	289·4	84·4	184	363·2
31·1	88	190·4	62·2	144	291·2	85·0	185	365·0
31·7	89	192·2	62·8	145	293·0	85·6	186	366·8
32·2	90	194·0	63·3	146	294·8	86·1	187	368·6
32·8	91	195·8	63·9	147	296·6	86·7	188	370·4
33·3	92	197·6	64·4	148	298·4	87·2	189	372·2
33·9	93	199·4	65·0	149	300·2	87·8	190	374·0
34·4	94	201·2	65·6	150	302·0	88·3	191	375·8
35·0	95	203·0	66·1	151	303·8	88·9	192	377·6
35·6	96	204·8	66·7	152	305·6	89·4	193	379·4
36·1	97	206·6	67·2	153	307·4	90·0	194	381·2
36·7	98	208·4	67·8	154	309·2	90·6	195	383·0
37·2	99	210·2	68·3	155	311·0	91·1	196	384·8
37·8	100	212·0	68·9	156	312·8	91·7	197	386·6
38·3	101	213·8	69·4	157	314·6	92·2	198	388·4
38·9	102	215·6	70·0	158	316·4	92·8	199	390·2
39·4	103	217·4	70·6	159	318·2	93·3	200	392·0
40·0	104	219·2	71·1	160	320·0	93·9	201	393·8
40·6	105	221·0	71·7	161	321·8	94·4	202	395·6
41·1	106	222·8	72·2	162	323·6	95·0	203	397·4
41·7	107	224·6	72·8	163	325·4	95·6	204	399·2
42·2	108	226·4	73·3	164	327·2	96·1	205	401·0
42·8	109	228·2	73·9	165	329·0	96·7	206	402·8
43·3	110	230·0	74·4	166	330·8	97·2	207	404·6
43·9	111	231·8	75·0	167	332·6	97·8	208	406·4
44·4	112	233·6	75·6	168	334·4	98·3	209	408·2
45·0	113	235·4	76·1	169	336·2	98·9	210	410·0
45·6	114	237·2	76·7	170	338·0	99·4	211	411·8
46·1	115	239·0	77·2	171	339·8	100·0	212	413·6
46·7	116	240·8	77·8	172	341·6			
47·2	117	242·6						
47·8	118	244·4						
48·3	119	246·2						
48·9	120	248·0						
49·4	121	249·8						
50·0	122	251·6						
50·6	123	253·4						
51·1	124	255·2						
51·7	125	257·0						
52·2	126	258·8						
52·8	127	260·6						
53·3	128	262·4						
53·9	129	264·2						
54·4	130	266·0						
55·0	131	267·8						

Values of single degrees

°C	°F		°F	°C	
1 =	1·8		1 =	0·56	
2 =	3·6		2 =	1·11	
3 =	5·4		3 =	1·67	
4 =	7·2		4 =	2·22	
5 =	9·0		5 =	2·78	
6 =	10·8		6 =	3·33	
7 =	12·6		7 =	3·89	
8 =	14·4		8 =	4·44	
9 =	16·2		9 =	5·0	

Appendix II The Classification and Nomenclature of Plants and Animals

Plants and animals are classified into groups which seem to have a close affinity or outward similarity. There is not always agreement on the details of classification but for general purposes these groups or *taxa* (sing. *taxon*) may be listed as below:

Divisions or phyla (sing. *phylum*): The major or highest grouping used in classification. In the plant kingdom the names usually end in -*phyta*; e.g. *Bryophyta*.

CLASSES

Orders: Each plant order has a name ending in -*ales*.
Families: Most plant families have names ending in -*acaea*.
Genera (sing. *genus*): The members of each genus usually look alike.
Species: The smallest unit of classification in general use. Its members have a broad similarity to each other.

The scientific name of each species is normally made up of 2 Latin or latinised words of which the first is the name of the genus, and the second its own epithet. The first or generic name is capitalised and that of the specific epithet is left in lower case. The scientific name is printed in italics or is underlined in manuscript or typescript; e.g. *Brassica campestris*—the turnip.

Chief Divisions of Plants

SCHIZOPHYTA: These are the bacteria.

THALLOPHYTA: These are the algae, fungi and moulds.

BRYOPHYTA: Includes mosses and liverworts. Unlike the thallophyta they show differentiation into root, stem and leaves, but not full roots.

PTERIDOPHYTA: Includes ferns, horsetails and club-mosses. Roots, stems and leaves all occur. Regeneration is by spores.

SPERMATOPHYTA: Includes all seed plants and has 2 subdivisions:

(1) *Gymnospermae* Cone-bearing plants. This subdivision includes the classes of cycads, ginkoales, conifers, and gnetales.
(2) *Angiospermae* These are true flowering plants with 2 classes: monocotyledons and dicotyledons. The monocotyledons have parallel-veined leaves and one seed leaf. They are mainly herbs but they include the palms, some trees, the lily family and grasses. Dicotyledons have 2 seed leaves and have net-veined leaves. They include herbs, shrubs and trees.

Divisions of Animals

VERTEBRATA: Animals with backbones.

Class Mammalia (Mammals)

> Sub-class I *Placentals:* Development occurs within the mother's womb.
>
> Sub-class II *Marsupials:* The young are born incompletely developed.
>
> Sub-class III *Monotremes:* Egg-laying mammals. There are only 3 living forms, all of which are confined to Australia.

Class Aves (Birds)

Class Reptilia (Reptiles)
There are 5 living orders:

> (1) Crocodiles
> (2) Turtles and tortoises
> (3) Snakes
> (4) Lizards
> (5) Beaked lizards—the only living form occurs in New Zealand.

Class Amphibia (Amphibians)
The eggs are fertilised and laid in water. The class includes frogs, toads, newts, salamanders and blindworms.

Class Pisces (Fishes)

INVERTEBRATA: This division of animals without backbones includes all mollusca or shellfish, slugs and cuttlefish; all arthropods, crabs, shrimps and barnacles, all insects, starfish, worms, jellyfish, sponges, etc.

Glossary

Abyssal Pertaining to the area from the margin of the continental shelf to the greatest depths and bounded by the pelagic zone.

Acclimatisation The capacity of a species to fulfil its life cycle in an environment to which it is not indigenous.

Accretion The growth of an ice particle by collision with water drops or, in a general sense, the growth of ice crystals or water drops by collision.

Accumulated temperature The integrated excess or deficiency of temperature measured with reference to a fixed temperature. If on a particular day the temperature is above the fixed level for 2 hours and the mean temperature during that period exceeds the fixed level by 10 degrees, the accumulated temperature for the day is 2×10 or 20 degree hours or 20/24 degree days.

Acid brown forest soils A great soil group occurring under cool perhumid climatic conditions; different from *brown forest soils* in that the (B) is weakly developed, subangular blocky, and has low base saturation of the clay.

Adaptation A measure of physiological fitness of an organism with respect to one or all of the conditions of its environment.

Adiabatic An adiabatic process is one in which heat does not enter or leave the system. In the expansion of a body of air its temperature falls, and in compression of a body of air its temperature rises. Such adiabatic processes in rising and descending air are largely responsible for the vertical temperature distribution in the atmosphere.

Advection The process of transfer by horizontal movement as in the process of transfer of heat by air masses.

Advection fog Fog formed by the passage of warm, moist, stable air over a cool surface.

Aerobic Adapted to conditions of the environment in which there is free oxygen. Particularly of bacteria.

Aerology The study of the atmosphere, particularly of the atmosphere above the surface layers.

Aerosol An aggregation of minute solid or liquid particles suspended in the atmosphere.

Aestivation (Estivation) A state or period of dormancy coincident with and probably induced by high temperatures and/or drought, e.g. desert snails which can only live where there is an occasional dew. They overcome the problem of lack of moisture for transpiration by spending hot, moistureless periods in a state of aestivation. This period may last as long as 4 years.

Aggregate A soil structural unit consisting of more than 1 primary particle.

A Horizon The uppermost layers of a mineral soil profile where accumulation of organic matter and eluviation commonly occur.

Air The mixture of gases which form the atmosphere.

Air mass A body of air with homogeneous characteristics of temperature and humidity. Air masses have horizontal dimensions of hundreds or thousands of kilometres.

Albedo A measure of the reflecting power of a surface, being the fraction of the incoming radiation which is reflected by that surface. Typical values are: forest 10% (or 0·1); sand 20–30% (or 0·2–0·3); fresh snow 80% (or 0·8).

Alfisols Soils that have no mollic epipedon, oxidic or spodic horizon, but do have argillic or natric horizons with high base saturation. (7th Approximation).

Allitic Describes soils from which silica has been largely removed, and which have a clay fraction in which Al and Fe compounds predominate.

Allitisation Synonymous with *laterisation*.

Allotrophic Of lakes which receive drainage containing organic matter in addition to that produced internally.

Alpine Of vegetation which is low, herbaceous or shrubby found above the timber-line on high mountains.

Alpine meadow soils A great soil group of the intrazonal order, comprised of dark soils of grassy meadows at altitudes above the timber-line.

Altocumulus ⎱
Altostratus ⎰ A type of cloud.

Amorphous Without shape or structure.

Anabatic wind A local wind which blows up a slope heated by the sun.

Anaerobic Adapted to environmental conditions containing no free oxygen.

Andosols Soils with an AC or A(B)C profile characterised by non-crystalline alumino-silica gels in the clay fraction. These soils have a high sorptive capacity; a relatively thick, friable dark A horizon with high organic matter content; they have low bulk density and low stickiness. They occur under humid and subhumid conditions.

Anemometer An instrument for measuring the speed of the wind.

Aneroid barometer An instrument which measures atmospheric pressure by the movement of a thin capsule of corrugated metal which contains a vacuum.

Anion An ion carrying a negative electrical charge.

Anomaly The difference between the value of a meteorological element, such as temperature, at a given time and place, and its average value at the same place.

Anticyclone An atmospheric pressure system in which the centre has a high pressure compared with its surroundings.

Anti-trades Upper winds which in some low latitudes prevail above the trade winds. They are more or less opposite in direction to the trade winds and move air polewards.

Aphelion That point of the orbit of a planet which is farthest from the sun. The aphelion for the earth occurs on 1 July.

Arctic The region of high northern latitudes generally limited southwards by the timber-line.

Arctic air An air mass originating over the snow and ice of the Arctic.

Arctic front The front which separates Arctic air from the mP or cP air to the south.

Argillisation The formation of secondary clay minerals within the soil.

Argilluviation A recently introduced term for the migration of clay from A to B horizons in soils. Equivalent to the French term *lessivage*.

Arid A climate in which rainfall is insufficient to support vegetation.

Aridisols Soils that are usually dry with a calcic horizon immediately underlying a light-coloured surface horizon (7th Approximation).

Aspect In vegetation the seasonal appearance.

Association A unit of vegetation dominated by 2 or more species.

Atmosphere The gaseous envelope of the earth.

Autecology The study of a species or individual plant in relation to its environment.

Autotrophs Animals and plants which are independent of the tissue of other organisms for their food supply. See *Heterotrophs*.

Avalanche wind A powerful wind produced in advance and at the sides of a descending mass of ice and snow. It is often very destructive.

Azimuth The horizontal angle between an observer's meridian and the line joining the observer and the object observed. It is normally measured in degrees (0–360°) from geographic north.

Azonal Of a soil maintained in a condition with a permanently immature or truncated profile by deposition or erosion.

Backing The change of the wind in an anticlockwise direction.

Banner cloud A stationary cloud attached to and extending downwind from an isolated mountain peak. It is probably caused by the rise of air in a lee eddy from a lower level than on the windward side of the peak.

Bar The unit of atmospheric pressure. It is equal to the pressure of 29·53 in or 750·05 mm of mercury at 0°C and at standard gravity.

Baroclinic A baroclinic atmosphere is one in which surfaces of equal density intersect surfaces of equal pressure so that temperature changes vertically and horizontally, and thermal winds result.

Barograph A recording barometer.

Barometer An instrument for measuring atmospheric pressure.

Base status A measure of the amount of replaceable basic ions present in the soil.

Beaufort scale A scale for classifying wind force. 0—calm; 2 to 6—light, gentle, moderate, fresh and strong breezes respectively; 7 to 9—near gale, gale and strong gale respectively; 10—storm; 11—violent storm; 12—hurricane.

Benthic The zone of the floor of an ocean or lake where there are rooted organisms.

Bergeron theory The theory that attributes the initiation of precipitation from a cloud to the presence of ice crystals and supercooled water droplets.

B Horizon Part of a soil profile below the A horizon; usually illuvial.

Biocenosis A term used by Russian ecologists to indicate the *biome*.

Biochore Defined by Dansereau as one of the main subdivisions of world environments. The four main ones are forest, savanna, grassland and desert.

Biocycle The biosphere is divided into 3 biocycles— the salt water, fresh water and land biocycles.

Biogeocenosis See *Geobiocenosis*.

Biogeography The study of the origin, distribution, adaptation and association of plants and animals.

Biomass The weight of living organisms in an ecosystem, expressed either as fresh weight or dry weight.

Biome A large primary biotic community in which the climax vegetation is more or less uniform.

Biosphere That part of the earth's crust and lower atmosphere inhabited by living organisms.

Biota Groups of living plants and animals occupying a place, e.g. the biota of the Galapagos Islands.

Biotic Pertaining to the complex of soils, plants and animals covering a part of the earth's surface.

Biotope The smallest subdivision of the habitat.

Black earth A term used by some as synonymous with *chernozem*; by others (in Australia) to describe self-mulching black clays.

Bleisand The bleached Ae horizon of a podzol.

Blizzard A high wind accompanied by snow and great cold. Originally used in North America.

Bog A wet area of impeded drainage covered by peat.

Bog soil See *Organic soil*.

Bora A cold, dry, often violent northeasterly wind which blows from the mountains east of the Adriatic. It is most frequent when pressure is high over central Europe and low over the Mediterranean.

Boreal A high latitude area more or less coincident with the needle-leaf forest formations.

Boreal climate A climate with snowy winter and warm summer with a large annual range of temperature, e.g. North America between about 40° and 60°N.

Brown earth Soils with a mull horizon but having no horizon of accumulation of clay or sesquioxides. In Europe this term is used as a synonym for *brown forest soils* but sometimes for similar soils acid in reaction.

Brown forest soils A great soil group occurring under temperate subhumid climates. They have A(B)C profiles with merging horizons. The A horizon is dark-coloured and the (B) horizon is granular. The silica/sesquioxide ratio is reasonably uniform through the profile and there is little or no evidence of clay illuviation.

Brown Mediterranean soils These soils have a similar ABC profile to those of *red Mediterranean* soils but the B horizon is browner in colour. They occur under warm temperate subhumid climates with a pronounced dry season.

Brown podzolic An acid soil, developed under forest, with little or no bleached A₂ horizon and only a weak textured profile, but with no illuvial accumulations of sesquioxides.

Brown soils A great soil group of the temperate to cool arid regions, composed of soils with a brown surface and a light-coloured transitional subsurface horizon over calcium carbonate accumulation.

Brunizems Soils with AC or ABC profiles with thick, dark-coloured granular A horizons and a high base saturation. A textural or colour B may be present. These soils are moderately calcareous, with $CaCO_3$ content increasing with depth. They occur in subhumid to semiarid climates in which the solum is only occasionally moistened throughout.

Bush (1) Usually a low shrub branching from the ground.
(2) In New Zealand it may refer to true forest communities.

Buys Ballot's Law States that in the northern hemisphere if an observer stands with his back to the wind, pressure is lower on his left hand than on his right; in the southern hemisphere the converse is true.

Caatinga Thorn scrub in northeast Brazil.

Calcification A soil-forming process in which lime is deposited as an illuvial layer. A characteristic of low rainfall climates.

Caliche Thick layer of calcium carbonate.

Calorie A unit of heat, being the heat required to raise the temperature of 1 gramme of water by 1°C.

Campo cerrado Brazilian savanna consisting of a dense seasonal grass layer and scattered trees.

Canopy The uppermost layer in a forest.

Capillary fringe That part of the soil immediately above the water table in which tension is linearly related to the distance from the water table and the soil is close to saturation.

Capillary water Water retained in pores primarily by surface tension.

Carbonation A soil process involving the combination of carbonate and bicarbonate ions with a mineral.

Carbon-nitrogen ratio Weight ratio of organic carbon to total nitrogen.

Carnivore A flesh-eating animal.

Carr A mound bearing trees in a peat bog.

Catena A sequence of soils of about the same age, derived from similar parent material, and occurring under similar climatic conditions, but having different characteristics due to variation in relief and in drainage. See *Clinosequence* and *Toposequence*.

Cation Ion carrying a positive electrical charge.

Cation exchange The interchange between a cation in solution and another cation on the surface of any surface-active material such as clay colloid or organic colloid.

Cauliflory The formation of flowers on the 'bare' wood of trees. A feature common only in tropical rain forests.

Celsius scale Also called Centigrade scale: in which temperature is so measured that the interval between the freezing and boiling points of pure water is divided into 100 degrees.

Centigrade See *Celsius*.

Chamaephyte A plant with a perennating bud near the ground, e.g. many shrubs. (Raunkiaer's system).

Chaparral A dense shrub vegetation mostly evergreen and growing in a Mediterranean type of climate, especially in California.

Chelation A soil process involving the combination of organic compounds in the soil with metallic cations.

Cheluviation A soil process in which ions, particularly iron and aluminium, are made soluble and leached from a soil horizon and redeposited at depth in the soil profile.

Chernozem A zonal great soil group consisting of soils with a thick, nearly black or black, organic matter-rich A horizon high in exchangeable calcium, underlain by a lighter coloured transitional horizon above a zone of calcium carbonate accumulation; occurs in a cool, subhumid climate under a vegetation of tall and mid-grass prairie.

Chestnut soil A zonal great soil group consisting of soils with a moderately thick, dark-brown A horizon over a lighter coloured horizon that is above a zone of calcium carbonate accumulation.

Chinook A warm, dry west wind, of the föhn type which occurs on the eastern side of the Rocky Mountains.

C Horizon Horizon of weathered rock material little affected by biological soil forming processes.

Chroma The purity or strength of colour. See *Munsell colour system*.

Chronosequence A sequence of related soils that differ, one from the other, in certain properties primarily as a result of time as a soil forming factor.

Cinnamon soils Characteristic soils of the summer-dry, winter-wet, subtropics, as in the Mediterranean area.

Circulation See *General circulation*.

Clay Soil mineral particles of diameter less than 0·002 mm.

Clay-humus complex The linked clay and humus molecules which make up the chemically active fraction of the soil.

Clay mineral Any mineral substance occurring in the clay fraction. Most such minerals are crystalline but some like allophane are amorphous.

Claypan A dense subsoil horizon having a high clay content compared with that of the upper horizons.

Clay skins Coatings of clay on soil aggregates. They imply illuviation.

Climate The climate of a place is the synthesis of the daily values of the meteorological elements which affect it.

Climax The plant community at the final stage of a succession which perpetuates itself indefinitely.

Climosequence A sequence of related soils that differ one from the other in certain properties, primarily as a result of the effect of climate as a soil forming factor.

Clinosequence A group of related soils that differ one from the other in certain properties, primarily as a result of the effect of the degree of slope on which they were formed. See *Toposequence*.

Clisere The replacement of one zone of vegetation by another as a result of climatic change.

Cloud An aggregate of very small water droplets and/or ice crystals.

Cloud classification In the *International Cloud Atlas* clouds are classified according to 10 basic forms, and 3 levels—high, middle and low.

Cloud forest The usually broadleaved evergreen montane forests which develop where condensation to form clouds is very frequent, as in parts of the Peruvian Andes.

Cloud genera The 10 characteristic cloud types are cirrus—Ci, cirrocumulus—Cc, cirrostratus—Cs, altocumulus—Ac, altostratus—As, nimbostratus—Ns, stratocumulus—Sc, stratus—St, cumulus—Cu, and cumulonimbus—Cb.

Cohesion The property of soil particles to stick together to form an aggregate.

Cold front A front whose movement is such that a colder air mass is replacing a warmer one.

Cold sector That part of a depression occupied by cold air on the earth's surface.

Colloidal complex The whole ion-adsorbing material of the soil.

Colloids Exceedingly small-sized substances which can either exist in a mobile (sol) or a stable (gel) state.

Commensalism The coaction in which 2 or more species are mutually associated in activities centring on food and 1 species at least derives benefit from the association, while the other associates are neither benefited nor harmed.

Commensals All organisms which exist in the same ecosystem.

Community Any group of plants growing together which has an individuality produced by a uniformity of structure and floristic composition.

Competition The struggle between individuals of the same or different species to obtain greater quantities of the available resources. Forest trees may compete for light and moisture.

Concretion A local concentration of a chemical compound, such as calcium carbonate or iron oxide, in the form of a grain or nodule of varying size, shape, hardness and colour.

Condensation The process of formation of a liquid from its vapour.

Conduction The process of heat transfer through matter.

Confluence The approach to each other of adjacent stream-lines in the direction of flow.

Conservation The protection of resources with an intention of sustaining the yield.

Consociation A plant community dominated by only 1 species.

Continental climate A type of climate characteristic of the interior of large land masses of middle latitudes. Large annual and daily ranges of air temperature and low precipitation often with a summer maximum are usual.

Continental drift A hypothesis that the continents were once joined and have drifted apart. It has been used to explain, among other things, climatic changes on a world scale.

Continentality A measure of the extent to which a place has a climate dominated by land as opposed to maritime influences.

Control (1) Biological control is obtained by the introduction and encouragement of the natural enemies to control a pest.

(2) Experimental controls have all conditions the same as those in an experiment except for the 1 or more factors being studied.

Convection A mode of heat transfer involving movements within the substance concerned, e.g. currents in air or water.

Convective rain Rain caused by vertical movement of air which is warmer than its environment. It is typically intensive and associated with cumulonimbus cloud and sometimes with thunder.

Convergence The horizontal motion of air into a central region.

Co-operation Co-actions that are beneficial to 1 or more of the participants.

Coprolite The fine-structured excreta of earthworms and other soil organisms.

Coriolis Effect or Coriolis Force A fictitious force used in meteorology to simplify calculations in which the rotation of the earth has to be taken into account.

Crotovina (Krotovina) Filled-in animal burrow in the soil.

Croute calcaire Hardened caliche, often found in thick masses or beds overlain by only some centimetres of earth. See *Caliche*.

Crumb Rounded, porous and soft soil aggregate up to 10 mm in diameter.

Cryophyte A plant with its perennating bud covered by soil or water.

Cryptopodzol A soil in which the podzolisation process is only detectible by chemical analysis. Such soils in Wales have a dark humic horizon overlying a pale brown B horizon and no Ae.

Crystal lattice See *Lattice structure*.

Cursorial Running.

Cushion plant A chamaephyte with its shoots densely massed together and with a short stature.

Cutin The cellulose body forming the skin or cuticle of plants.

Cyclogenesis The initiation of cyclonic circulation.

Cyclone A pressure system with a low pressure centre relative to its surroundings. Cyclonic circulation is anticlockwise round the centre in the northern hemisphere and clockwise in the southern hemisphere. Cyclones of middle and high latitudes are called depressions. A tropical cyclone is an intense storm. In the western Pacific it would be called a typhoon and in most other tropical latitudes a hurricane.

Cyclonic rain Rain caused by depressions and fronts.

Decalcification The removal of calcium carbonate or calcium ions from the soil by leaching.

Decomposer An organism which utilises dead plant or animal matter for food and thus releases component elements back into the nutrient cycle.

Degradation The changing of a soil to a more highly leached and more highly weathered condition, usually accompanied by morphological changes such as development of an A_2 horizon.

Degraded chernozem A zonal great soil group consisting of soils with a very dark brown or black A_1 horizon underlain by a dark grey, weakly expressed A_2 horizon and a brown B (?) horizon; formed in the forest-prairie transition of cool climates.

Denitrification The biochemical reduction of nitrate or nitrite to gaseous nitrogen either as molecular nitrogen or as an oxide of nitrogen.

Depression A low pressure system of middle and high latitudes. See *Cyclone*.

Desert A region in which rainfall is insufficient, in relation to evaporation, to support vegetation.

Desert soil A zonal great soil group consisting of soils with a very thin, pale-coloured surface horizon, which may be vesicular and is ordinarily underlain by calcareous material; formed in arid regions under sparse shrub vegetation.

Desert varnish A glossy sheen or coating on stones and gravel in arid regions.

Desiccation The disappearance of water from an area due to change of climate.

Dew The moisture deposited in drops upon a cool surface by the condensation of water vapour in the atmosphere.

Dew point The temperature to which air must be cooled before it becomes saturated with water vapour.

D Horizon Unweathered rock below the C horizon, synonymous with *R horizon*.

Disclimax A plant community or an ecosystem whose structure and composition is controlled by the interference of humans or domestic animals. This term is the equivalent for the *plagioclimax* of British usage.

Dispersal Spread of individuals away from their home sites.

Disperse (1) To break up compound particles, such as aggregates, into the individual component particles.

(2) To distribute or suspend fine particles, such as clay, in or throughout a dispersion medium, such as water.

Disseminule A seed, fruit or spore adapted for dispersion.

Divergence Motion of air away from a centre.

Doldrums The equatorial oceanic regions of light winds and calms. They are variable in extent and position, and discontinuous.

Dominant (1) An organism which controls the habitat.

(2) An organism which gives the community its characteristic appearance.

Dormancy A slowing down of metabolic processes which reduces the requirements of an organism for environmental resources and increases its tolerance to environmental adversities. Estivating and hibernating plants enter periods of dormancy.

Drought A shortage of ground water or precipitation which visibly affects the behaviour of plants and animals.

Duff See *Mor*.

Duripan Indurated soil horizon that is usually cemented by an agent that is soluble in concentrated alkali. A synonym for *hardpan*.

Dy Sedimentary peat formed from humic substances deposited in a colloidal condition from water. See *Gyttja*.

Easterly wave A shallow trough in the tropical easterlies. It brings increased cloud and showers as it moves westward.

Ecesis The successful establishment of plants or animals on a new site as a result of migration.

Ecological niche Particular position in a community and habitat occupied by an animal as the result of its peculiar structural adaptations, its physiological adjustments and the special behaviour patterns that have evolved to make best use of these potentialities.

Ecology The study of the interrelations of organisms with each other and with their immediate environment.

Ecosphere The whole terrestrial *ecosystem*.

Ecosystem The dynamic whole formed by the habitat and the association of living beings that occupy it, e.g. a bog. Ecosystem = habitat + biotic community. The idea of energy exchange is fundamental in the ecosystem concept.

Ecotone A transition area between 2 clearly defined communities.

Edaphic (1) Of or pertaining to the soil.

(2) Resulting from or influenced by factors inherent in the soil or other substrate, rather than by climatic factors.

Eddy A turbulent motion in a fluid.

Effective precipitation That part of total precipitation which is of use to plants.

Egg-cup podzol A local podzol of cup shape formed beneath a tree releasing strongly podzolising solutions from its litter.

Eluvial Of a soil horizon from which material has been removed.

Eluvial horizon Soil horizon from which material has been removed in solution or suspension.

Endemic Indigenous or native to a locality.

Energy balance The difference between the amounts of radiation which are absorbed by and emitted from the earth's surface or by the atmosphere at a specified level.

Entisol Soil without natural horizons or only the beginning of horizons (7th Approximation).

Entropy The name given to one of the quantitative elements which determine the thermodynamic condition of a portion of matter: e.g. a portion of matter at uniform temperature retains its entropy unchanged so long as no heat passes to or from it, but if it receives a quantity of heat without change of temperature the entropy is increased.

Environment The resultant of all the external conditions which act upon an organism. The area of a particular set of conditions, e.g. a salt-marsh environment.

Epiphyte A plant which grows upon another for support but is not dependent on it for nutrients.

Equator The earth's equator is the great circle whose plane is perpendicular to the earth's polar axis.

Equinox The period of the year when the day and night are of equal duration—about 21 March and 22 September.

Erosion (1) The wearing away of the land surface by running water, wind, ice, or other geological agents, including such processes as gravitational creep.

(2) Detachment and movement of soil or rock by water, wind, ice, or gravity. The following terms are used to describe different types of water erosion:

Accelerated erosion Erosion much more rapid than normal, natural, geological erosion, primarily as a result of the influence of the activities of man or, in some cases, of animals.

Geological erosion The normal or natural erosion caused by geological processes acting over long geologic periods and resulting in the wearing away of mountains, the building up of flood plains, coastal plains, etc. Synonymous with *natural erosion*.

Gully erosion The erosion process whereby water accumulates in narrow channels and, over short periods, removes the soil from this narrow area to considerable depths, ranging from 0·3–0·6 metre to as much as 25–30 metres.

Natural erosion Wearing away of the earth's surface by water, ice, or other natural agents under natural environmental conditions of climate, vegetation, etc., undisturbed by man. Synonymous with *geological erosion*.

Normal erosion The gradual erosion of land used by man which does not greatly exceed natural erosion. See *Natural erosion*.

Rill erosion An erosion process in which numerous small channels of only some centimetres in depth are formed; occurs mainly on recently cultivated soils.

Sheet erosion The removal of a fairly uniform layer of soil from the land surface by runoff water.

Splash erosion The spattering of small soil particles caused by the impact of raindrops on very wet soils. The loosened and spattered particles may or may not be subsequently removed by surface runoff.

Estival Pertaining to summer.

Estivation (Aestivation) A period of dormancy induced by high temperatures and/or drought.

Ethiopian Region The animal realm which includes Africa south of the Sahara.

Eutrophic (1) Of soils with a high nutrient content.
(2) Of a lake, usually shallow with rich flora and fauna.
(3) Accumulating in ground-water rich in bases.

Evaporation The change of liquid water or ice to water vapour.

Evaporation Fog Fog which is formed by evaporation of relatively warm water into cool air.

Evapotranspiration The combined loss of water from a given area, and during a specified period of time, by evaporation from the soil surface and by transpiration from plants.

Evergreen A plant which does not shed its leaves.

Exchangeable Describes ions capable of replacement in the adsorption complex of the soil.

Exchange capacity Milliequivalents of ions that can be adsorbed by 100 grammes of material at a specific pH.

Exotic A plant or animal introduced into an area. Not native.

Fahrenheit A scale of temperature on which the freezing point of water is at 32° and the boiling point is at 212°.

Fallow Condition of soil left without a crop for some time.

Family A group of plant or animal genera which are thought to have evolved from a common ancestor.

Fauna The sum total of animal species in an area.

Feedback The more or less circular passage of a resource through several points; e.g. a nutrient from the soil to the plant to the litter to the soil.

Fen A plant formation growing on wet organic soil dominated by shrubs and graminoids. The soil is usually alkaline in reaction—often as a result of seawater.

Ferrallitic soil A soil of the humid tropics which is rich in iron and aluminium sesquioxides. A synonym for *latosol*, *lateritic soil* and *oxisol*.

Ferrallitisation A modern term for *laterisation*.

Ferritic soil A soil of the humid tropics rich in iron alone.

Fetch The length of the traverse of moving air across a sea or ocean.

Field capacity Water held in a well drained soil after excess has drained away and the rate of downward movement has materially decreased.

Flark An elongated stretch of open shallow water with a long axis which is at right angles to the slope. Flarks occur in boreal peatlands.

F layer A layer of partially decomposed litter with portions of plant structures still recognisable. Occurs below the L layer on the forest floor in forest soils. It is the fermentation layer.

Flocculate To group together or aggregate, as of colloids.

Flora The sum of all plant species occupying an area at a given time.

Foehn An alternative for *föhn*.

Fog Obscurity in the lower layers of the atmosphere caused by a suspension of water droplets and/or smoke particles. By international agreement a fog is deemed to exist when visibility is less than 1000 metres.

Föhn A warm, dry wind occurring in the lee of a mountain range. Air which rises at the saturated adiabatic lapse rate (about 0·5°C per 100 metres) and looses water by precipitation on the windward side of the mountain, descends at the dry adiabatic lapse rate (1°C per 100 metres) and becomes a warm dry wind.

Föhn wall A mass of precipitating clouds on the windward side of a range of mountains which descend part way down the lee slope and terminate along a line parallel to the main range.

Food chain The series of organisms through which food energy moves before it is expended.

Food web The multiple-pathed interlocking system through which food energy moves.

Forb A broadleaf herb; e.g. clover.

Forest A stand of close-growing trees of considerable extent.

Formation A geographically distinct part of a formation-type: e.g. west European broadleaf deciduous forest.

Formation-type The largest and most comprehensive kind of plant community with a distinctive structure and physiognomy; e.g. a broadleaf deciduous forest.

Fossorial Burrowing.

Fragipan Compact, massive horizons, rich in silt, sand or both, and low in clay. When dry they are brittle, but cementation disappears on wetting.

Free atmosphere The atmosphere above about 600 metres in which surface friction does not influence air motion.

Freezing point The constant temperature at which the solid and liquid forms of a pure substance are in equilibrium at standard atmospheric pressure. For pure water at standard pressure the point is o°C.

Front The sloping transition zone separating air masses of contrasting temperature and humidity.

Frontal fog The fog formed at a front when warm rain falls from above a front into cold air beneath it to evaporate and saturate the cold air.

Frontogenesis The development or intensification of a front; mainly by convergence.

Frontolysis The weakening of a front, mainly by divergence.

Frost The icy deposits formed on the ground and on objects when the temperature is below o°C.

Frost-hollow A basin-shaped topographic feature in which cold air is trapped so that the basin experiences more and greater frosts than surrounding areas.

Fulvic acid A term of varied usage but usually referring to the mixture of organic substances remaining in solution upon acidification of a dilute alkali extract from the soil.

Gale A mean wind-speed, over a period of at least 10 minutes, of 34–40 knots (63–74 k.p.h.), at a free exposure (10 metres above the ground).

Galleria A strip or gallery forest along a watercourse in savannas or grasslands.

Garrigue A low open scrub with many evergreen species in the Mediterranean. It usually occurs on limestones or other pervious rocks.

Gel A colloidal substance in its stable state.

General circulation The global atmospheric flow which results from the major temperature and pressure gradients in the atmosphere which result from the differences between radiation received in equatorial and polar regions.

Genus A group of species which are thought to have evolved from a common ancestor.

Geobiocenosis A term used by Russian ecologists as an equivalent for *ecosystem*.

Geocenosis A term used by Russian ecologists to indicate the physical habitat.

Geophyte The life-form of a plant in which the perennating bud is below ground level; e.g. a tuber.

Geostrophic wind The horizontal equilibrium wind which flows parallel to the isobars and maintains a balance between the Coriolis Effect/Force and the pressure gradient. It approximates to the actual wind in the free atmosphere.

G Horizon The soil horizon in which gleying occurs.

Gilgai The micro-relief of soils produced by expansion and contraction with changes in moisture. Found in soils that contain large amounts of clay which swells and shrinks considerably with wetting and drying. Usually a succession of micro-basins and micro-knolls in nearly level areas or of micro-valleys and micro-ridges parallel to the direction of the slope. See *Micro-relief*.

Gleying A soil forming process resulting in the development of gley soils. The term *gleysation* is sometimes used in the USA as a synonym.

Gleysation A soil forming process resulting in the development of gley soils.

Gley soil A soil developed under conditions of poor drainage resulting in reduction of iron and other elements and in grey colours and mottles. (This term is obsolete in USA.)

Gradient The rate of change of an element, such as temperature, pressure or velocity, with distance. The pressure gradient is the rate of change of pressure per unit distance measured perpendicular to the isobars. Vertical temperature gradients are called lapse rates. The gradient of wind change is called a shear.

Graminoid A grass-like herb with long narrow leaves; e.g. sedges.

Granule Rounded soil aggregate of irregular shape, up to 10 mm in diameter, which is hard and relatively non-porous.

Grassland An area vegetated by perennial grasses; meadow—wet conditions; prairie—North America; steppe—Eurasia; pampa—South America; puszta —Hungary; veld—South Africa.

Gravel Particles between 20 and 2 mm in diameter.

Great soil group A taxonomic group of soils similar in kind and arrangement of horizons.

Greenhouse effect The effect by which incoming radiation passes through the atmosphere and so reaches the earth's surface from which it is re-radiated as long wave radiation, which is trapped in the atmosphere and so warms it.

Grey-brown podzolic soil A zonal great soil group consisting of soils with a thin, moderately dark A_1 horizon and with a greyish-brown A_2 horizon underlain by a B horizon containing a high percentage of bases and an appreciable quantity of illuviated silicate clay; formed on relatively young land surfaces, mostly glacial deposits, from material relatively rich in calcium, under deciduous forests in humid temperate regions.

Grey desert soil A term used in Russia, and frequently in the United States, synonymously with *desert soil*. See *Desert soil*.

Grey forest soil There is no unanimous opinion on the origin of these soils. They may develop as a result of the degradation of steppe chernozems with the encroachment of forest vegetation, but some workers believe that they formed with the replacement of needle-leaf coniferous forest by deciduous forest with a rich grass ground layer.

Grey wooded soil Soil with a thin A_1, thick grey A_2 and greyish or yellowish-brown B horizons, commonly over calcareous formations.

Ground-water gley soil A soil with a permanently high water table. A thin A horizon occurs over a G horizon.

Ground-water laterite soil A great soil group of the intrazonal order and hydromorphic suborder, consisting of soils characterised by hardpans or concretional horizons rich in iron and aluminium (and sometimes manganese) that have formed immediately above the water table.

Ground-water podzol soil A great soil group of the intrazonal order and hydromorphic suborder, consisting of soils with an organic mat on the surface over a very thin layer of acid humus material underlain by a whitish-grey leached layer, which may be as much as 0·6–0·9 metre in thickness, and is underlain by a brown, or very dark-brown, cemented hardpan layer; formed under various types of forest vegetation in cool to tropical, humid climates under conditions of poor drainage.

Growing season The period of the year during which vegetation grows. For most plants of north-western Europe 5·6°C is regarded as a critical temperature.

Growth-form The habit of a plant as determined by its branching system; e.g. rosette herbs, cushion plants.

Grumusols Soils with AC profiles and a thick A_1 horizon even though the organic matter content may be low. These soils have a heavy texture and clay minerals which swell and shrink greatly, with changing moisture conditions with the result that they show strong churning action. Grumusols have a high base saturation, are often calcareous and form in climates with distinct wet and dry seasons.

Gulf Stream A warm ocean current which originates in the Gulf of Mexico, flows through the Straits of Florida, northwards along the coast of the United States to about 40°N, where it turns eastwards and crosses the Atlantic as the North Atlantic Drift.

Gumbotil Soil, developed upon glacial deposits, with an impervious subsoil.

Gyttja Sedimentary peat consisting of plant material in a finely divided condition deposited from water.

Haar A local name in eastern Scotland and England for a wet sea fog.

Habitat (1) The sum total of the environmental conditions which are effective in determining the existence of a community in a certain place.

(2) The part of the environment at which exchanges occur between organisms and the resources they use; e.g. a tree's habitat is that part of the forest at which it obtains its nutrients.

Hadley cell A simple thermal circulation carrying warm rising equatorial air polewards, usually to sink at about 30° latitude and return equatorwards as a surface flow.

Hail Precipitation in the form of ice pellets.

Half-bog soil A great soil group, of the intrazonal order and hydromorphic suborder consisting of soil with dark brown or black peaty material over greyish and rust mottled mineral soil; formed under conditions of poor drainage under forest, sedge, or grass vegetation in cool to tropical humid climates.

Halomorphic soil A suborder of the intrazonal soil order, consisting of saline and alkali soils formed under imperfect drainage in arid regions and including the great soil groups solonchak or saline soils, solonetz soils, and solod soils. See *Saline soil*.

Halophyte A plant which grows on a salt soil.

Halophytic vegetation Salt-adapted or salt-tolerant vegetation, usually having fleshy leaves or thorns and resembling desert vegetation.

Halosere A series of communities which in turn occupy an originally bare salt marsh.

Hardpan See *Duripan*.

Harmattan A dry cool easterly or northeasterly wind which reaches northwest Africa after crossing the Sahara. It is often laden with dust, but it does bring relief from humid weather.

Haze A suspension of very small, non-aqueous, solid particles of dust, smoke, etc. which give the sky a milky appearance.

Heath Originally a shrubby formation dominated by heather but now extended to include other shrubs.

Heliophyte A plant tolerant to existence in full sunlight.

Hemicryptophyte The life-form of a plant in which the perennating bud is situated at ground level, as in many herbs.

Herb A non-woody plant which dies down after flowering. This includes grasses.

Herbivores Animals which feed on living plants.

Heterotrophic Fed by others, as of animals which require other animals to synthesise nutrients for them.

Heterotrophs Animals and plants which obtain their food supply from the tissues of other organisms. See *Autotrophs*.

High An anticyclone or high pressure system.

Histosol Soil that is characterised throughout by a high organic content (7th Approximation).

H layer A layer occurring in mor humus consisting of well decomposed organic matter of unrecognisable origin. The O_2 horizon.

Hoar frost Thin ice crystals deposited on surfaces cooled below $0°C$ by radiation.

Hochmoor A synonym for *raised bog*, and highmoor.

Holocenosis The living community considered as a whole, without undue emphasis on any of its parts and without necessary reference to any causal factors, e.g. the bog as an ecosystem reveals the mutual interdependence of substratum, plants and animals.

Horizon A soil layer with clearly defined and characteristic features.

Horse latitudes The belts of light, variable winds associated with the subtropical high pressure systems.

Hue One of the 3 variables of colour. It is caused by light of certain wavelengths and changes with the wavelength. See *Munsell colour system*, *Chroma*, and *Value*.

Humic acid A mixture of variable or indefinite composition of dark-coloured organic substances, precipitated upon acidification of a dilute alkali extract from soil. (Used by some workers to designate only the alcohol-soluble portion of this precipitate.) (In chemical literature, it is sometimes used to designate a preparation obtained by the treatment of sugars with mineral acids.)

Humic gley soil Soil of the intrazonal order and hydromorphic suborder that includes wiesenboden and related soils, such as half-bog soils, which have a thin muck or peat O_2 horizon and an A_1 horizon. Developed in wet meadow and in forested swamps.

Humification Transformation of fresh organic residues into humus.

Humus (1) That more or less stable fraction of the soil organic matter remaining after the major portion of added plant and animal residues have decomposed. Usually it is dark coloured.

(2) Includes the F and H layers in undisturbed forest soils.

Hurricane An intense tropical low pressure system, especially in the Caribbean. See *Cyclone*.

Hydrarch A succession which originates in wet conditions and progresses towards drier conditions.

Hydrological cycle The chain of events through which water passes from the atmosphere to the earth's surface and back to the atmosphere.

Hydromorphic soils A suborder of intrazonal soils, consisting of 7 great soil groups, all formed under conditions of poor drainage in marshes, swamps, seepage areas, or flats.

Hydrophyte A plant growing in water and dependent upon being partially immersed at all times.

Hydrosere The total number of plant communities which successively occupy the same site in its progression from a wet to a drier state. Especially fresh water successions. See also *Halosere*.

Hydrosphere That part of the earth's surface covered by water.

Hydrous mica A silicate clay with $2:1$ lattice structure, but of indefinite chemical composition since usually part of the silicon in the silica tetrahedral layer has been replaced by aluminium, and containing a considerable amount of potassium which serves as an additional bonding between the crystal units, resulting in particles larger than normal in montmorillonite and, consequently, in a lower cation-exchange capacity. Sometimes referred to as illite.

Hyphae White filamentous tubes making up mycelium.

Hypolimnion The thermic strata of water below the thermocline. In a deep lake this is usually at $4°C$.

Ice Water in solid form.

Iceberg A mass of ice broken from a glacier (glacier berg) or an ice shelf (tabular berg), which floats in the sea.

Illite A hydrous mica. See *Hydrous mica*.

Illuvial Of a soil horizon in which leached materials accumulate.

Illuvial horizon A soil layer or horizon in which material carried from an overlying layer has been precipitated from solution or deposited from suspension. The layer of accumulation. See *Eluvial horizon*.

Illuviation The process of deposition of soil material removed from one horizon to another in the soil; usually from an upper to a lower horizon in the soil profile.

Immature soil A soil with indistinct or only slightly developed horizons because of the relatively short time it has been subjected to the various soil forming processes. A soil that has not reached equilibrium with its environment.

Impeded drainage A condition which hinders the movement of water through soils under the influence of gravity.

Inceptisol Soil that forms quickly and shows little evidence of significant eluviation or illuviation (7th Approximation).

Indigenous Native to: a plant or animal belonging to the original biota of an area.

Infiltration Movement of water into the soil.

Infra-red radiation Electromagnetic radiation with a wave length range of 0·7–1000 microns. 52% of solar radiation is of this type. In meteorology 'long wave radiation' is used as a synonym.

Insectivores Animals that feed largely on insects.

Insolation The intensity or amount of solar radiation received on a unit area of the earth's surface in a specified time.

Intertidal That part of the littoral zone which lies between mean high and mean low tide marks.

Intertropical convergence zone The ITC is the narrow zone in which air masses from the northern and southern hemispheres converge.

Intrazonal Of a soil which does not yet have a mature profile, e.g. floodplain soils.

Intrazonal soil Well developed soil whose morphology reflects the influence of some local factor of relief, parent material or age rather than that of climate or vegetation.

Inversion A temperature inversion is the condition in which atmospheric temperature increases with altitude.

Ion An electrically-charged particle which arises in solution by the dissociation of a dissolved chemical substance.

Ionisation The splitting of a dissolved chemical substance into positive and negative ions.

Iron pan Layer cemented with iron oxides.

Iso A prefix meaning 'equal'.

Isobar A line of constant atmospheric pressure.

Isogon A line of constant wind direction.

Isohyet A line of constant precipitation amount.

Isomorphous substitution The replacement of one atom by another of similar size in a crystal lattice without disrupting or changing the crystal structure of the mineral.

Isophene A line drawn on a map to connect all neighbouring points of similar biological periodicity, e.g. date of first flowering of a particular plant.

Isotach A line of constant wind-speed.

Isotherm A line of constant temperature.

Jet stream A strong narrow current of air, generally near the tropopause, characterised by strong vertical and lateral wind shears. The subtropical and polar jet streams are the 2 main types.

Joule A unit of energy: the work done when the point of application of a force of 1 newton is displaced 1 metre in the direction of the force.

Jungle Second-growth tropical vegetation, usually very dense in its lower layers.

Kaolin (1) An aluminosilicate mineral of the 1 : 1 crystal lattice group; that is, consisting of one silicon tetrahedral layer and one aluminium oxide-hydroxide octahedral layer.
(2) The 1 : 1 group or family of aluminosilicates.

Kaolisol A soil of the humid tropics rich in kaolinite.

Kastanozem *Chestnut* soil.

Katabatic wind Otherwise known as a mountain breeze, it is a downslope flow of cold air which displaces warm lighter air.

Khamsin The hot dry southerly wind which blows over Egypt from the south towards low pressure areas in the Mediterranean.

Kilogrammecalorie (kcal) One thousand (10^3) calories.

Krasnozem Red soil formed on red-coloured clayey weathering products of the primary crystalline rocks and crystalline conglomerates beneath deciduous forest in hot, humid climates. Particularly in areas on the eastern seaboard of the Black Sea.

Krummholz Stunted growth-form of trees, caused by wind action.

Kuro shio A warm ocean current which flows northeastwards along the south coast of Japan before merging into the general drift of the north Pacific.

Lag deposits Residual deposits.

Lagg The depressed margin of a raised bog.

Land and sea breezes Local winds caused by unequal heating and cooling of adjacent land and sea surfaces. During the day the land is warmest and the sea breeze is onshore. At night the sea is warmest and the breeze offshore.

Lapse rate The decrease of temperature per unit increase in altitude.

Latent heat The quantity of heat absorbed or emitted, without change of temperature, during the change of state of unit mass of a material, e.g. water changing to water vapour, ice to water and the reverse.

Laterisation A soil-forming process characteristic of the humid tropics in which organic materials do not accumulate on the surface, silica is leached downwards, and iron oxides remain near the surface.

Laterite Buchanan's name for a red subsoil which hardens permanently on exposure or has already hardened under natural conditions.

Lateritic soil A suborder of zonal soils formed in warm, temperate, and tropical regions and including the following great soil groups: yellow podzolic, red podzolic, yellowish-brown lateritic, and lateritic.

Latosol A suborder of zonal soils including soils formed under forested, tropical, humid conditions and characterised by low silica-sesquioxide ratios of the clay fractions, low base-exchange capacity, low activity of the clay, low content of most primary minerals, low content of soluble constituents, a high degree of aggregate stability, and usually having a red colour.

Lattice structure The orderly arrangement of atoms in a crystalline material.

Leaching The removal of substances in solution or suspension by downward percolation. A soil process.

Lee waves A system of standing waves formed in the lee of a hill. The existence of such waves is shown by the presence of clouds near the wave crests.

Lentic Environment; standing waters, e.g. ponds, lakes. See *Lotic*.

Lessivage See *Argilluviation*.

Leucinisation A biochemical process which leads to the destruction of the black pigment of organic compounds. This process is responsible for the pale colour of the humus in many tropical soils.

Liana A climbing and twining woody plant or vine.

Liebig's Law of Minimum To occur and thrive in a given situation, an organism must have essential materials which are necessary for growth and reproduction. These basic requirements vary with the species and the situation. The essential material available in amounts most closely approaching the critical minimum needed will tend to be the limiting one.

Life-form (1) The morphological features of a plant which are associated with its life history, e.g. bulbous plants, evergreen needle-leaves.

(2) The vegetative size, structure, and appearance of a plant, e.g. tree, shrub.

Lightning A visible electrical discharge in the atmosphere.

Liman Dry lacustrine or lagoon-like basin of the dry Russian steppe.

Limiting factor Environmental factor limiting the growth or reproduction of a plant or a community.

Liquid limit/Upper plastic limit Minimum moisture content at which soil will barely flow under a standard stress.

Lithosequence A group of related soils that differ, one from the other, in certain properties primarily as a result of differences in the parent rock as a soil forming factor.

Lithosere A sere having its origin on rock.

Lithosols A great soil group of azonal soils characterised by an incomplete solum or no clearly expressed soil morphology and consisting of freshly and imperfectly weathered rock or rock fragments.

Lithosphere That part of the earth which is solid.

Litter Leaves and other undecomposed residues lying loosely on the soil.

Littoral (1) The upper part of the benthic zone.

(2) The region lying along the shore.

L layer (litter) The surface layer of the forest floor consisting of freshly fallen leaves, needles, twigs, stems, bark, and fruits. This layer may be very thin or absent during the growing season. The O_1 horizon.

Loam Soil having clay and coarser particles in proportions which usually form a permeable friable mixture.

Loess A uniform deposit of silt-sized material deposited by the wind.

Long wave Also termed *Rossby wave*. A smooth wave-shaped isobaric pattern in the middle or high troposphere with a wave length of the order of 2000 km.

Long wave radiation Terrestrial radiation. See *Infra-red radiation*.

Lotic Environment; running waters, e.g. streams and rivers. See *Lentic*.

Low A depression or low pressure system.

Lowmoor A synonym for *fen*.

Macropore The large soil pores occurring between soil aggregates.

Maquis A formation of evergreen and spiny shrubs of the Mediterranean area. They are closely spaced and usually occur on siliceous and impermeable rocks. See *Garrigue*.

Margalitic soil Refers to tropical black soils especially of Indonesia.

Maritime climate A climate dominated by marine influence. Characterised by small diurnal and annual temperature changes.

Marsh A tract of poor, but not permanently impeded, drainage usually characterised by herbaceous vegetation.

Mature soil A soil with well developed soil horizons produced by the natural processes of soil formation and essentially in equilibrium with its present environment.

Mechanical analysis The analysis of the particle sizes of soils.

Melanisation Darkening of the soil horizons by the incorporation of humus into the mineral soil.

Mesic Quality of an organism requiring constant water availability, or of the site that provides it.

Mesophyte A plant adapted to a moderately moist habitat.

Mesothermal Of a climate with warm-temperate temperatures, e.g. Köppen's C-climates.

Mesothermal climate A climate with a moderate temperature regime. In the Köppen classification a climate in which the mean temperature of the coldest month lies between $-3°C$ and $+18°C$. Such climates occur mainly between latitudes 30° and 45°.

Meteorology The science of the atmosphere. It embraces weather and climate and is concerned with the dynamic, physical and chemical state of the atmosphere.

Micelle A clay, humus or clay-humus soil particle.

Micro-climate The climate of a small area close to the earth's surface. It is often regarded as the climate beneath the head of the vegetation.

Micro-habitat The ultimate division of the biosphere. The most intimately local and immediate set of conditions surrounding an organism, e.g. burrow of a rodent, a decaying log.

Micropore The small soil pores occurring within soil aggregates.

Micro-relief Small-scale, local differences in topography, including mounds, swales, or pits that are only a few metres in diameter and with elevation differences of up to 2 metres.

Microthermal A climate with low temperatures. In the Köppen classification a climate of long cold winters and short cool summers, with a mean temperature for the coldest month of less than $-3°C$ and for the warmest month of greater than $+10°C$. Such climates occur mostly between latitudes 40° and 65°.

Migration Like *dispersal*, involves movements and the invasion of new areas but migration differs from dispersal in that it is a periodic movement back and forth between two areas.

Millibar The thousandth part of a bar. The millibar (mb) is the standard unit of atmospheric pressure.

Mineralisation Release of mineral matter of soil from organic combination by decomposition.

Minerotrophic Of peats which obtain their minerals from ground-water.

Mist Obscurity in the atmosphere caused by a suspension of fine water droplets and hygroscopic particles. Visibility exceeds 1 km.

Mistral A cold dry northerly or northwesterly which blows down the Rhône Valley and over the Mediterranean coast from Ebro to Genoa.

Moder Humus form transitional between mor and mull.

Mollisols Soils characterised by thick A horizons and underlying ca or cs horizons. Includes chernozem, brunizem and chestnut soils. (7th Approximation).

Monolith A vertical section taken from the soil.

Monsoon The term is derived from an Arabic word for 'season' and referred to the seasonal winds of the Arabian Sea which blow for about 6 months from the northeast and 6 months from the southwest. It is now used to denote other markedly seasonal winds as in south and east Asia, northern Australia and western and eastern Africa.

Montane Of a type of vegetation developed on mountains which resembles that of the lowlands of higher latitudes.

Montmorillonite An aluminosilicate clay mineral with a 2 : 1 expanding crystal lattice; that is, with two silicon tetrahedral layers enclosing an aluminium octahedral layer. Considerable expansion may be caused along the C axis by water moving between silica layers of contiguous units. See *Montmorillonite group*.

Montmorillonite group Clay minerals with 2 : 1 crystal lattice structure; that is, two silicon tetrahedral layers enclosing an aluminium octahedral layer. Consists of montmorillonite, beidellite, nontronite, saponite and others.

Mor A surface soil horizon of partly decomposed organic remains resting on the mineral soil. A synonym for *duff*.

Motile Capable of motion.

Mottled zone A layer that is marked with spots or blotches of different colour or shades of colour. The pattern of mottling and the size, abundance, and colour contrast of the mottles may vary considerably and should be specified in soil description.

Mottling Spots or blotches of different colour or shades of colour interspersed with the dominant colour.

Mountain breeze See *Katabatic wind*.

Mountain soil Soil, usually skeletal, formed in mountain regions mainly by physical weathering.

Mountain wave See *Lee wave*.

Muck An organic soil which develops on periodically flooded land, e.g. on a flood plain.

Mull A mineral-rich humus form which is friable, porous and has a diffused lower boundary.

413

Munsell colour system A colour designation system that specifies the relative degrees of the three simple variables of colour; hue, value, and chroma. For example: 10YR 6/4 is a colour (of soil) with a hue = 10YR, value = 6, and chroma = 4. These notations can be translated into several different systems of colour names as desired. See *Chroma*, *Hue* and *Value*.

Muskeg Peat bogs within the general area of the North American boreal forests.

Mutation A genetic change in a species. It may cause noticeable changes in morphology.

Mutualism An association between two or more species in which all derive benefit in feeding or in some other way. Mutualism may be *facultative* when the species involved are capable of existence independent of one another, or *obligative* when the relationship is imperative to the existence of one or both species.

Mycelium The vegetative part of the thallus of fungi, consisting of white filamentous tubes (*hyphae*).

Mycorrhiza Symbiotic association of fungi and roots.

Nearctic region North America as an animal realm.

Neotropical region South and central America as an animal realm.

Neritic Subzone; that part of the *pelagic* zone that lies above the *littoral* subzone.

Newton Denotes the force which, when applied to a mass of 1 kg, gives it an acceleration of 1 metre per second per second.

Niche The status of an animal in its community; e.g. an eagle and an owl occupy niches as avian predators in some communities.

Nitrification Biological oxidation of nitrogen to nitrate.

Nitrogen fixation Conversion of atmospheric nitrogen to a combined form.

Nodule bacteria Nitrogen fixing micro-organisms living in nodules on roots of legumes.

Norte A strong, cold, northerly wind which blows in winter on the shores of the Gulf of Mexico.

Nucleus In meteorology, a minute solid particle suspended in the atmosphere.

Nutrient Substance required for plant growth.

Occlusion A front which develops during the later stages in the life cycle of a depression. It shuts off, or occludes, the warm air from the ground surface.

Oceanic Subzone; that part of the *pelagic* zone that lies above the *abyssal* subzone.

O Horizon The surface layers of organic matter above the mineral soil.

Oligotrophic (1) Of soils with a low nutrient content.
(2) Of a clear-water lake, often deep and cold, with little plankton.
(3) Of peat developed entirely in downward percolating rainwater and hence having few nutrients.

Ombrotrophic Of peats which derive their minerals from the atmosphere.

Omnivore An eater of both plants and animals.

Organic soil Peaty soil with surface peat deeper than about 16 cm formed under poorly drained conditions. A synonym for *bog soil*.

Oriental region India and southeast Asia as an animal realm.

Orographic cloud Cloud which is formed by forced uplift of air over high ground. The reduction of pressure within the rising air produces adiabatic cooling and condensation.

Orographic rain Precipitation caused by the forced uplift of moist air over high ground.

Orterde A deep soft iron-rich B horizon of a podzol.

Ortstein Hard, cemented B horizon of a podzol.

Oxisols Soils that have an oxic horizon; an altered subsurface horizon consisting of a mixture of hydrated oxides of iron or aluminium together with variable amounts of non-expanding lattice clays and accessory diluents like quartz, that are highly insoluble (7th Approximation).

Palaearctic region Eurasia with north Africa, but excluding southeast Asia, as an animal realm.

Palaeoclimatology The study of climates during geological time.

Palaeoecology The study of the relationship of past floras and faunas to their environment.

Palaeomagnetism The study of the earth's magnetic field during geological time.

Palsa A mound formed in boreal peatlands. Palsas have a height of up to 7·6 metres. Their interiors consist of alternating layers of peat and ice.

Palynology The study of pollen, spores and micro-fossils.

Pampas South American continuous grassland.

Parasite An organism living upon another from which it derives sustenance, e.g. lice upon birds.

Parasitism A coaction between two individuals where the parasite receives benefit at the expense of the host. *Ectoparasites* live on the outside of the host; *endoparasites* live in the alimentary tract, blood, or various organs of the host. See *Parasitoid*, *Social parasitism*.

Parasitoid Larvae that infect the host body and kill it slowly as they grow; a type of parasite that kills its host.

Parent material The rock or mineral matter which forms the material which is weathered to form soil.

Parkland A formation of widely and regularly spaced trees.

Particle size The effective diameter of a particle measured by sedimentation, sieving, or micrometric methods.

Peat A deposit of organic material decomposed under anaerobic conditions.

Peat soil An organic soil containing more than 50% organic matter. Used in the United States to refer to the stage of decomposition of the organic matter, *peat* referring to the slightly decomposed or undecomposed deposits and *muck* to the highly decomposed materials.

Ped A unit of soil structure such as an aggregate, crumb, prism, block, or granule, formed by natural processes (in contrast with a clod, which is formed artificially).

Pedalfer Soil containing accumulations of iron and aluminium compounds.

Pedocal Soil containing an accumulation of calcium carbonate.

Pedogenesis The formation of soil from parent material.

Pedology The study of soils.

Pedon The smallest volume that can be called a soil.

Pedosphere That part of the earth in which soil forming processes occur.

Pelagic Zone; the open water of the ocean (see *Benthic*). Whales and sharks are pelagic organisms.

Pelosol A term used in Germany for soils with characteristics similar to those of *grumusols*.

Peptisation The more or less complete dispersion of colloidal particles in liquids. It is the opposite term to flocculation.

Perennial Of plants which live for a number of years, although they may die down after each flowering.

Permafrost Ground which is permanently frozen. The limit of permafrost is approximately indicated by the $-5°C$ annual mean temperature isotherm.

Permeability The readiness with which air or water can pass through soil.

pF A measure of the energy with which water is held by soils.

pH The logarithm of the reciprocal of the hydrogen ion concentration in a solution. Neutrality is pH 7; acidity is below pH 7; alkalinity is above pH 7. The scale extends from pH 0–14.

Phanerophyte The life-form of a plant in which the perennating bud is situated well above the ground, e.g. in tall shrubs and trees.

Phenology The study of the periodicity in plants and animals, e.g. the time of flowering.

Photoperiodism The conditioning in a plant or animal which causes an activity to occur in response to a certain amount of daylight, e.g. some plants flower in response to short days.

Photosynthesis The mode of nutrition characteristic of all green plants. It is the synthesis, in the presence of chlorophyll and sunlight, of complex organic compounds from simple inorganic ones existing free in the environment, e.g. in the air or soil.

Phreatophyte A plant which derives its water supply from ground-water and is more or less independent of rainfall.

Phytogeography The study of the geographical distribution of plants.

Phytoplankton Plant plankton, e.g. blue-green algae.

Pioneer Plants or communities which occupy newly available sites.

Plagioclimax A climax community which is maintained by continuous human activity, e.g. mowing of grass.

Plagiosere A series of communities which in turn occupy a site after the climatic climax vegetation is removed by human activity.

Plankton Microscopic plants and animals in suspension in water, e.g. algae and crustaceans. Divisible into phytoplankton and zooplankton.

Planosol A great soil group of the intrazonal order and hydromorphic suborder consisting of soils with eluviated surface horizons underlain by B horizons more strongly illuviated, cemented, or compacted than associated normal soil.

Plastic Of soil which can be deformed without rupture.

Plastic limit/Lower plastic limit Minimum moisture content permitting deformation of a small soil sample without rupture.

Platy Consisting of soil aggregates that are developed predominantly along the horizontal axes; laminated; flaky.

Plinthite A modern term for *laterite*—the red tropical subsoil which hardens permanently on exposure.

Plough sole Layer of soil compacted by passage of a plough.

Pluvial period A geological period characterised by large rainfalls compared with preceding and succeeding periods.

Pneumatophore An adventitious root in a plant, whose habitat is wet, which protrudes above the surface of the water and soil, e.g. as in the swamp cypress (*Taxodium distichum*) and some mangroves (*Avicennia spp.*).

Podzol A great soil group of the zonal order consisting of soils formed mainly in cool temperate to temperate, humid climates, under coniferous or mixed coniferous and deciduous forest, and characterised particularly by a highly leached, whitish-grey (podzol) A_2 horizon, and illuvial B horizon.

Podzolisation A process of soil formation resulting in the genesis of podzols and podzolic soils.

Poikilotherm An animal whose body temperature tends to stabilise at the level of its habitat, and likewise fluctuates; e.g. reptiles.

Polar air Air originating in high latitudes. It is normally divided into maritime polar (mP) and continental polar (cP).

Polar front A front which divides polar and tropical air masses, and along which many travelling low pressure systems form.

Polar wandering Hypothetical movement of the earth's axis of rotation relative to the earth's surface, during geological time.

Pollination The act of shedding pollen upon the stigma, etc. in order to fertilise.

Pollution Atmospheric pollution is the contamination of the air by gases and solids, particularly the products of combustion.

Pore space ⎱ Fraction of the total soil volume not
Porosity ⎰ occupied by solid particles.

Prairie A continuous grassland in North America.

Prairie soil (1) A zonal great soil group developed under grass in humid temperate regions and resembling chernozem, but dark brown on the surface, ordinarily with some textural profile and without a prominent horizon of accumulated calcium carbonate.
(2) A general term for all dark soils of grassland plains of temperate climates.

Precipitation A deposit of water from the atmosphere in solid or liquid form, e.g. snow, hail, rain.

Predation A form of disoperation whereby one animal kills another for food.

Pressure In meteorology, the force per unit area exerted on a surface by the liquid or gas in contact with it. The atmospheric pressure at any point is the weight of the air which lies vertically above it. On average this is $1·03$ kgf/cm^2 or $10·132$ N/cm^2.

Prevailing wind The direction of the wind which occurs more frequently than any other during a specified period.

Primary mineral A crystal constituent of unweathered rock.

Prisere A series of plant communities which occupy in turn a fresh untenanted site.

Prismatic soil structure A soil structure type with prismlike aggregates that have a vertical axis much longer than the horizontal axes.

Profile, soil A vertical section of the soil through all its horizons and extending into the parent material.

Psammosere A series of communities which occupy in turn an area of originally unconsolidated sand.

Pseudogley soils Soils with an ABC profile with a compact textural B horizon. Both A and B horizons show strong mottling caused by temporary waterlogging above or in the B horizon. The soils are very acid and destruction of clay minerals occurs in the upper part of the B horizon. The A horizon may penetrate into the B horizon in the form of tongues. These soils occur under humid climatic conditions.

Puszta The Hungarian grassland.

Quadrat A measured area of any shape or size used to sample vegetation.

Radiation The transmission of energy by electromagnetic waves.

Radiation fog A type of fog which forms over land and inland waters at night when the sky is clear and the air calm. Condensation occurs in the cooled air layers in contact with the surface which has been cooled by outgoing radiation.

Rain Liquid precipitation in the form of drops with a diameter greater than 500 microns (0·5 mm).

Rainbow A bow or arch exhibiting the prismatic colours in their order, formed in the sky opposite the sun by the reflection, double refraction, and dispersion of the sun's rays in falling drops or rain.

Rain forest A plant formation of humid environments, characterised by tall, usually evergreen broadleaf trees. It usually has many lianas and epiphytes.

Rainshadow An area receiving a low average precipitation because it is sheltered from rain-bearing winds by a range of hills.

Raised bog Synonymous with *highmoor* and *hochmoor*. A peat accumulation in which the central part of the bog is raised and is surrounded by a lower *moat*. Many European raised bogs are composed of sphagnum but elsewhere other plants form the peat.

Ranker A thin soil with an AC profile developed on silicate rocks.

Reaction, soil The degrees of acidity or alkalinity of a soil, usually expressed as a pH value. Descriptive terms commonly associated with certain ranges in pH are: extremely acid, $<4·5$; very strongly acid, $4·5–5·0$; strongly acid, $5·1–5·5$; moderately acid, $5·6–6·0$; slightly acid, $6·1–6·5$; neutral, $6·6–7·3$;

slightly alkaline, 7·4–7·8; moderately alkaline, 7·9–8·4; strongly alkaline, 8·5–9·0; and very strongly alkaline, >9·1.

Red desert soil A zonal great soil group. Pinkish grey to light reddish brown soil over a somewhat more clayey yellowish red or red subsoil in desert or semidesert regions.

Red earth Highly leached, red clayey soils of the humid tropics, usually with very deep profiles that are low in silica and high in sesquioxides. The term is often used as a synonym for *krasnozem*.

Red Mediterranean soils These soils have an ABC profile. The B horizon is textural and shows strong illuviation of clay as reflected in the presence of thick continuous clay coatings on ped surfaces and clay linings to pores and cavities. The B horizon has strong blocky or prismatic structure and is red or red-brown in colour. A ca horizon may occur in the lower part of the B horizon. These soils occur under warm temperate subhumid climates with a pronounced dry season.

Red podzolic A zonal great soil group. Soil with thin O_2 and A_1, yellowish-brown to nearly white leached A_2 and red B horizons.

Red-yellow podzolic soils A great soil group with an ABC profile and strongly illuvial B horizon as shown by clay coatings on peds and in pores. They have a light-coloured but not bleached, A_2 horizon. They usually occur on old land surfaces under warm temperate humid climates.

Regolith The unconsolidated mantle of weathered rock and soil material on the earth's surface; loose earth materials above solid rock. (Approximately equivalent to the term 'soil' as used by many engineers.)

Regosol Any soil of the azonal order without definite genetic horizons and developing from or on deep, unconsolidated, soft mineral deposits such as sands, loess, or glacial drift.

Regressive Soils whose surface horizons are progressively being lowered.

Regur An intrazonal group of dark calcareous soils high in clay, which is mainly montmorillonitic, and formed mainly from rocks low in quartz; occurring extensively on the Deccan plateau of India.

Relative humidity The ratio between the amount of water vapour actually in the air and the amount of water vapour the same volume of air could hold at the same temperature and pressure. The ratio is usually expressed by a percentage.

Rendzina A great soil group of the intrazonal order and calcimorphic suborder consisting of soils with brown or black friable surface horizons underlain by light grey to pale yellow calcareous material; developed from soft, highly calcareous parent material under grass vegetation or mixed grasses

and forest in humid and semiarid climates.

Rhizome A laterally growing underground stem by which vegetative reproduction is effected.

Rhizosphere That portion of the soil directly affected by plant roots.

R Horizon Underlying consolidated bedrock, such as granite, sandstone, or limestone. If presumed to be like the parent rock from which the adjacent overlying layer or horizon was formed, the symbol R is used alone. If presumed to be unlike the overlying material, the R is preceded by a Roman numeral denoting lithologic discontinuity. (USA System).

Ridge A ridge of high pressure is an elongated extension of an anticyclone.

Ringelmann shades A scale of shades, varying in degree of blackness, which is used by an observer to estimate the blackness of smoke and so estimate the percentage of solids the smoke contains.

Roaring forties The prevailing westerly winds of temperate latitudes in the southern hemisphere.

Rossby wave See *Long wave*.

Rubification Reddening of the soil by precipitated crystalline iron. A feature common to soils of the humid and subhumid tropics.

Rubruzem Reddish-black soils of the subtropical grasslands, especially of La Plata lowland and parts of the South African High Veld.

Saline soil A soil whose properties have been largely determined by the presence of salts. A synonym for *halomorphic soil*.

Saltatorial Leaping.

Sand Particles of diameter 2–0·02 mm (UK).
Particles of diameter 2–0·05 mm (US).

Saprovores Animals that feed on dead plants, animals and excreta.

Saprophyte A plant which lives upon dead organic matter.

Sastruga A wave pattern on the surface of snow caused by persistent winds.

Savanna Also spelt savannah, savana. A xerophilous grassland containing isolated trees.

Scavenger An animal that lives upon dead organic matter, e.g. vultures that prey upon carrion.

Sciophyte A plant which tolerates and thrives in shade.

Scirocco A southerly wind blowing from Africa across the Mediterranean to southern Europe. Over Africa it is hot and dry, over Europe warm and moist.

Sclerophyte, sclerophyllous A shrub, or of a shrub, having evergreen leaves heavily cutinised, and resistant to summer drought.

Sea breeze See *Land and sea breeze*.

Secondary mineral A crystalline compound formed from the weathering products of primary minerals.

Selva Tropical rain forest as in the low plain of the Amazon basin.

Sere A group of plant communities which successively occupy the same site.

Sesquioxide An oxide which contains two metallic atoms to every three of oxygen.

Sessile Animals fixed to the substratum; sedentary; having no means of self-propulsion.

Shade temperature The temperature of the air indicated by a thermometer sheltered from precipitation and from radiation, and around which the air can circulate. A Stevenson screen satisfies these conditions.

Shrub A woody perennial of smaller structure than a tree.

Siallite (Obsolete) Weathered rock material consisting largely of alumino-silicate clay minerals and being highly leached of the alkalis and alkaline earths.

Sierozem A zonal great soil group consisting of soils with pale greyish A horizons grading into calcareous material at a depth of 0·3 metre or less, and formed in temperate to cool, arid climates under a vegetation of desert plants, short grass, and scattered brush. (Synonymous with *grey desert soil*).

Silica-alumina ratio The molecules of silicon dioxide (SiO_2) per molecule of aluminium oxide (Al_2O_3) in clay minerals or in soils.

Silica-sesquioxide ratio The molecules of silicon dioxide (SiO_2) per molecule of aluminium oxide (Al_2O_3) plus ferric oxide (Fe_2O_3) in clay minerals or in soils.

Silt Soil particles of diameter 0·02–0·002 mm (UK). Soil particles of diameter 0·05–0·002 mm (US).

Simoom A hot dry suffocating wind or whirlwind which occurs in the deserts of Arabia and North Africa. It carries much sand and usually lasts for less than 20 minutes.

Sirocco See *Scirocco*.

Sleet Precipitation of rain and snow together (European terminology). In North America it is used to indicate ice pellets.

Smog A fog in which smoke and other pollutants occur.

Snow Solid precipitation of ice crystals.

Snow-line The lower altitudinal limit of perpetual snow.

Social parasitism Describes the exploitation of one species by another, e.g. cuckoo abandons eggs and young to the care of foster parents.

Sod-podzolic soil A podzolised soil formed beneath deciduous forest with a ground layer of grasses. The A_1 is granular. Synonymous with *turf-podzolic*.

Soil (1) The unconsolidated mineral material on the immediate surface of the earth that serves as a natural medium for the growth of land plants.

(2) The unconsolidated mineral matter on the surface of the earth that has been subjected to and influenced by genetic and environmental factors of: parent material, climate (including moisture and temperature effects), macro and micro-organisms, and topography, all acting over a period of time and producing a product—soil— that differs from the material from which it is derived in many physical, chemical, biological and morphological properties, and characteristics.

Soil body The lifeless part of the soil composed of inorganic and organic materials in which the living organisms operate.

Soil horizon A layer of soil or soil material approximately parallel to the land surface and differing from adjacent genetically related layers in physical, chemical, and biological properties or characteristics such as colour, structure, texture, consistency, kinds and numbers of organisms present, degree of acidity or alkalinity, etc.

Soil skeleton The chemically inert fraction of the soil, usually composed of quartz grains.

Sol A colloidal substance in its mobile state.

Solod/Soloth Leached saline soil (degraded solonetz) having a pale A_2 horizon and a degraded fine-textured B horizon.

Solonchak A great soil group of the intrazonal order and halomorphic suborder, consisting of soils with grey, thin, salty crust on the surface, and with fine granular mulch immediately below being underlain with greyish, friable, salty soil; formed under subhumid to arid, hot or cool climate, under conditions of poor drainage, and under a sparse growth of halophytic grasses, shrubs and some trees.

Solonetz A great soil group of the intrazonal order and halomorphic suborder, consisting of soil with a very thin, friable, surface soil underlain by a dark, hard columnar layer usually highly alkaline; formed under subhumid to arid, hot to cool climates, under better drainage than solonchaks, and under a native vegetation of halophytic plants.

Soloth See *Solod*.

Solstice The time of maximum or minimum declination of the sun. It occurs on about 22 June and 22 December.

Solum The genetic soil developed by soil forming processes.

Sor or Shor A central Asian term for a drying lake or saltpan.

Species A type of plant or animal which 'breeds true' from generation to generation.

Spodosols Soils that have a bleached A₂ horizon that is demonstrable after ploughing and cultivation (7th Approximation).

Squall A strong wind, sometimes accompanied by precipitation, which lasts for only a few minutes.

Squall line The term originally used to denote a cold front. It is now usually used to denote a violent convective phenomena extending along a non-frontal line.

Stagnogley See *Surface water gley*.

Standing wave See *Lee wave*.

Steady state The state in which inflow of energy and materials in a system is just sufficient to maintain forms or biomass at a steady level.

Sten- A prefix indicating narrowness.

Stevenson screen The standard housing for meteorological thermometers.

Stratification Of vegetation it is the arrangement of layers within a stand, e.g. tree, shrub, moss.

Streamline A line or curve parallel to the wind direction at all points along it.

Structure (1) Of vegetation it is the distribution in space of the plants in a stand, e.g. broadleaf trees with lianas, epiphytes and shrubs with ground mosses.
(2) Arrangement of primary soil particles in aggregates—prisms, blocks, crumbs, etc.

Subalpine The zone immediately below the timber line.

Subarctic brown forest soils Soils similar to brown forest soils except having more shallow sola and average temperatures of $<5°C$ at 45·5 cm or more below the surface.

Subclimax A phase of succession occupied by associations which are not of climatic climax status but which persist because of some arresting factor.

Subdominant A modifier of the community composition in a secondary manner. See *Dominant*.

Suberin Corky tissue forming the bark of some trees.

Sublittoral Division of the littoral subzone that lies between the *eulittoral* and the *archibenthal*.

Subsere A sere which is initiated when an arresting factor is removed, e.g. as in an old pasture.

Subsidence In meteorology, the slow downward motion of air over a large area.

Subtropical high One of the anticyclones of the semi-permanent belt of high pressure which exists between the equator and 40°N and S.

Succession The process through which one plant community invades and replaces another.

Sucker A laterally growing, subterranean offshoot from the base of the main stem of a plant, by which vegetative reproduction is effected.

Surface water gley soil Soil with an impermeable horizon which impedes drainage and causes gleying. The gleyed horizon may be an Ag or a Bg horizon. When surface wetting is prolonged such soils are called *stagnogleys*.

Swamp A more or less permanently flooded area. Often having a tree cover.

Swell A wave motion in the ocean with a long regular undulation caused by a distant disturbance.

Symbiosis The living together of two organisms which derive mutual benefits from the association; e.g. a lichen is composed of a chlorophyll-making alga and a sugar-storing fungus.

Synecology The study of living communities.

Synoptic chart A map or chart showing the distribution of selected meteorological elements at a specified instant of time.

Synusia (1) A horizontal layer of definite depth in a stand of vegetation, e.g. the shrub layer in a forest.
(2) The collective term for similar ecological niches within a formation, e.g. of epiphytes.

Taiga The formation occurring between the boreal forest and the tundra composed of scattered low trees with an abundant ground layer of lichens and mosses.

Takyr Soils formed in deserts on clayey plains under the influence of stagnant alkaline surface waters which evaporate to leave compact clay crusts which crack on drying. The lower parts of the profile are usually strongly saline.

Taxon A systematic unit (plant or animal) of unspecified rank.

Taxonomy The classification of plants and animals.

Temperature The condition which determines the flow of heat from one substance to another.

Terra rossa Clayey red soils of the Mediterranean lands, found particularly in the karst areas of the Balkans.

Terra roxa A base-rich, low-humic, clayey latosol of the São Paulo area of Brazil.

Texture Composition of soil in respect of particle size distribution.

Thermocline The layer in a body of water where the decrease in the temperature exceeds 1°C per metre in depth.

Thermosequence A sequence of related soils that differ, one from the other, primarily as a result of temperature as a soil forming factor.

Therophyte Life-form of a plant in which the perennating bud is in the seed, i.e. an annual.

Thunder The noise which follows a flash of lightning, caused by a sudden heating and explosive expansion of air along the path of the lightning.

Tilth State of aggregation of soil after cultivation.

Timber-line The altitudinal or latitudinal limit of tree growth on mountains or in the Arctic.

Tirs Black clay soil of North Africa, resembling *regur*.

Tolerance The ability of an organism to live within an environment in which one or more conditions are unfavourable to its growth.

Toposequence A sequence of related soils that differ, one from the other, primarily because of topography as a soil formation factor. See *Clinosequence*.

Topsoil The layer of soil moved in cultivation.

Tornado A violent whirl, cyclonic in sense, averaging a few hundred metres in diameter and with an intense vertical current capable of lifting heavy objects into the air. The winds within a tornado are estimated to reach over 200 knots (370 k.p.h.) but no instrument has survived to record this accurately. They are extremely destructive local phenomena. They are most frequent and intense in the United States, east of the Rocky Mountains.

Trace element Nutrient required by a plant in very small amounts.

Trade winds The winds which diverge from the subtropical highs centred between 30°–40°N and S. Also known as the tropical easterlies, they do not form a continuous belt but occur most frequently in the eastern half of the tropical oceans.

Translocation The leaching and redeposition of clay or sesquioxides in the soil profile.

Transpiration The process whereby water is removed from the soil by plants and discharged as vapour into the atmosphere.

Tree-line The latitudinal and altitudinal limit of tree growth.

Trophic level An energy level in a food chain of an ecosystem.

Tropical air An air mass originating in low latitudes, normally in the subtropical highs.

Tropical cyclone A low pressure system occurring within tropical regions. Terminology is not clear, but distinction between a depression and the more severe storms—cyclone, hurricane or typhoon—should be made.

Tropics The region of the earth's surface lying between the tropics of Cancer and Capricorn at 23°27′N and S respectively.

Tropopause The atmospheric boundary between the troposphere and the stratosphere.

Troposphere The lower layers of the atmosphere, extending from the surface of the earth to about 18 km above sea level at the equator and 8 km at the poles.

Trough An elongated low pressure feature.

Truncated Of a soil profile from which all or part of the upper horizons have been removed.

Tundra Arctic or alpine vegetation beyond or above the timber-line characterised by low shrubs, graminoids, mosses and lichens.

Tundra soils (1) Soils characteristic of tundra regions.

(2) A great soil group consisting of soils with dark brown peaty layers over greyish horizons mottled with rust and having continually frozen substrata; formed under frigid, humid climates, with poor drainage, and native vegetation of lichens, moss, flowering plants, and shrubs.

Turbulence The state of fluid motion in which the velocity exhibits large and apparently random fluctuations.

Turf-podzolic soil See *Sod-podzolic soil*.

Typhoon An intense tropical low pressure system occurring in the western Pacific Ocean.

Ultisol Soils that have no oxic or natric horizon, but do have an argillic horizon (7th Approximation).

Ultra-violet radiation Electromagnetic radiation of wavelength 10–4000 angströms and therefore not visible.

Upslope fog Fog formed on the windward side of high ground by the forced uplift of stable moist air till saturation is reached by adiabatic expansion.

Upwelling The movement of cold deep ocean water to the ocean surface.

Valley breeze An upslope (anabatic) wind which blows during the day because of the warming of valley slopes.

Value, colour The relative lightness or intensity of colour and approximately a function of the square root of the total amount of light. One of the 3 variables of colour. See *Munsell colour system*, *Hue*, and *Chroma*.

Van't Hoff Formula Chemical and biological processes, within favourable temperature limits, are increased by an approximate constant (2–3 times) for each 10°C rise in temperature.

Vector A physical quantity with direction and magnitude.

Vegetation The total plant cover of an area.

Veld Temperate grasslands of the Union of South Africa.

Vernal Pertaining to the spring.

Vertisol Soil with large amounts of expanding lattice clays in climatic areas with pronounced dry seasons. (7th Approximation).

Visibility In meteorology, the greatest distance at which an object of specified characteristics can be identified by the unaided eye in any particular circumstances.

Warm front A front whose movement is such that the warmer air mass is replacing the colder.

Warm sector The surface zone of warm air (usually T air) within a depression. It is usually occluded as the depression gets older.

Waterspout A funnel-shaped tornado cloud which extends from the surface of the sea to the base of a cumulonimbus cloud.

Weather Changing atmospheric conditions. The synthesis of weather conditions is climate.

Weed A plant which grows where man does not want it.

Wet meadow soil/wiesenboden Poorly drained soil with humus rich A_1 horizon grading into grey mineral soil. Recent classifications include this in *humic gley soil*.

Wilting point The maximum moisture percentage of soil that will induce permanent wilting of plants.

Xeric Of an organism adapted to dry conditions; dry.

Xerophyte A plant adapted to withstand drought.

Xerophytic Of a plant which is capable of withstanding drought.

Xerosere The successive communities which occupy the same site in the course of its amelioration from a dry to a mesic state.

Yellow podzolic A group of well drained, well developed acid soils with thin O_2 and organic mineral A_1 horizons over a light coloured, bleached A_2 horizon over a red, yellowish-red, or yellow and more clayey B horizon. Coarse streaks or mottles of red, yellow, brown and light grey characterise deep horizons where the siliceous parent materials are thick.

Zheltozem Yellow soils of the subtropics occurring particularly beneath a monsoon climatic regime.

Zonal Of a soil which develops on a freely drained site and reflects the influence of climate and vegetation.

Zonal soil Soil having a profile which shows a dominant influence of climate and vegetation on its development, and occupying a large area or zone. One of the 3 primary orders in soil classification.

Zoogeography That part of biogeography that deals with the animal cover of the world; its composition, local productivity and distribution.

Zooplankton Animal plankton.

Selected Bibliography

In the following list of references an attempt has been made to include only the more readily accessible publications, and summaries of some technical studies have been included rather than the original research reports.

1. Introduction

BUDYKO, M. I., 'On the Causes of the Extinction of Some Animals at the End of the Pleistocene.' *Soviet Geography: Review and Translation*, 8(10), 1967, pp. 783–93.

BUTZER, K. W., *Environment and Archaeology*. Methuen, London, 1964.

EYRE, S. R., AND JONES, G. R. J., *Geography as Human Ecology* (particularly Chapter 1). Arnold, London, 1966.

EYRE, S. R., 'Determinism and the Ecological Approach to Geography.' *Geography*, 49, 1964, pp. 369–76.

MACDONALD, G. J. F., 'Weather Modification.' *Science Journal*, 4(1), 1968, pp. 39–44.

Climate
General. Chapters 2–6

BATTAN, L. J., *The Nature of Violent Storms*. Doubleday, New York, 1961.

BOLIN, B., *The Atmosphere and the Sea in Motion* (Rossby Memorial Volume). Rockefeller and Oxford UP, New York, 1959.

CRITCHFIELD, H. J., *General Climatology*. Prentice-Hall, Englewood Cliffs, New Jersey, 1965.

DOBSON, G. M. B., *Exploring the Atmosphere*. Clarendon Press, Oxford, 1963.

HARE, F. K., *The Restless Atmosphere*. Hutchinson, London, 1958.

HAURWITZ, B., AND AUSTIN, J. M., *Climatology*. McGraw-Hill, New York, 1944.

MALONE, T. K. (Editor), *Compendium of Meteorology*. American Meteorological Society, Boston, 1951.

PETTERSSEN, S., *Introduction to Meteorology*. McGraw-Hill, New York, 1958.

Weather Analysis and Forecasting, II. McGraw-Hill, New York, 1956.

SUTTON, O. G., *The Challenge of the Atmosphere*. Hutchinson, London, 1962.

Understanding Weather. Penguin, London, 1960.

TREWARTHA, G. T., *An Introduction to Climate*. McGraw-Hill, New York, 1954.

The Earth's Problem Climates. Wisconsin UP, Madison, 1961.

WILLETT, H. C., AND SANDERS, F., *Descriptive Meteorology*. Academic Press, New York, 1959.

7. Primary Circulation

BECKINSALE, R. P., 'Some Recent Trends in Climatology.' In R. J. Chorley and P. Haggett, *Frontiers in Geographical Teaching*. Methuen, London, 1965.

HARE, F. K., 'The Westerlies.' *Geographical Review*, 50(3), 1960, pp. 354–67.

'The Stratosphere.' *Geographical Review*, 52(4), 1962, pp. 525–47.

'Energy Exchanges and the General Circulation.' *Geography*, 50(3), 1965, pp. 229–41.

MCDONALD, J. E., 'The Coriolis Effect.' *Scientific American*, 186(5), 1952, pp. 72–8.

NAMIAS, J., 'The Index Cycle and its Role in the General Circulation.' *Journal of Meteorology*, 7, 1950, pp. 130–9.

'The Jet Stream.' *Scientific American*, 187(4), 1952, pp. 26–31.

NEWELL, R. E., 'The Circulation of the Upper Atmosphere.' *Scientific American*, 210(3), 1964, pp. 62–74.

PFEFFER, R. L., 'The Global Atmospheric Circulation.' *Transactions of the New York Academy of Sciences*, Series II, 26(8), 1964, pp. 984–97.

STARR, V. P., 'The General Circulation of the Atmosphere.' *Scientific American*, 195(6), 1956, pp. 40–5.

SUTTON, G., 'The Energy of the Atmosphere.' *Science Journal*, 1(i), 1965, pp. 76–81.

8. Pressure Systems and Monsoons

CHANG, J-H., 'The Indian Summer Monsoon.' *Geographical Review*, 57(3), 1967, pp. 373–96.

LOCKWOOD, J. G., '700 mb Contour Charts for Southeast Asia and Neighbouring Areas.' *Weather*, 21(9), 1966, pp. 325–34.

MALKUS, J. S., 'The Origin of Hurricanes.' *Scientific American*, 197(2), 1957, pp. 33–9.

RIEHL, H., *Introduction to the Atmosphere*, McGraw-Hill, New York, 1965.

Tropical Meteorology, McGraw-Hill, New York, 1954.

(Section 8 continued overleaf.)

SCORER, R. S., 'Vorticity.' *Weather*, 12(3), 1957, pp. 72–84.

'Origin of Cyclones.' *Science Journal*, 2(3), 1966, pp. 46–52.

TREWARTHA, G., 'Climate as related to the Jet Stream.' *Erdkunde*, 12(2), 1958, pp. 205–14.

9. Tertiary Circulation

PEDGLEY, D. E., 'Weather in the Mountains.' *Weather*, 22(7), 1967, pp. 266–75.

SCORER, R. S., 'Lee Waves in the Atmosphere.' *Scientific American*, 204(3), 1961, pp. 124–34.

TEPPER, M., 'Tornadoes.' *Scientific American*, 198(5), 1958, pp. 31–7.

10. Human Activity

CRITCHFIELD, H. J., 1965 (see *Climate, general*)

EDHOLM, O. G., 'Problems of Acclimatization in Man.' *Weather*, 21(10), 1966, pp. 340–50.

GRIFFITHS, J. F., *Applied Climatology*. Oxford UP, London, 1966.

PRIESTLEY, C. H. B., 'Microclimates of Life.' *Science Journal*, 3(4), 1967, pp, 67–73.

10. Air Pollution

HAAGEN-SMIT, A. J., 'The Control of Air Pollution.' *Scientific American*, 210(1), 1964, pp. 25–31.

LEIGHTON, P. A., 'Geographical Aspects of Air Pollution.' *Geographical Review*, 54(2), 1966, pp. 151–74.

MCDERMOTT, W., 'Air Pollution and Public Health.' *Scientific American*, 205(4), 1961, pp. 49–57.

WILLETT, H. C., 'Meteorology as a Factor in Air Pollution.' *Industrial Medicine & Surgery*, 19(3), 1950, pp. 116–20.

Air Pollution. World Health Organisation, Geneva, 1961.

10. Microclimates

GEIGER, R., *The Climate Near the Ground*. Harvard UP, Cambridge, Mass., 1957.

THORNTHWAITE, C. W., 'Modification of Rural Microclimates.' In W. L. Thomas (Editor), *Man's Role in Changing the Face of the Earth*. Chicago UP, Chicago, 1956.

10. Urban Climates

CHANDLER, T. J., 'London's Urban Climate.' *Geographical Journal*, 128, 1964, pp. 279–302.

LOWRY, W. P., 'The Climate of Cities.' *Scientific American*, 217(2), 1967, pp. 15–23.

PARRY, M., 'The Climates of Towns.' *Weather*, 5, 1950, pp. 351–6.

11. Weather Forecasting and Satellites

BARRETT, E. C., *Viewing Weather from Space*. Longmans, London, 1967.

BUSHBY, F. H., 'Reckoning with the Rain.' *New Scientist*, 34(542), 1967.

FRITZ, S., 'Pictures from Meteorological Satellites and Their Interpretation.' *Space Science Reviews*, 3, 1964, pp. 541–80.

METEOROLOGICAL OFFICE, *Weather Map*. HMSO, London, 1956.

SUTCLIFFE, R. C., 'Weather Forecasting by Electronic Calculation.' *Endeavour*, 88, 1964, pp. 27–32.

SUTTON. O. G., 1960 (see *Climate, general*)

12. Climatic Change

BROOKS, C. E. P., *Climate Through the Ages*. Benn, London, 1949.

DORF, E., 'Climatic Changes of the Past and Present.' *American Scientist*, 48, 1960, pp. 341–64.

EMELIANI, C., 'Ancient Temperatures.' *Scientific American*, 198(2), 1958, pp. 54–63.

LAMB, H. H., *The Changing Climate*. Methuen, London, 1966.

MANLEY, G., 'Climatic Variation.' *Quarterly Journal Royal Meteorological Society*, 79, 1953, pp. 185–208.

OPIK, E. J., 'Climate and the Changing Sun.' *Scientific American*, 198(6), 1958, pp. 85–92.

SCHWARZBACH, M., *Climates of the Past*. Van Nostrand, London, 1963.

SHAPLEY, H., *Climatic Change*. Harvard UP, Cambridge, Mass., 1953.

TUCKER, G. B., 'Solar Influences on the Weather.' *Weather*, 19(10), 1964, pp. 302–11.

WEXLER, H., 'Volcanoes and World Climate.' *Scientific American*, 186(4), 1952, pp. 74–80.

13. Classification

HARE, F. K., 'Climatic Classification.' In L. D. Stamp and S. W. Wooldridge, *London Essays in Geography*. Longmans, London, 1951.

STRAHLER, A. N., *Physical Geography*. Wiley, New York, 1960.

THORNTHWAITE, C. W., 'Problems in the Classification of Climates.' *Geographical Review*, 33(2), 1943, pp. 233–55.
    'An approach Toward a Rational Classification of Climate.' *Geographical Review*, 38(1), 1948, pp. 55–94.

14–18. Climate Types

HAURWITZ, B., AND AUSTIN, J. M., 1944 (see *Climate, general*)

TREWARTHA, G. T., 1954 and 1961 (see *Climate, general*)

KENDREW, W. G., *The Climates of the Continents*. Clarendon Press, Oxford, 1953.

Soils

General

BAVER, L. D., *Soil Physics*. Wiley, New York, 1956.

BEAR, F. E. (Editor), *Chemistry of the Soil (American Chemical Society Monograph No. 160)*. Reinhold, New York, 1964.

BRADE-BIRKS, S. G., *Good Soil*. EUP, London, 1959.

BUCKMAN, H. O., AND BRADY, N. C., *The Nature and Properties of Soils*. Macmillan, New York, 1960.

BUNTING, B. T., *The Geography of Soil*. Hutchinson, London, 1965.

BURGES, A., *Micro-Organisms in the Soil*. Hutchinson, London, 1958.

COMBER, N. M., AND TOWNSEND, W. N., *An Introduction to the Scientific Study of the Soil*. Arnold, London, 1960.

DUCHAUFOUR, P. *Précis de Pédologie*. Masson, Paris, 1960.

EDEN, T., *Elements of Tropical Soil Science*. Macmillan, New York, 1964.

GANSSEN, R., AND HÄDRICH, F., *Atlas zur Bodenkunde*. Bibliographisches Institut, Mannheim, 1965.

GERASIMOV, I. P., AND GLAZOVSKAYA, M. A., *Fundamentals of Soil Science and Soil Geography*. Israel Programme for Scientific Translations, Jerusalem, 1960.

GLINKA, K. D., *Treatise on Soil Science*. Israel Programme for Scientific Translations, Jerusalem, 1931. Translated by A. Gourevitch, 1963.

HALL, A. D., AND ROBINSON, G. W., *The Soil*. Murray, London, 1945.

HALLSWORTH, E. G., AND CRAWFORD, D. V. (Editors), *Experimental Pedology*. Butterworth, London, 1965.

JACKS, G. V., *Soil*. Nelson, London, 1954.

JENNY, H., *Factors of Soil Formation*. McGraw-Hill, New York, 1941.

JOFFE, J. S., *Pedology*. Pedology Publications, New Jersey, 1949.

KELLOGG, C. E., *The Soils That Support Us*. Macmillan, New York, 1951.

KUBIENA, W. L., *The Soils of Europe*. Murby, London, 1953.

LEEPER, G. W., *Introduction to Soil Science*. Melbourne UP, Parkville, 1961.

ROBINSON, G. W., *Mother Earth*. Murby, London, 1937.
    Soils—Their Origin, Constitution and Classification. Murby, London, 1949.

RODE, A. A., *Soil Science*. Israel Programme for Scientific Translations, Jerusalem, 1955. Translated by A. Gourevitch, 1962.

RUSSELL, E. J., *The World of the Soil*. Collins, London, 1963.

RUSSELL, E. W., *Soil Conditions and Plant Growth*. Longmans, London, 1961.

SOIL SURVEY STAFF, *Soil Survey Manual*. US Dept of Agriculture, Washington, 1951.
    Soil Classification; A Comprehensive System, 7th Approximation. US Government Printing Office, Washington, 1960.

TAMM, O., *Northern Coniferous Soils*. Scrivener Press, Oxford, 1950.

THOMPSON, L. M., *Soils and Soil Fertility*. McGraw-Hill, New York, 1957.

US DEPT OF AGRICULTURE, *Soils and Men, Yearbook of Agriculture 1938*. Government Printing Office, Washington, 1938.
    Soil, The Yearbook of Agriculture 1957. Government Printing Office, Washington, 1957.

WILDE, S. A., *Forest Soils*. Ronald Press, New York, 1958.

20–21. Soil Texture, Structure, Moisture and Atmosphere

BUCKMAN, H. O., AND BRADY, N. C., 1960 (see *Soils, general*)

RUSSELL, E. W., 1961 (see *Soils, general*)

RUSSELL, E. J., 1963 (see *Soils, general*)

22. Organic Matter

BROADBENT, F. E., 'The Soil Organic Fraction.' *Advances in Agronomy*, 5, 1953, pp. 153–81.

BURGES, A., 1958 (see *Soils, general*)

HANDLEY, W. R. C., *Mull and Mor Formation in Relation to Forest Soils*. HMSO, London, 1954.

KONONOVA, M. M., *Soil Organic Matter*. Pergamon, London, 1961.

RUSSELL, E. W., 1961 (see *Soils, general*)

WAKSMAN, S. A., *Soil Microbiology*. Wiley, New York, 1952.

23. Clays

BEAR, F. E., 1964 (see *Soils, general*)
GRIM, R. E., *Applied Clay Mineralogy*. McGraw-Hill, New York, 1962.
MACKENZIE, R. O., AND MITCHELL, B. D., 'Clay Mineralogy.' *Earth Science Reviews*, 2, 1966, pp. 47–91.

24. Nutrients

BUCKMAN, H. O., AND BRADY, N. C., 1960 (see *Soils, general*)
RUSSELL, E. W., 1961 (see *Soils, general*)
RUSSELL, E. J., 1963 (see *Soils, general*)
JACKS, G. V., 1954 (see *Soils, general*)
THOMPSON, L. M., 1957 (see *Soils, general*)

25. Weathering

ROBINSON, G. W., 1949 (see *Soils, general*)
BEAR, F. E., 1964 (see *Soils, general*)
JACKSON, M. L., AND SHERMAN, G. D., 'Chemical Weathering of Minerals in Soils.' *Advances in Agronomy*, 5, 1953, pp. 219–318.

26. Soil Profile

CLARKE, G. R., *The Study of the Soil in the Field*. Clarendon Press, Oxford, 1957.
ROBINSON, G. W., 1949 (see *Soils, general*)
SOIL SURVEY STAFF, 1960 (see *Soils, general*)
WINTERS, E., AND SIMONSON, R. W., 'The Subsoil.' *Advances in Agronomy*, 3, 1951, pp. 1–92.

27. Factors

BEAR, F. E., 1964 (see *Soils, general*)
CLINE, M. G., 'The Changing Model of Soil.' *Soil Science Society of America, Proceedings*, 25, 1961, pp. 442–6.
DUCHAUFOUR, P., 1960 (see *Soils, general*)
JACKSON, M. L., 'Clay Transformations in Soil Genesis During the Quaternary.' *Soil Science*, 99(1), 1965, pp. 15–21.
JENNY, H., 1941 (see *Soils, general*)
NIKIFOROFF, C. C., 'Weathering and Soil Formation.' *Soil Science*, 67, 1949, pp. 219–30.
SIMONSON, R. W., 'Outline of a Generalised Theory of Soil Genesis.' *Soil Science Society of America, Proceedings*, 23(2), 1959, pp. 152–6.
STEPHENS, C. G., 'Climate as a Factor of Soil Formation through the Quaternary.' *Soil Science*, 99(1), 1965, pp. 9–14.
WILDE, S. A., 1958 (see *Soils, general*)

28. Soil in the Field

CLARKE, G. R., 1957 (see *Soil Profile*)
SOIL SURVEY STAFF, 1951 (see *Soils, general*)
TAYLOR, J. A., 'Methods of Soil Study.' *Geography*, 45, (1–2), 1960, pp. 52–67.
TAYLOR, N. H., AND POHLEN, I. J., 'Soil Survey Method.' *NZDSIR Soil Bureau Bulletin*, No. 25, Wellington, 1962.

29. Soil Classification

'A New System of Soil Classification.' *Soil Science* (Special Issue), 96, 1963, pp. 1–74.
KUBIENA, W. L., 1953 (see *Soils, general*)
SIMONSON, R. W., 'Soil Classification in the United States.' *Science*, 137, 1962, pp. 1027–34.
SMITH, G. D., 'Lectures on Soil Classification.' *Pedologie*, Special No. 4, 1965, pp. 3–134.
'Soil Classification.' *Soil Science* (Special Issue), 67, 1949, pp. 77–191.
SOIL SURVEY STAFF, 1960 (see *Soils, general*)
US DEPT OF AGRICULTURE, 1938 (see *Soils, general*)

30. Latosolic Soils

AUBERT, G., 'Observations on Pedological Factors That May Limit the Productivity of Soils of the Humid Tropics.' *Transactions International Soil Conference*, New Zealand, 1962, 15p.
  'Soil With Ferruginous or Ferrallitic Crusts of Tropical Regions.' *Soil Science*, 95, 1963, pp. 235–42.
KELLOGG, C. E., 'Preliminary Suggestions For the Classification and Nomenclature of Great Soil Groups in Tropical and Equatorial Regions.' *Commonwealth Bureau of Soil Science Technical Communication*, 46, 1949, 10p.
  'Tropical Soils.' *Transactions, International Congress of Soil Science*, Amsterdam, 1950, 11p.
MOHR, E. C. J., AND VAN BAREN, F. A., *Tropical Soils*. Van Hoeve, The Hague, 1959.
NYE, P. H., 'Some Soil-Forming Processes in the Humid Tropics.' Parts 1 and 2. *Journal of Soil Science*, 5(1), 1954, pp. 7–21; 6(1), 1954, pp. 51–83.
PRESCOTT, J. A., AND PENDLETON, R. L., 'Laterite and Lateritic Soils.' *Commonwealth Bureau of Soil Science, Tech. Comm.*, 47, 1952.
SIMONSON, R. W., 'Morphology and Classification of the Regur Soils of India.' *Journal of Soil Science*, 5(2), 1954, pp. 275–88.
SIVARAJASINGHAM, S., ALEXANDER, L. T., CADY, J. G., AND CLINE, M. G., 'Laterite.' *Advances in Agronomy*, 14, 1962, pp. 1–60.
'Symposium on Tropical Soil Resources.' *Soil Science* (Special Issue) 95(4), 1963, pp. 219–82.

31. Desertic, Saline and Alkaline Soils

BUOL, S. W., 'Present Soil-Forming Factors and Processes in Arid and Semiarid Regions.' *Soil Science*, 99(1), 1965, pp. 45–9.

JEWITT, T. N., 'Soils of Arid Lands.' Ch. 6 in E. S. Hills, *Arid Lands*. Methuen, London, 1966.

KELLEY, W. P., *Alkali Soils; Their Formation, Properties and Reclamation.* New York, Reinhold, 1951.

KELLOGG, C. E., 'Potentialities and Problems of Arid Soils.' *Proceedings, International Symposium on Desert Research, Jerusalem*, 1953, pp. 19–42.

ROBINSON, G. W., 1949 (see *Soils, general*)

32. Chernozemic Soils

DUCHAUFOUR, P., 1960 (see *Soils, general*)

GERASIMOV, I. P., AND GLAZOVSKAYA, M. A., 1960 (see *Soils, general*)

JOFFE, J. S., 1949 (see *Soils, general*)

ROBINSON, G. W., 1949 (see *Soils, general*)

33. Brown and Podzolised Soils

BLOOMFIELD, C., 'Mobilisation Phenomena in Soils.' *Report of the Rothamsted Experimental Station*, 1963, pp. 226–39.

CARTER, G. F., AND PENDLETON, R. L., 'The Humid Soil, Process and Time.' *Geographical Review*, 46(4), 1956, pp. 488–507.

DUCHAUFOUR, P., 1960 (see *Soils, general*)

GERASIMOV, I. P., AND GLAZOVSKAYA, M. A., 1960 (see *Soils, general*)

MUIR, A., 'The Podzol and Podzolic Soils.' *Advances in Agronomy*, 13, 1961, pp. 1–56.

RUSSELL, E. W., 1961 (see *Soils, general*)

TAVERNIER, R., AND SMITH, G. D., 'The Concept of Braunerde (Brown Forest Soil) in Europe and the United States.' *Advances in Agronomy*, 9, 1957, pp. 217–89.

33. Arctic Soils

HILL, D. E., AND TEDROW, J. C. F., 'Weathering and Soil Formation in the Arctic Environment.' *American Journal of Science*, 259, 1961, pp. 84–101.

TEDROW, J. C. F., AND CANTLON, J. E., 'Concepts of Soil Formation and Classification in Arctic Regions.' *Arctic*, 11, 1958, pp. 166–79.

TEDROW, J. C. F., DREW, J. V., HILL, D. E., AND DOUGLAS, L. A., 'Major Genetic Soils of the Arctic Slope of Alaska.' *Journal of Soil Science*, 9(1), 1958, pp. 33–45.

TEDROW, J. C. F., AND HILL, D. E., 'Arctic Brown Soil.' *Soil Science*, 80, 1955, pp. 265–75.

33. Antarctic Soils

MCCRAW, J. D., 'Soils of the Ross Dependency, Antarctica.' *New Zealand Society of Soil Science, Proceedings*, 4, 1960, pp. 30–5.

'Soils of Taylor Dry Valley, Victoria Land, Antarctica, With Notes on Soils from Other Localities in Victoria Land.' *New Zealand Journal of Geology and Geophysics*, 10(2), 1967, pp. 498–539.

TEDROW, J. C. F. (Editor), 'Antarctic Soils and Soil Forming Processes.' *American Geophysical Union*, Washington, DC, 1966.

'Polar Desert Soils.' *Soil Science Society of America, Proceedings*, 30(3), 1966, pp. 381–7.

34. Azonal Soils

COSTIN, A. B., 'Alpine Soils in Australia.' *Journal of Soil Science*, 6(1), 1955, pp. 35–50.

DUCHAUFOUR, P., 1960 (see *Soils, general*)

GANSSEN, R., AND HÄDRICH, F., 1965 (see *Soils, general*)

34. Intrazonal Soils

DAWSON, J. E., 'Organic Soils.' *Advances in Agronomy*, 8, 1956, pp. 378–401.

DUCHAUFOUR, P., 1960 (see *Soils, general*)

FARNHAM, R. S., AND FINNEY, H. R., 'Classification and Properties of Organic Soils.' *Advances in Agronomy*, 17, 1965, pp. 115–62.

SJÖRS, H., 'Surface Patterns in Boreal Peatland.' *Endeavour*, 20(80), 1961, pp. 217–24.

TAYLOR, N. H., AND POHLEN, I. J., 1962 (see *Soil in the Field*)

WILDE, S. A., 1958 (see *Soils, general*)

35. Soils and Man

BENNETT, H. H., *Soil Conservation*. McGraw-Hill, New York, 1939.

BIDWELL, O. W., AND HOLE, F. D., 'Man as a Factor of Soil Formation.' *Soil Science*, 99(1), 1965, pp. 65–72.

BOON, W. R., 'The Quaternary Salts of Bipyridyl—A New Agricultural Tool.' *Endeavour*, 26(97), 1967, pp. 27–32.

BUTLER, M. D., *Conserving Soil*. Van Nostrand, Princeton, New Jersey, 1955.

FAO, *Soil Conservation—An International Study*, FAO Agricultural Studies 4, Rome, 1948.

HAW, R. C., *The Conservation of Natural Resources*. Faber, London, 1959.

GREENE, H., *Using Salty Land*. FAO Agricultural Studies 3, FAO, Rome, 1948.

JACKS, G. V., 'The Human Factor in Soil Formation.' *Imperial Bureau of Soil Science, Monthly Letter*, 67, 1937.

JACKS, G. V., AND WHYTE, R. O., *The Rape of the Earth*, Faber, London, 1939.

NYE, P. H., AND GREENLAND, D. J., *The Soil Under Shifting Cultivation*. Commonwealth Agricultural Bureaux, Farnham Royal, Bucks, 1960.

TAYLOR, N. H., *Soils and Mankind*. Thomas Cawthron Memorial Lecture 36, Nelson, 1961.

THORNE, D. W., AND PETERSON, H. B., *Irrigated Soils*. Blakiston, New York, 1954

Biogeography

General

ANDERSON, M. S., *Geography of Living Things*. EUP, London, 1955.

CAIN, S. A., *Foundations of Plant Geography*. Harper, New York, 1944.

CAIN, S. A., AND CASTRO, G. M. DE O., *Manual of Vegetation Analysis*. Harper, New York, 1959.

CARPENTER, J. R., *An Ecological Glossary*, Hafner, New York, 1962.

CROIZAT, L., *Manual of Phytogeography*. Junk, The Hague, 1952.

DANSEREAU, P., *Biogeography, an Ecological Perspective*. Ronald Press, New York, 1957.

EYRE, S. R., *Vegetation and Soils, a World Picture*. Arnold, London, 1963.

GLEASON, H., AND CRONQUIST, A., *The Natural Geography of Plants*. Columbia UP, New York, 1964.

GOOD, R., *The Geography of the Flowering Plants*, Longmans, London, 1964.

NEWBIGIN, M. I., *Plant and Animal Geography*, Methuen, London, 1936.

POLUNIN, N., *Introduction to Plant Geography*. Longmans, London, 1960.

RILEY, D., AND YOUNG, A., *World Vegetation*, Cambridge UP, Cambridge, 1966.

SHELFORD, V. E., *The Ecology of North America*. Illinois UP, Urbana, 1963.

SHIMPER, A. F. W., *Plant-geography upon a Physiological Basis* (translated by W. R. Fisher, revised and edited by P. Groom and I. B. Balfour). Clarendon Press, Oxford, 1903.

TANSLEY, A., *Introduction to Plant Ecology*. Allen and Unwin, London, 1954.

WULFF, E. V., *An Introduction to Historical Plant Geography* (translated by E. Brissenden). Chronica Botanica, Waltham, Mass., 1943.

36. Introduction

DANSEREAU, P., 1957 (see *Biogeography, general*)

WACE, N. M., 'The Units and Uses of Biogeography.' *Australian Geographical Studies*, 5(1), 1967, pp. 15–29.

37. Environment

DAUBENMIRE, R. F., *Plants and Environment*. Wiley, New York, 1959.

38. Communities

BRAUN-BLANQUET, J., *Plant Sociology*, Stechert-Hafner, New York, 1932.

DU RIETZ, G. E., 'Life Forms of Terrestrial Flowering Plants.' *Acta Phytogeographica Suecica*, 3(1), 1931, 95p.

OOSTING, H. J., *The Study of Plant Communities*. Freeman, San Francisco, 1956.

RAUNKIAER, C., *The Life Forms of Plants and Statistical Plant Geography*. Oxford UP, Oxford, 1934.

39. Ecosystems

BILLINGS, W. D., *Plants and the Ecosystem*. Macmillan, London, 1964.

EYRE, S. R., 'Determinism and the Ecological Approach to Geography.' *Geography*, 49, 1964, pp. 369–76.

MORGAN, W. B., AND MOSS, R. P., 'Geography and Ecology: The Concept of the Community and its Relationship to Environment.' *Annals of the Association of American Geographers*, 55(2), 1965, pp. 339–50.

ODUM, E. P., *Fundamentals of Ecology*. Saunders, Philadelphia, 1959.
    Ecology. Holt, Rinehart, New York, 1963.

PHILLIPSON, J., *Ecological Energetics*. Arnold, London, 1966.

SIMMONS, I. G., 'Ecology and Land Use.' *Institute of British Geographers, Transactions*, 38, 1966, pp. 59–72.

STODDART, D. R., 'Geography and the Ecological Approach. The Ecosystem as a Geographic principle and method.' *Geography*, 50(3), 1965, pp. 242–51.

40. Dispersal and Migration

DARLINGTON, P. J., *Biogeography of the Southern End of the World*. Harvard UP, Cambridge, Mass., 1965.

GOOD, R., 1964 (see *Biogeography, general*)

POLUNIN, N., 1960 (see *Biogeography, general*)

41. Southern Continents

DARLINGTON, P. J., 1965 (see *Dispersal and Migration*)

FALLA, R. A., 'Oceanic birds as Dispersal Agents.' *Proc. Roy. Soc. London, Ser. B.*, 152, 1960, pp. 655–9.

FLEMING, C. A., 1963 (in GRESSITT, see below)

FLORIN, R., 'The Tertiary Fossil Conifers of South Chile and their Phytogeographical significance, with a Review of the Fossil Conifers of Southern Lands.' *Svenska Vet.-Akad.*, 19, Pt 2, 1940.

GRESSITT, J. L., *Pacific Basin Biogeography, A Symposium*. Bishop Museum Press, Hawaii, 1963.

HOLDGATE, M., 'Biological Routes between the Southern Continents.' *New Scientist*, 239, 1961, pp. 636–8.

HOLLOWAY, J. T., 'Forests and Climates in the South Island of New Zealand.' *Trans. Roy. Soc. N.Z.*, 82, 1954, pp. 329–410.

42. Origins and Distribution

CAIN, S. A., 1944 (see *Biogeography, general*)

DANSEREAU, P., 1957 (see *Biogeography, general*)

GOOD, R., 1964 (see *Biogeography, general*)

POLUNIN, N., 1960 (see *Biogeography, general*)

WULFF, E. V., 1943 (see *Biogeography, general*)

43 and 44. Animals

ALLEE, W. C., AND SCHMIDT, K. P., *Ecological Animal Geography*. Wiley, New York, 1951.

ANDREWARTHA, H. G., AND BIRCH, L. C., *The Distribution and Abundance of Animals*. Chicago UP, Chicago, 1954.

BROWNING, T. O., *Animal Populations*. Hutchinson, London, 1963.

BURTON, M., *Systematic Dictionary of Mammals of the World*. Museum Press, London, 1962.

DARLINGTON, P. J., *Zoogeography: The Geographical Distribution of Animals*. Wiley, New York, 1957.

DOWDESWELL, W. H., *Animal Ecology*. Methuen, London, 1959.

ELTON, C. S., *The Ecology of Invasions by Animals and Plants*. Methuen, London, 1958.
    The Pattern of Animal Communities. Methuen, London, 1966.

GEORGE, W., *Animal Geography*. Heinemann, London, 1962.

KENDEIGH, S. C., *Animal Ecology*. Prentice-Hall, Englewood Cliffs, New Jersey, 1961.

LEOPOLD, A., *Game Management*. Charles Scribner, New York, 1933.

MACFADYEN, A., *Animal Ecology*. Pitman, London, 1963.

PEARSALL, W. H., 'Biological Invasions.' *Penguin Science Survey 1964 B*, 1964.

WODZICKI, K. A., *Introduced Mammals of New Zealand*. Dept of Scientific and Industrial Research, Wellington, 1950.

45. Field Study of Plants

BROWN, D., *Methods of Surveying and Measuring Vegetation*, Commonwealth Agricultural Bureaux, Farnham Royal, Bucks, 1954.

GRIEG-SMITH, P., *Quantitative Plant Ecology*. Butterworth, London, 1964.

KERSHAW, K. A., *Quantitative and Dynamic Ecology*. Arnold, London, 1964.

KÜCHLER, A. W., 'Analyzing the Physiognomy and Structure of Vegetation.' *Annals of the Association of American Geographers*, 56(1), 1966, pp. 112–25.
    Vegetation Mapping. Ronald Press, New York, 1967.

SANKEY, J., *A Guide to Field Biology*. Longmans, London, 1958.

46. Humid Tropics

AUBERT DE LA RÜE, E., BOURLIÈRE, F., AND HARROY, J. P., *The Tropics.* George Harrap, London, 1958.

BEARD, J. S., 'The Savanna Vegetation of Northern Tropical America.' *Ecological Monographs*, 23(2), 1953, pp. 149–215.
    'The Classification of Tropical American Vegetation-Types.' *Ecology*, 36(1), 1955, pp. 89–100.

COLE, M. M., 'Cerrado, Caatinga and Pantanal: The Distribution and Origin of the Savanna Vegetation of Brazil.' *Geographical Journal*, 126(2), 1960, pp. 168–179.

DAVIS, T. A. W., AND RICHARDS, P. W., 'The Vegetation of Morabilli Creek, British Guiana, an Ecological Study of a Limited Area of Tropical Rain Forest.' Parts I and II. Pt I, *Journal of Ecology*, 21, 1933, pp. 350–84. Pt II, *Journal of Ecology*, 22, 1934, pp. 106–55.

HOPKINS, B., *Forest and Savanna.* Heinemann, London, 1965.

RICHARDS, P. W., *The Tropical Rain Forest.* Cambridge UP, Cambridge, 1957.

STEENIS, C. G. G. J. VAN., 'The Mountain Flora of the Malaysian Tropics.' *Endeavour*, 21 (83–84), 1962, pp. 183–94.

TROLL, C., 'Tropical Mountain Vegetation.' *Proceedings Ninth Pacific Science Congress*, 20, 1958, pp. 37–46.

47. Dry Tropics

CLOUDSLEY-THOMSON, J. L., 'The Ecology of Oases.' *Science Journal*, 3(8), 1967, pp. 47–53.

HILLS, E. S., *Arid Lands.* Methuen, London, 1966.

HUNT, C. B., 'Plant Ecology of Death Valley, California.' *US Geological Survey Professional Paper* 509, 1966.
    Physiography of the United States. Freeman, San Francisco, 1967.

KIRMIZ, J. P., *Adaptation to Desert Environment.* Butterworth, London, 1962.

POLUNIN, N., 1960 (see *Biogeography, general*)

SCHMIDT-NIELSEN, K., *Desert Animals.* Clarendon Press, Oxford, 1964.

SHELFORD, V. E., 1963 (see *Biogeography, general*)

WENT, F. W., 'The Ecology of Desert Plants.' *Scientific American*, 192(4), 1955, pp. 68–75.

48. Temperate Regions

COCKAYNE, L., *The Vegetation of New Zealand.* 3rd edition, Wheldon and Wesley, London, 1958.

COCKAYNE, L., AND TURNER, E. P., *The Trees of New Zealand.* Government Printer, Wellington, 1958.

EYRE, S. R., 1963 (see *Biogeography, general*)

PEARSALL, W. H., *Mountains and Moorlands.* Collins, London, 1950.

POLUNIN, N., 1960 (see *Biogeography, general*)

POOLE, A. L., AND ADAMS, N. M., *Trees and Shrubs of New Zealand.* Government Printer, Wellington, 1963.

ROBBINS, R. G., 'The Podocarp-Broadleaf Forests of New Zealand.' *Trans. Roy. Soc. of New Zealand, Botany*, 1(5), 1962, pp. 33–75.

SHELFORD, V. E., 1963 (see *Biogeography, general*)

TANSLEY, A. G., *The British Isles and Their Vegetation.* 2 vols, Cambridge UP, Cambridge, 1953.

49. Cold Regions

BAIRD, P. D., *The Polar World*, Longmans, London, 1964.

HATHERTON, T. (Editor), *Antarctica.* A. H. and A. W. Reed, Wellington, 1965.

KIMBLE, G. H. T., AND GOOD, D., *Geography of the Northlands.* Chapman and Hall, London, 1955.

VAN MIEGHEM, J., AND VAN OYE, P. (Editors), *Biogeography and Ecology In Antarctica.* Junk, The Hague, 1965.

50. Ecology and Man

DEBACH, P., *Biological Control of Insect Pests and Weeds.* Chapman and Hall, London, 1964.

CARSON, R., *Silent Spring.* Houghton Mifflin, Boston, 1962, and Hamish Hamilton, London, 1963.

EGLER, F. E., 'Pesticides in Our Ecosystem.' *American Scientist*, 52(1), 1964, pp. 110–36.

ELLIOT, M., 'Synthesizing Pyrethrin-like Insecticides.' *Science Journal*, 3(3), 1967, pp. 61–6.

MELLANBY, K., *Pesticides and Pollution.* Collins, London, 1967.

ORDISH, G., *Biological Methods in Crop Pest Control.* Constable, London, 1967.

RATCLIFFE, F. N., 'Biological Control.' *Australian Journal of Science*, 28(6), 1965, pp. 237–40.

SWEETMAN, H. L., *The Principles of Biological Control.* Brown, Iowa, 1958.

VARLEY, G. C., 'The Biological Control of Agricultural Pests.' *Journal of the Royal Society of Arts*, 107, 1959, pp. 475–90.

WATERHOUSE, D. F., 'The Use of Sterile Insects For Their Own Destruction.' *Australian Journal of Science*, 28(6), 1965, pp. 235–7.

WILLIAMS, H., 'Bikini nine years later.' *Science Journal*, 3(4), 1967, pp. 48–53.

WILLIAMS, C. M., 'Third Generation Pesticides.' *Scientific American*, 217(1), 1967, pp. 13–17.

WOODWELL, G. M., 'The Ecological Effects of Radiation.' *Scientific American*, 208(6), 1963, pp. 40–9.

Index

Plants: adaptation, *see* Adaptation, plant; classification, 243–4, 281–4, 303–4, 310; climaxes, 258–60; communities, 252–6, 387; composition, 132; dependence, 250–1; distribution, 278, 281–4; effect of animals, 250, 294, 295, 364; effect of pollution, 66, 246, 247, 384, 385; environmental factors, 240, 241–51; life-forms, 254–6, 304; limiting factors, 151; migration and dispersal, *see* Migration and dispersal; nutrients, 145–147; origin, 278; productivity, 265; recording, 305–6; sampling, 306–9; stratification, 304; successions, 256–8, 375; survival of species, 279–81, 284. *See also* Vegetation

Plate river, 350

Platypus, 300

Pleistocene period, 3, 79, 180; ice age, *see* Glaciations

Plinthite, 199

Pneumatophores, 247

Podocarps, 355, 356, 357–8, 359, 361

Podzolic soils: agriculture on, 215; composition and structure, 134–5, 141, 142, 162, 214–15; egg-cup podzol, 175; formation, 143, 167, 168, 169–70, 173, 192, 357, 365, *see also* Podzolisation; gleying, 224; types, 167, 192, 202–3, 212, 221, 320, 373; zonal, 212, 220

Podzolisation, 205; caused by forest clearance, 230; caused by soil, 169, 215; caused by vegetation, 174–5, 186, 213, 215, 366, 373; defined, 211, 213; in different soils, 202, 217, 223; variations, 212, 215

Poikilotherms, 61

Pollen, 248, 274, 277; analysis, 82, 362–3

Pollination methods, 249–50, 251, 341

Pollution: atmospheric, 7, 23, 65–6, 246, 247; chemical, 385–7; effect on animals, 3, 286, 383, 384, 385–7; effect on vegetation, 3, 66, 246, 247, 384, 385; land, 3, 382, 383; radioactive, 384–5

Polunin, N., 271

Populations, ecological: animal, 290–1, 383; defined, 237; forest, 309

Portugal, 102

Post, F. von, 226

Potatoes, 63; blight, 252

Pott, Sir Percival, 385

Prairies: agriculture, 209, 344; soil, 163, 209; vegetation, 248, 304, 344, 348–9, 372

Prairie soils, 209–10, 212, 230

Prasolov, N., 201

Precipitation: causes, 22–4, 25; convectional, 25; frontal, 26, 55; man-made, 24–5, 67; orographic, 26, 97; effectiveness (P/E), 86; soil formation, 175–6

Prescott, J. A., 201

Pressure, atmospheric: distribution, 19–21; measurement, 20; systems, 55–7; wind, 20–1. *See also* Anticyclones; Depressions

Pressure gradients, 20–1, 58, 60

Productivity: animal, 265–7, 394; plant, 265

Protoplasm, world production of, 267

Protozoa, 128

Puna grassland (western Andes), 335

Punjab, 48

Pygmies, 287

Pyrenees mountains, 280, 283

Quail, 288

Quartz: in laterite, 163; in podzol, 214, 215; soil formation, 118, 158, 195; weathering, 142, 153

Quaternary period: animal migration, 296; effect on soil, 177, 193; ice ages, *see* Glaciations; vegetation, 312, 325, 344, 355

Quito (Ecuador), 332

Rabbits: classification, 285; dispersal, 293; effect on vegetation, 127–8, 250, 294, 364; myxomatosis, 250, 293, 389; productivity, 265

Radiation curves (Milankovitch), 79, 80

Radiation, solar. *See* Insolation

Radioactivity, 384–5

Rain. *See* Precipitation

Rainbows, 23

Rain forest, 283, 333; agriculture, 316; characteristics, 312; classifications, 237, 238, 239, 310; climate and distribution, 311–12, 320, 344, 353; composition, 251, 268, 313–16, 320, 353–4; insolation, 246–7; stratification, 313

Ramann, E., 176

Rats, 294, 300, 384

Raunkiaer, Professor C., vegetation classification, 255–6, 339

Red Sea, 321, 343

Redwood trees (*Sequoia sempervirens*), 280, 313, 353

Regolith, 148, 150, 151

Regosols, 220, 221

Reindeer, 376, 378

Rendzina, 170, 202, 223

Rhine river, 82

Rhinoceroses, 299, 300

Rhizobium (soil bacterium), 242

Rhizosphere, 129

Rice, 63

Richards, P. W., 306

Richardson, L. F., 72

Robin, 290

Robinson, G. W., 178

Rock: bedrock, 158, 168; classification, 148, 151; flour, 117; inselberg, 150, 320; plant growth on, 241; weathering, *see* Weathering

Rocky Mountains, 30, 41, 174, 283, 332

Rossby waves (Long waves), 41, 56, 71

Roth, E. S., 150

Rubber tree (*Hevea brasiliensis*), 271

Russia. *See* USSR

Ruwenzori mountains, 335, 336

Sage-brush, 249, 348

Sahara Desert, 281, 322; animals, 299; boundary of zoogeographical realms, 297; climate, 60, 99, 109; extent, 98; oases, 340, 341; vegetation, 81, 242, 335, 337, 339

St Lawrence Seaway, 64

Saline soils, 205–7

Salt: atmospheric, 7, 23, 248; plant tolerance, 146, 175, 242; soil, 75, 119, 125, 137, 159, 219, 229, *see also* Saline soils; weathering, 150

Salt flats (playas), 242

Sand, 113, 116, 117, 118, 124, 182

Sandstone, 75, 154, 168, 223, 319

Saprophytes, 252–3

Satellites, weather, 7, 19, 52, 53, 73, 93

Savanna, 1, 335; agriculture, 232, 249, 327; animals, 265–7, 299–300, 331–2; distribution, 81, 312, 322, 328–30; origin, 324–8; soil, 135, 194, 200, 231, 325–6; vegetation, 238, 304, 310, 328, 348

Schimper, A. F. W., 310, 311, 324

Sciophytes, 247, 249

Sclater, P. L., 297

Sclerophyllous vegetation, 335, 345–7

Scotland, 292, 361, 367, 381

Scrub vegetation, 193, 337–9, 345; in New Zealand, 356, 359–60

Seals, 302, 380

Seaweed, 270, 274

Separation of land masses, and effect on animals and plants, 268, 271, 275; Asia, 281; Britain, 301; New Zealand, 277, 301, 353–4; southern continents, 272, 274, 281

Seres, 290; primary, 256, 271; types, 257, 259, 370

Sesquioxides: in calcimorphic soil, 223; in podzolisation, 213; in prairie soil, 209; in soil horizons, 158, 161, 162; in tropical soils, 193, 194, 197, 198

Shanghai, 102

Shantz, H. L., 173

Shaw, Sir Napier, 53

Sheep, 231, 250, 294, 341, 350, 392

Siberia, 274; climate, 15, 50, 100, 106, 107, 109; forest, 303, 372, 373; soil, 215, 225; tundra marsh, 378

Sibirtsev, N. M., 187

Sierozem (serozem), 134, 204, 210

Silt, 113, 116, 117, 118, 182

Simpson, Sir George, 80

Singapore, 316

Smog, 23, 66

Smoke, 7, 23, 63, 65, 66

Snakes, 301, 330

Snow, 26, 88; albedo, 77; disrupts communications, 65; effect on vegetation, 246, 369, 370, 376; formation and precipitation, 23, 27, 55; in cold climates, 15, 105, 106, 107; in polar climates, 108, 109, 216, 219; in temperate climates, 101; mountain, 16, 75, 110

Soil: atmosphere, 122, 125–6; azonal, 187, 188, 220–2; classification, 113, 186–92; colour, 163–4, 181, 184; composition, 113, 117–18; consistence, 121, 182–3; effect on vegetation, 241–2; equilibrium concepts, 192; formation, *see* Pedogenesis; intrazonal, 169, 187, 188, 222–6; modified by man, 227–34; moisture, 122–5; organic matter, 114, 132–5; organisms, 127–32; pH, 146–7, 183–4; profile, 156–63, 194–5; salt, *see* Salt; structure, 114, 118–21, 183; study of, 114–15, 179–85; texture, 113, 116–17, 181–2, *see also* Clay, Sand, Silt; zonal, 187, 188, 192, 216, 220, 222

Solberg, H., 54

Solod (soloth, soloti, solodi), 206